A UNIVERSE OF PHYSICS

A UNIVERSE OF PHYSICS
A Book of Readings

Jerry B. Marion
University of Maryland
College Park, Maryland

JOHN WILEY & SONS, INC.   *New York · London · Sydney · Toronto*

Library of Congress Catalogue Card Number: 79-115651

ISBN   0-471-56913-5

Printed in the United States of America

10 9 8 7 6 5 4

PREFACE

The physics student who attempts to locate reading material on the subject or who wishes to supplement his textbook finds himself confronted with a vast and confusing literature. The usual result is that a reader does not spend the necessary time in locating appropriate material and, therefore, loses an opportunity to become acquainted with the technical, general, or historical aspects of the subject. This collection of articles was prepared with the objective of providing such reading matter. Here the student or casual reader will find a wide variety of essays, general articles, and historical documents dealing with the physics of the Universe in which we live. Some of the material consists of original papers that were important in the development of particular concepts and ideas; some of the articles are historical reviews; some are popular expositions by noted lecturers and teachers; some are personal accounts of important events or discoveries; and some are just *fun*. Material from such noted educators and lecturers as Richard Feynman and Norman Feather has been liberally drawn upon.

This is not a book of equations and problems—it is meant to be browsed through, not studied as a textbook. The reader of these articles should gain some insight into the way that physics was built and the way in which the building continues; he should catch a glimpse of the way in which great minds have approached the mysteries of the Universe; and he should find some indications of the scientific philosophy that has guided men to discoveries of the ways in which Nature works.

These articles provide only a taste of the essence of physics. Hopefully, the taste is sufficiently pleasant to induce the reader to gain an even wider acquaintance with the literature of science.

Jerry B. Marion

CONTENTS

A UNIVERSE OF PHYSICS

CHAPTER 1

SCIENCE, PHYSICS, AND THE WORLD

1.1

The Value of Science
Richard P. Feynman

From time to time, people suggest to me that scientists ought to give more consideration to social problems—especially that they should be more responsible in considering the impact of science upon society. This same suggestion must be made to many other scientists, and it seems to be generally believed that if the scientists would only look at these very difficult social problems and not spend so much time fooling with the less vital scientific ones, great success would come of it.

It seems to me that we do think about these problems from time to time, but we don't put full-time effort into them—the reason being that we know we don't have any magic formula for solving problems, that social problems are very much harder than scientific ones, and that we usually don't get anywhere when we do think about them.

I believe that a scientist looking at nonscientific problems is just as dumb as the next guy—and when he talks about a nonscientific matter, he will sound as naive as anyone untrained in the matter. Since the question of the value of science is not a scientific subject, this discussion is dedicated to proving my point—by example.

The first way in which science is of value is familiar to everyone. It is that scientific knowledge enables us to do all kinds of things and to make all kinds of things. Of course if we make good things, it is not only to the credit of science; it is also to the credit of the moral choice which led us to good work. Scientific knowledge is an enabling power to do either good or bad—but it does not carry instructions on how to use it. Such power has evident value—even though the power may be negated by what one does.

I learned a way of expressing this common human problem on a trip to Honolulu. In a Buddhist temple there, the man in charge explained a little bit about the Buddhist religion for tourists, and then ended his talk by telling them he had something to say to them that they would *never* forget—and I have never forgotten it. It was a proverb of the Buddhist religion:

"To every man is given the key to the gates of heaven; the same key opens the gates of hell."

What then, is the value of the key to heaven? It is true that if we lack clear instructions that determine which is the gate to heaven and which the gate to hell, the key may be a dangerous object to use, but it obviously has value. How can we enter heaven without it?

SOURCE. From *Frontiers in Science—A Survey*, Edward Hutchings, Jr., (ed.) (New York: Basic Books, 1958).

1

The instructions, also, would be of no value without the key. So it is evident that, in spite of the fact that science could produce enormous horror in the world, it is of value because it *can* produce *something*.

Another value of science is the fun called intellectual enjoyment which some people get from reading and learning and thinking about it, and which others get from working in it. This is a very real and important point and one which is not considered enough by those who tell us it is our social responsibility to reflect on the impact of science on society.

Is this mere personal enjoyment of value to society as a whole? No! But it is also a responsibility to consider the value of society itself. Is it, in the last analysis, to arrange things so that people can enjoy things? If so, the enjoyment of science is as important as anything else.

But I would like *not* to underestimate the value of the world view which is the result of scientific effort. We have been led to imagine all sorts of things infinitely more marvelous than the imaginings of poets and dreamers of the past. It shows that the imagination of nature is far, far greater than the imagination of man. For instance, how much more remarkable it is for us all to be stuck—half of us upside down—by a mysterious attraction, to a spinning ball that has been swinging in space for billions of years, than to be carried on the back of an elephant supported on a tortoise swimming in a bottomless sea.

I have thought about these things so many times alone that I hope you will excuse me if I remind you of some thoughts that I am sure you have all had—or this type of thought— which no one could ever have had in the past, because people then didn't have the information we have about the world today.

For instance, I stand at the seashore, alone, and start to think. There are the rushing waves ... mountains of molecules, each stupidly minding its own business ... trillions apart ... yet forming white surf in unison.

Ages on ages ... before any eyes could see ... year after year ... thunderously pounding the shore as now. For whom, for

what? ... on a dead planet, with no life to entertain.

Never at rest ... tortured by energy ... wasted prodigiously by the sun ... poured into space. A mite makes the sea roar.

Deep in the sea, all molecules repeat the patterns of one another till complex new ones are formed. They make others like themselves ... and a new dance starts.

Growing in size and complexity ... living things, masses of atoms, DNA, protein ... dancing a pattern ever more intricate.

Out of the cradle onto the dry land ... here it is standing ... atoms with consciousness ... matter with curiosity.

Stands at the sea ... wonders at wondering ... I ... a universe of atoms ... an atom in the universe.

THE GRAND ADVENTURE

The same thrill, the same awe and mystery, come again and again when we look at any problem deeply enough. With more knowledge comes deeper, more wonderful mystery, luring one on to penetrate deeper still. Never concerned that the answer may prove disappointing, but with pleasure and confidence we turn over each new stone to find unimagined strangeness leading on to more wonderful questions and mysteries—certainly a grand adventure!

It is true that few unscientific people have this particular type of religious experience. Our poets do not write about it; our artists do not try to portray this remarkable thing. I don't know why. Is nobody inspired by our present picture of the universe? The value of science remains unsung by singers, so you are reduced to hearing—not a song or a poem, but an evening lecture about it. This is not yet a scientific age.

Perhaps one of the reasons is that you have to know how to read the music. For instance, the scientific article says, perhaps, something like this: "The radioactive phosphorous content of the cerebrum of the rat decreases to one-half in a period of two weeks." Now, what does that mean?

It means that phosphorus that is in the brain of a rat (and also in mine, and yours) is not the same phosphorus as it was two weeks ago, but that all of the atoms that are in the brain are being replaced, and the ones that were there before have gone away.

So what is this mind, what are these atoms with consciousness? Last week's potatoes! That is what now can *remember* what was going on in my mind a year ago—a mind which has long ago been replaced.

That is what it means when one discovers how long it takes for the atoms of the brain to be replaced by other atoms, to note that the thing which I call my individuality is only a pattern or dance. The atoms come into my brain, dance a dance, then go out; always new atoms but always doing the same dance, remembering what the dance was yesterday.

THE REMARKABLE IDEA

When we read about this in the newspaper, it says, "The scientist says that this discovery may have importance in the cure of cancer." The paper is only interested in the use of the idea, not the idea itself. Hardly anyone can understand the importance of an idea, it is so remarkable. Except that, possibly, some children catch on. And when a child catches on to an idea like that, we have a scientist. These ideas do filter down (in spite of all the conversation about TV replacing thinking), and lots of kids get the spirit—and when they have the spirit you have a scientist. It's too late for them to get the spirit when they are in our universities, so we must attempt to explain these ideas to children.

I would now like to turn to a third value that science has. It is a little more indirect, but not much. The scientist has a lot of experience with ignorance and doubt and uncertainty, and this experience is of very great importance, I think. When a scientist doesn't know the answer to a problem, he is ignorant. When he has a hunch as to what the result is, he is uncertain. And when he is pretty darn sure of what the result is going to be, he is in some doubt. We have found it of paramount importance that in order to progress we must recognize the ignorance and leave room for doubt. Scientific knowledge is a body of statements of varying degrees of certainty—some most unsure, some nearly sure, none *absolutely* certain.

Now, we scientists are used to this, and we take it for granted that it is perfectly consistent to be unsure—that it is possible to live and *not* know. But I don't know whether everyone realizes that this is true. Our freedom to doubt was born of a struggle against authority in the early days of science. It was a very deep and strong struggle. Permit us to question—to doubt, that's all—not to be sure. And I think it is important that we do not forget the importance of this struggle and thus perhaps lose what we have gained. Here lies a responsibility to society.

We are all sad when we think of the wondrous potentialities human beings seem to have, as contrasted with their small accomplishments. Again and again people have thought that we could do much better. They of the past saw in the nightmare of their times a dream for the future. We, of their future, see that their dreams, in certain ways surpassed, have in many ways remained dreams. The hopes for the future today are, in good share, those of yesterday.

EDUCATION, FOR GOOD AND EVIL

Once some thought that the possibilities people had were not developed because most of those people were ignorant. With education universal, could all men be Voltaires? Bad can be taught at least as efficiently as good. Education is a strong force, but for either good or evil.

Communications between nations must promote understanding: so went another dream. But the machines of communication can be channeled or choked. What is communicated can be truth or lie. Communication is a strong force also, but for either good or bad.

The applied sciences should free men of material problems at least. Medicine controls diseases. And the record here seems all to the good. Yet there are men patiently working to create great plagues and poisons. They are to be used in warfare tomorrow.

Nearly everybody dislikes war. Our dream today is peace. In peace, man can develop best the enormous possibilities he seems to have. But maybe future men will find that peace, too, can be good and bad. Perhaps peaceful men will drink out of boredom. Then perhaps drink will become the great problem which seems to keep man from getting all he thinks he should out of his abilities.

Clearly, peace is a great force, as is sobriety, as are material power, communication, education, honesty and the ideals of many dreamers.

We have more of these forces to control than did the ancients. And maybe we are doing a little better than most of them could do. But what we ought to be able to do seems gigantic compared with our confused accomplishments.

Why is this? Why can't we conquer ourselves?

Because we find that even great forces and abilities do not seem to carry with them clear instructions on how to use them. As an example, the great accumulation of understanding as to how the physical world behaves only convinces one that this behavior seems to have a kind of meaninglessness. The sciences do not directly teach good and bad.

Through all ages men have tried to fathom the meaning of life. They have realized that if some direction or meaning could be given to our actions, great human forces would be unleashed. So, very many answers must have been given to the question of the meaning of it all. But they have been of all different sorts, and the proponents of one answer have looked with horror at the actions of the believers in another. Horror, because from a disagreeing point of view all the great potentialities of the race were being channeled into a false and confining blind alley. In fact, it is from the history of the enormous monstrosities created by false belief that philosophers have realized the apparently infinite and wondrous capacities of human beings. The dream is to find the open channel.

What, then, is the meaning of it all? What can we say to dispel the mystery of existence?

If we take everything into account, not only what the ancients knew, but all of what we know today that they didn't know, then I think that we must frankly admit that *we do not know.*

But, in admitting this, we have probably found the open channel.

This is not a new idea; this is the idea of the age of reason. This is the philosophy that guided the men who made the democracy that we live under. The idea that no one really knew how to run a government led to the idea that we should arrange a system by which new ideas could be developed, tried out, tossed out, more new ideas brought in; a trial and error system. This method was a result of the fact that science was already showing itself to be a successful venture at the end of the eighteenth century. Even then it was clear to socially-minded people that the openness of the possibilities was an opportunity, and that doubt and discussion were essential to progress into the unknown. If we want to solve a problem that we have never solved before, we must leave the door to the unknown ajar.

OUR RESPONSIBILITY AS SCIENTISTS

We are at the very beginning of time for the human race. It is not unreasonable that we grapple with problems. There are tens of thousands of years in the future. Our responsibility is to do what we can, learn what we can, improve the solutions and pass them on. It is our responsibility to leave the men of the future a free hand. In the impetuous youth of humanity, we can make grave errors that can stunt our growth for a long time. This we will do if we say we have the answers now, so young and ignorant; if we suppress all discussion, all criticism, saying, "This is it, boys, man is saved!" and thus doom man for a long time to the chains of authority, confined to the limits of our present imagination. It has been done so many times before.

It is our responsibility as scientists, knowing the great progress and great value of a satisfactory philosphy of ignorance, the great progress that is the fruit of freedom of thought, to proclaim the value of this freedom, to teach how doubt is not to be feared but welcomed and discussed, and to demand this freedom as our duty to all coming generations.

I.2   *The Privilege of being a Physicist*
Victor F. Weisskopf

There are certain obvious privileges that a physicist enjoys in our society. He is reasonably paid; he is given instruments, laboratories, complicated and expensive machines, and he is asked not to make money with these tools, like most other people, but to spend money. Furthermore he is supposed to do what he himself finds most interesting, and he accounts for what he spends to the money givers in the form of progress reports and scientific papers that are much too specialized to be understood or evaluated by those who give the money—the federal authorities and, in the last analysis, the taxpayer. Still, we believe that the pursuit of science by the physicist is important and should be supported by the public. In order to prove this point, we will have to look deeper into the question of the relevance of science to society as a whole. We will not restrict ourselves to physics only; we will consider the relevance of all the natural sciences, but we will focus our attention on basic sciences, that is to those scientific activities that are performed without a clear practical application in mind.

The question of the relevance of scientific research is particularly important today, when society is confronted with a number of immediate urgent problems. The world is facing threats of nuclear war, the dangers of overpopulation, of a world famine, mounting social and racial conflicts, and the destruction of our natural environment by the byproducts of ever-increasing applications of technology. Can we afford to continue scientific research in view of these problems?

I will try to answer this question affirmatively. It will be the trend of my comments to emphasize the diversity in the relations between science and society; there are many sides and many aspects, each of different character, but of equal importance. We can divide these aspects into two distinct groups. On the one hand, science is important in shaping our *physical* environment; on the other, in shaping our *mental* environment. The first refers to the influence of science on technology, the second to the influence on philosophy, on our way of thinking.

TECHNOLOGY

The importance of science as a basis of technology is commonplace. Obviously, knowledge as to how nature works can be used to obtain power over nature. Knowledge acquired by basic science yielded a vast technical return. There is not a single industry today that does not make use of the results of atomic physics or of modern chemistry. The vastness of the return is illustrated by the fact that the total cost of all basic research, from Archimedes to the present, is less than the value of ten days of the world's present industrial production.

We are very much aware today of some of the detrimental effects of the ever increasing pace of technological development. These effects begin to encroach upon us in environmental pollution of all kinds, in mounting social tensions caused by the stresses and dislocations of a fast changing way of life and, last but not least, in the use of modern technology to invent and construct more and more powerful weapons of destruction.

In many instances, scientific knowledge has been and should continue to be applied to counteract these effects. Certainly, physics and chemistry are useful to combat many forms of pollution and to improve public transportation. Biological research could and must be used to find more effective means of birth control and new methods to increase our food resources. It has been pointed out

SOURCE. From *Physics Today*, August 1969.

many times that our exploitation of the sea for food gathering is still in the hunting stage; we have not yet reached the neolithic age of agriculture and animal breeding in relation to the oceans.

Many of the problems that technology has created cannot be solved by natural science. They are social and political problems, dealing with the behavior of man in complicated and rapidly evolving situations. In particular, the questions arise: "What technical possibilities should or should not be realized? How far should they be developed?" A systematic investigation of the positive and negative social effects of technical innovations is necessary. But it is only partly a problem for natural sciences; to a greater extent, it is a problem of human behavior and human reaction. I am thinking here of the supersonic transport, of space travel, of the effects of the steadily increasing automobile traffic and again, last but not least, of the effects of the development of weapons of mass destruction.

PHYSICAL ENVIRONMENT

What role does basic science have in shaping our physical environment? It is often said that modern basic physical science is so advanced that its problems have little to do with our terrestrial environment. It is interested in nuclear and subnuclear phenomena and in the physics of extreme temperatures. These are objectives relating to cosmic environments, far away from our own lives. Hence, the problems are not relevant for society; they are too far removed; they are studied for pure curiosity only. We will return later to the value of pure curiosity.

Let us first discuss how human environment is defined. Ten thousand years ago, metals were not part of human environment; pure metals are found only very rarely on earth. When man started to produce them, they were first considered as most esoteric and irrelevant materials and were used only for decoration purposes during thousands of years. Now they are an essential part of our environment. Electricity went through the same development, only much faster. It is observed naturally only in a few freak phe-nomena, such as lightning or friction electricity, but today it is an essential feature of our lives.

This shift from periphery to center was most dramatically exhibited in nuclear physics. Nuclear phenomena are certainly far removed from our terrestrial world. Their place in nature is found rather in the center of stars or of exploding supernovae, apart from a few naturally radioactive materials, which are the last embers of the cosmic explosion in which terrestrial matter was formed. This is why Ernest Rutherford remarked in 1927, "Anyone who expects a source of power from transformations of atoms is talking moonshine." It is indeed a remarkable feat to recreate cosmic phenomena on earth as we do with our accelerators and reactors, a fact often overlooked by the layman, who is more impressed by rocket trips to the moon. That these cosmic processes can be used for destructive as for constructive purposes is more proof of their relevance in our environment.

Even phenomena as far removed from daily life as those discovered by high-energy physicists may some day be of technical significance. Mesons and hyperons are odd and rare particles today, but they have interactions with ordinary matter. Who knows what these interactions may be used for at the end of this century? Scientific research not only investigates our natural environment, it also creates new artificial environments, which play an ever-increasing role in our lives.

MENTAL ENVIRONMENT

The second and most important aspect of the relevance of science is its influence on our thinking, its shaping of our mental environment. One frequently hears the following views as to the effect of science on our thought: "Science is materialistic, it reduces all human experience to material processes, it undermines moral, ethical and aesthetic values because it does not recognize them, as they cannot be expressed in numbers. The world of nature is dehumanized, relativized; there are no absolutes any more; nature is regarded as an abstract formula; things and objects are nothing but vibrations of an abstract mathematical concept . . ." (Science is accused at the same time of being materialistic and of negating matter.)

Actually science gives us a unified, rational view of nature; it is an eminently successful search for fundamental laws with universal validity; it is an unfolding of the basic processes and principles from which all natural happenings are derived, a search for the absolutes, for the invariants that govern natural processes. It finds law and order— if I am permitted to use that expression in this context—in a seemingly arbitrary flow of events. There is a great fascination in recognizing the essential features of nature's structure, and a great intellectual beauty in the compact and all-embracing formulation of a physical law. Science is a search for meaning in what is going on in the natural world, in the history of the universe, its beginnings and its possible future.

PUBLIC AWARENESS

These growing insights into the workings of nature are not only open to the scientific expert, they are also relevant to the non-scientist. Science did create an awareness among people of all ways of life that universal natural laws exist, that the universe is not run by magic, that we are not at the mercy of a capricious universe, that the structure of matter is largely known, that life has developed slowly from inorganic matter by evolution in a period of several thousand million years, that this evolution is a unique experiment of nature here on earth, which leaves us humans with a responsibility not to spoil it. Certainly the ideas of cosmology, biology, paleontology and anthropology changed the ideas of the average man in respect to future and past. The concept of an unchanging world or a world subject to arbitrary cycles of changes is replaced by a world that continuously develops from more primitive to more sophisticated organization.

Although there is a general awareness of the public in all these aspects of science, much more could be and must be done to bring the fundamental ideas nearer to the intelligent layman. Popularization of science should be one of the prime duties of a scientist and not a secondary one as it is now. A much closer collaboration of scientists and science writers is necessary. Seminars, summer schools, direct participation in research should be the rule for science writers, in order to obtain a free and informal contact of minds between science reporters and scientists on an equal level, instead of an undirected flow of undigested information.

EDUCATION

Science also shapes our thinking by means of its role in education. The study of open scientific frontiers where unsolved fundamental problems are faced is, and should be, a part of higher education. It fosters a spirit of inquiry; it lets the student participate in the joy of a new insight, in the inspiration of new understanding. The questioning of routine methods, the search for new and untried ways to accomplish things, are important elements to bring to any problem, be it one of science or otherwise. Basic research must be an essential part of higher education. In elementary education, too, science should and does play an increasing role. Intelligent play with simple, natural phenomena, the joys of discovery of unexpected experiences, are much better ways of learning to think than any teaching by rote.

A UNIVERSAL LANGUAGE ...

The international aspect of science should not be forgotten as an important part of its influence on our mental environment. Science is a truly human concern; its concepts and its language are the same for all human beings. It transcends any cultural and political boundaries. Scientists understand each other immediately when they talk about their scientific problems, and it is thus easier for them to speak to each other on political or cultural questions and problems about which they may have divergent opinions. The scientific community serves as a bridge across boundaries, as a spearhead of international understanding.

As an example, we quote the Pugwash meetings, where scientists from the East and West met and tried to clarify some of the divergences regarding political questions that are connected with science and technology. These meetings have contributed to a few steps that were taken towards peace, such as the stopping of bomb tests, and they prepared the ground for more rational discussions of arms control. Another example is the western European laboratory for nuclear research in Geneva—CERN—in which twelve nations collaborate successfully in running a most active center for fundamental research. They have created a working model of the United States of Europe as far as high-energy physics is concerned. It is significant that this laboratory has very close ties with the laboratories in the east European countries; CERN is also equipping and participating in experiments carried out together with Russian physicists at the new giant accelerator in Serpukhov near Moscow.

... OCCASIONALLY INADEQUATE

The influence of science on our thinking is not always favorable. There are dangers stemming from an uncritical application of a method of thinking, so incredibly successful in natural science, to problems for which this method is inadequate. The great success of the quantitative approach in the exploration of nature may well lead to an overstressing of this method to other problems. A remark by M. Fierz in Zürich is incisive: He said that science illuminates part of our experience with such glaring intensity that the rest remains in even deeper darkness. The part in darkness has to do with the irrational and the affective in human behavior, the realm of the emotional, the instinctive world. There are aspects of human experience to which the methods of natural science are not applicable. Seen within the framework of that science, these phenomena exhibit a degree of instability, a multidimensionality for which our present scientific thinking is inadequate and, if applied, may become dangerously misleading.

DEEP INVOLVEMENT, DEEP CONCERN

The foregoing should have served to illustrate the multilateral character of science in its relation to society. The numerous and widely differing aspects of relevance emphasize the central position of science in our civilization. Here we find a real privilege of being a scientist. He is in the midst of things; his work is deeply involved in what happens in our time. This is why it is also his privilege to be deeply concerned with the involvement of science in the events of the day.

In most instances he cannot avoid being drawn in one form or another into the decision-making process regarding the applications of science, be it on the military or on the industrial scene. He may have to help, to advise or to protest, whatever the case may be. There are different ways in which the scientist will get involved in public affairs; he may address himself to the public when he feels that science has been misused or falsely applied; he may work with his government on the manner of application of scientific results to military or social problems.

In all these activities he will be involved with controversies that are not purely scientific but political. In facing such problems and dilemmas, he will miss the sense of agreement that prevails in scientific discussions, where there is an unspoken understanding of the criteria of truth and falsehood even in the most heated controversies. Mistakes in science can easily be corrected; mistakes in public life are much harder to undo because of the highly unstable and nonlinear character of human relations.

HOW MUCH EMPHASIS?

Let us return to the different aspects of relevance in science. In times past, the emphasis has often shifted from one aspect to the other. For example at the end of the last century there was a strong overemphasis on the practical application of science in the US. Henry A. Rowland, who was the first president of the American Physical Society, fought very hard against the underemphasis of science as is seen in the following quotation from his address to the American Association for the Advancement of Science in 1883:

American science is a thing of the future, and not of the present or past; and the proper course of one in my position is to consider what must be done to create a science of physics in this country, rather than to call telegraphs, electric lights, and such conveniences by the name of science. I do not wish to underrate the value of all these things; the progress of the world depends on them, and he is to be honored who cultivates them successfully. So also the cook, who invents a new and palatable dish for the table, benefits the world to a certain degree; yet we do not signify him by the name of a chemist. And yet it is not an uncommon thing, especially in American newspapers, to have the applications of science confounded with pure science; and some obscure character who steals the ideas of some great mind of the past, and enriches himself by the application of the same to domestic uses, is often lauded above the great originator of the idea, who might have worked out hundreds of such applications, had his mind possessed the necessary element of vulgarity.

Rowland did succeed in his aim, although posthumously. He should have lived to see the US as the leading country in basic science for the last four decades. His statement—notwithstanding its forceful prose—appears to us today inordinately strong in its contempt of the applied physicists. The great success of this country in basic science derives to a large extent from the close cooperation of basic science with applied science. This close relation—often within the same person—provided tools of high quality, without which many fundamental discoveries could not have been made. There are a healthy equilibrium between basic and applied science during the last decades and thus also between the different aspects of the relevance of science.

Lately, however, the emphasis is changing again. There is a trend among the public, and also among scientists, away from basic science towards the application of science to immediate problems and technological shortcomings, revealed by the crisis of the day. Basic science is considered to be a luxury by the public; many students and researchers feel restless in pursuing science for its own sake.

PERSPECTIVE

The feeling that something should be done about the pressing social needs is very healthy. "We are in the midst of things," and scientists must face their responsibilities by using their knowledge and influence to rectify the detrimental effects of the misuse of science and technology. But we must not lose our perspective in respect to other aspects of science. We have built this great edifice of knowledge; let us not neglect it during a time of crisis. The scientist who today devotes his time to the solution of our social and environmental problems does an important job. But so does his colleague who goes on in the pursuit of basic science. We need basic science not only for the solution of practical problems but also to keep alive the spirit of this great human endeavor. If our students are no longer attracted by the sheer interest and excitement of the subject, we were delinquent in our duty as teachers. We must make this world into a decent and livable world, but we also must create values and ideas for people to live and to strive for. Arts and sciences must not be neglected in times of crisis; on the contrary, more weight should be given to the creation of aims and values. It is a great human value to study the world in which we live and to broaden the horizon of knowledge.

These are the privileges of being a scientist: We are participating in a most exhilarating enterprise right at the center of our culture. What we do is essential in shaping our physical and mental environment. We, therefore, carry a responsibility to take part in the improvement of the human lot and to be concerned about the consequences of our ideas and their applications. This burden makes our lives difficult and complicated and puts us in the midst of social and political life and strife.

But there are compensations. We are all working for a common and well defined aim: to get more insight into the workings of nature. It is a constructive endeavor, where we build upon the achievements of the past; we improve but never destroy the ideas of our predecessors.

This is why we are perhaps less prone to the feeling of aimlessness and instability that is observed in so many segments of our society. The growing insight into nature is not only a source of satisfaction for us, it also gives our lives a deeper meaning. We are a "happy breed of men" in a world of uncertainty and bewilderment.

1.3   *The Progress of Science*
Kenneth W. Ford

Thrilling insights and awesome power. These are the fruits of science. In this chapter, we shall examine the nature and progress of man's most successful enterprise, scientific inquiry, and hazard some guesses about its future.

To science we owe most of our comforts, our leisure, our health and longevity, our ability to mold the environment, to communicate instantly, and to move swiftly over the earth. To science we also owe our ability to wipe out populations with devices of mass destruction and to numb populations with devices of mass communication. What are the features that set science apart from other human activity? What are the reasons for its remarkable growth? What does the future hold for scientific progress? We can give at best partial answers to these questions, but one thing seems clear: Future scientific understanding can be no better legislated and controlled than can any other creative activity of man.

What might be called the defining feature of science is its quantitative character. By "quantitative" we mean not merely numerical, but mathematical in a general sense—subject to rules of logic and order, and reproducible. A heckler could find some examples of nonquantitative science and of quantitative nonscience, but in the main it is this quantitative aspect that distinguishes science from other avenues of search for understanding. Because of this feature, science gives to man the power to predict and to control, not merely the power to describe.

Paradoxically, it is the modest goals of science that are most responsible for its mushrooming growth. It is commonplace to refer to the vast scope and generality of natural laws, which are indeed magnificent and impressive. Yet it has to be borne in mind—and the scientist is perhaps particularly conscious of this fact—that of all the important questions man has posed to himself in recorded history, science has so far provided answers only to the easiest ones. Science encompasses a limited range of human experience, yet with no other range of experience has human endeavor been so successful.

Besides the self-imposed limitation to "easy" questions, another key to understanding the progress of science is the generality of application that springs from an economically few fundamentals. This "amplification factor" (or "leverage" as it is called in the business world) could as well be called a quality of nature as a property of science. Man has found himself able not only to describe nature quantitatively, but to do so with relatively few basic ideas and relationships. From each of the few important theories and laws in physics flows a wealth of application. The blossoming of a whole area of technology in a few years' time can spring from the discovery of a single fundamental fact. What is just one significant advance in

SOURCE. From *Basic Physics* (Waltham, Mass.: Ginn-Blaisdell, 1968) Chapter 28.

science may appear as an incredible rapid-fire series of advances because of the latent power of each forward step.

Has the growth of science actually been "explosive"? It is commonly said to be so. In the long view of human history it certainly seems explosive. In this century both the accumulation of scientific facts and the growth of technology merit the adjective "explosive." But the pace of scientific progress can be overemphasized. It is important to distinguish between merely *rapid* growth and *accelerating* growth. The expansion of the world's population is an example of accelerating growth. Every decade the increase is greater than in the previous decade. There is indeed a population "explosion." Without question, the most evident aspects of science—its catalogue of facts and its application for practical goals—have also shown accelerated growth. So have the number of scientific workers and the amount of money spent on science. Nevertheless, there is little evidence in recent history for an accelerated growth of fundamental understanding of nature. It is more accurate to think of science as having undergone a metamorphosis in the seventeenth century which has led to its continual steady (and rapid) growth since then, rather than to think of an ever accelerating pace of progress.

The true landmarks of scientific progress are the occasional revolutions in our view of the workings of nature and the enlargements of our horizons in the physical world. Such key points of progress in the twentieth century have been the new view of space, time, and gravitation initiated by the theory of relativity; the elucidation of atomic structure through the theory of quantum mechanics; the discovery of the subatomic world of transitory particles; the elevation of principles of invariance to a primary position among the laws of nature; and the discovery of the molecular structure of the basic units of living matter. This last advance, representing the union of physical science and biological science at the submicroscopic level, may prove to be as important as any in this century. Looking back to the nineteenth century,

one finds that revolutions of scientific thought occurred nearly as frequently as in the present century. In the nineteenth century the theory of electromagnetism revealed the nature of light and the unity of electricity and magnetism; the kinetic theory of matter, besides explaining the nature of heat and temperature, revealed the turmoil of random molecular motion underlying our apparently solid material world; the theory of thermodynamics brought scientific precision to the concepts of order and disorder and provided a basis for understanding why we experience only a one-way flow of time; the analysis of all matter into a small number of elements and the orderly arrangement of these elements into the periodic table brought a new simplicity to the view of the physical world; the universal scope of the law of conservation of energy brought hope (and some overoptimism) for the construction of a single overriding theory of natural phenomena; the theory of natural selection revolutionized man's view of human history; the discovery of laws of genetics showed the existence of quantitative simplicity in the living world as well as in the inanimate world.

It is, of course, impossible to measure exactly the relative importance of different advances in science or to arrive at any unique list of "truly fundamental" advances. The purpose of the brief catalogue of scientific progress above is to emphasize two facts about modern science. First, the technological marvels of the present day and the ever greater human resources poured into scientific research should not be confused with true progress in understanding the design of nature. The list is notably deficient in inventions (radio) or technical feats (space flight) or in reference to mere accumulation of data. Second, the rate of generation of fundamental ideas in science has shown no marked trend of acceleration, at least over the past 150 years. Progress in both the nineteenth and the twentieth centuries has been rich, but scarcely any richer at the present time than 100 years ago. It seems that any important new idea requires some time to be thoroughly assimilated and appreciated before it can serve as the basis of further advance. Perhaps the instant communication and high-speed travel of the present era serve to stimulate scientific progress through cross-fertilization of ideas. However, it is clear—and it is important to keep in mind—that the pace of fundamental scientific progress is by no means comparable either to the rate of technological development or to the input of human effort.

Can history teach us the best way to push forward the frontiers of understanding in science? Probably very little. In discussing the future of science, scientists agree only that the future is unpredictable. Past progress has revealed no single scientific method. Idealized versions of scientific induction or of the sequence of experimental and theoretical steps in the evolution of a theory have borne little resemblance to actual progress so far. But two of the aspects of past scientific progress stand out as features likely to persist. (1) Theory and experiment have both been necessary. On the one hand, the mere accumulation of data has been barren without the insight provided by hypothesis and theory. On the other hand, flights of fancy untempered by experiment have not been fruitful. (2) All of the great advances have been but small forward steps in the overall view, resting heavily on what has gone before. In discussing an important advance in science, there is a natural human tendency to emphasize its novelty or the creative genius which it reflects. The less heralded developments which preceded and made possible the giant stride tend to be forgotten.

The modern world abounds in scientific cranks, men who ignore one or the other of these paramount aspects of science, seeking radical new departures in science without reference to experiment, or without reference to the past stream of scientific thought. Notwithstanding the oft-repeated charge by the cranks that the "high priests of science" have closed minds, most scientists try to keep an open mind about the possible direction from which future progress might come. Anyone with an idea in science can get a hearing. However, past history makes it seem likely that the fruitful ideas will be generated by men who build upon their own deep comprehension of past achievement, not by men who attempt the great leap from no firm foundation.

It is easy to predict one near-certainty about the future, that scientific progress will continue. This sounds like an obvious statement, yet there have been times in the past when both mathematicians and physicists thought that their fields were drying up. In contemporary science, complacency about the state of our understanding is completely absent. Every practicing scientist is painfully aware of his areas of ignorance at the frontiers of science. Although a future unrolling of scientific progress accompanying ever deeper understanding of nature is easy enough to foresee, it is more fascinating to consider how future science might differ from past science. Is science getting more difficult to comprehend, and may progress therefore be slowed or stopped? Will the scope of science increase to make it encompass a larger fraction of man's experience?

Fundamental science is undoubtedly becoming more difficult. The new concepts are further removed from everyday experience and are less easy to visualize. The world of man's senses changes very little. The world of the fundamental theories of nature has been changing rapidly. Albert Einstein and Leopold Infeld expressed this trend in these words: "The simpler and more fundamental our assumptions become, the more intricate is our mathematical tool of reasoning; the way from theory to observation becomes longer, more subtle, and more complicated."[1] A warning is sounded for the future. Will the progress of science be slowed by the ever longer route which men must follow from the world of his senses to the elementary world of nature's design? So far, we humans have shown ourselves adept at learning to think in terms foreign to our direct sense experience. We believe in the annihilation

[1] *The Evolution of Physics* (New York: Simon and Schuster, 1950).

and creation of matter, the wave nature of particles, the relativity of time, the quantum of energy, and the curvature of space-time. How far this adaptability will carry us no one can predict.

About the future *scope* of science there can be little doubt. The range of experience encompassed by science has increased gradually, and a further enlargement of what comprises science is almost certain. Although the expansion of biology and the incorporation into science of parts of psychology will have the greatest impact on man, growth has occurred and will continue within physical science itself. A few centuries ago, the question, Why does the sun give out light? had no answer outside of religion. A hundred years ago, the question, Why is energy conserved? belonged to philosophy. Both questions are now answered within the framework of physics. Other scientific-sounding questions are easy to pose which in fact lie outside of science today but may be incorporated in the future: Is the universe boundless? What was going on fifty billion years ago? What is the true nature of time? How did the matter in the universe come into existence?

In the realm of biology, consider disease, which has moved from anger of the gods to a very scientific matter of germs and viruses. What the future undoubtedly holds is scientific understanding of some patterns of human behavior. The past problems raised by science have been mainly technological—industrial automation, and implements of war. The future problems are likely to be more directly concerned with man's behavior and are sure to be even more vexing. It is probably wiser to prepare for an expanding scope of scientific understanding—which means the power to predict and control—than to cry out for the bliss of past ignorance. Fortunately the human being is a mechanism of fantastic complexity. It is unlikely that science will ever impinge on the pleasurable uncertainties that characterize some human relations.

Upon the principles of physics rests the whole of science—biological as well as physical—whose growth provides ever deepening understanding and therefore ever broadening responsibility, and ever increasing opportunity for betterment of the human condition.

1.4   *Physics and Reality*

Albert Einstein

GENERAL CONSIDERATION CONCERNING THE METHOD OF SCIENCE

It has often been said, and certainly not without justification, that the man of science is a poor philosopher. Why then should it not be the right thing for the physicist to let the philosopher do the philosophizing? Such might indeed be the right thing at a time when the physicist believes he has at his disposal a rigid system of fundamental concepts and fundamental laws which are so well established that waves of doubt can not reach them; but, it can not be right at a time when the very foundations of physics itself have

become problematic as they are now. At a time like the present, when experience forces us to seek a newer and more solid foundation, the physicist cannot simply surrender to the philosopher the critical contemplation of the theoretical foundations; for, he himself knows best, and feels more surely where the shoe pinches. In looking for a new foundation, he must try to make clear in his own mind just how far the concepts which he uses are justified, and are necessities.

The whole of science is nothing more than a refinement of every day thinking. It is for this reason that the critical thinking of the

SOURCE. From *Out of My Later Years* (New York: Philosophical Library, 1950).

physicist cannot possibly be restricted to the examination of the concepts of his own specific field. He cannot proceed without considering critically a much more difficult problem, the problem of analyzing the nature of everyday thinking.

On the stage of our subconscious mind appear in colorful succession sense experiences, memory pictures of them, representations and feelings. In contrast to psychology, physics treats directly only of sense experiences and of the "understanding" of their connection. But even the concept of the "real external world" of everyday thinking rests exclusively on sense impressions.

Now we must first remark that the differentiation between sense impressions and representations is not possible; or, at least it is not possible with absolute certainty. With the discussion of this problem, which affects also the notion of reality, we will not concern ourselves but we shall take the existence of sense experiences as given, that is to say as psychic experiences of special kind.

I believe that the first step in the setting of a "real external world" is the formation of the concept of bodily objects and of bodily objects of various kinds. Out of the multitude of our sense experiences we take, mentally and arbitrarily, certain repeatedly occurring complexes of sense impression (partly in conjunction with sense impressions which are interpreted as signs for sense experiences of others), and we attribute to them a meaning—the meaning of the bodily object. Considered logically this concept is not identical with the totality of sense impressions referred to; but it is an arbitrary creation of the human (or animal) mind. On the other hand, the concept owes its meaning and its justification exclusively to the totality of the sense impressions which we associate with it.

The second step is to be found in the fact that, in our thinking (which determines our expectation), we attribute to this concept of the bodily object a significance, which is to a high degree independent of the sense impression which originally gives rise to it. This is what we mean when we attribute to the bodily object "a real existence." The jus-

tification of such a setting rests exclusively on the fact that, by means of such concepts and mental relations between them, we are able to orient ourselves in the labyrinth of sense impressions. These notions and relations, although free statements of our thoughts, appear to us as stronger and more unalterable than the individual sense experience itself, the character of which as anything other than the result of an illusion or hallucination is never completely guaranteed. On the other hand, these concepts and relations, and indeed the setting of real objects and, generally speaking, the existence of "the real world," have justification only in so far as they are connected with sense impressions between which they form a mental connection.

The very fact that the totality of our sense experiences is such that by means of thinking (operations with concepts, and the creation and use of definite functional relations between them, and the coordination of sense experiences to these concepts) it can be put in order, this fact is one which leaves us in awe, but which we shall never understand. One may say "the eternal mystery of the world is its comprehensibility." It is one of the great realizations of Immanuel Kant that the setting up of a real external world would be senseless without this comprehensibility.

In speaking here concerning "comprehensibility," the expression is used in its most modest sense. It implies: the production of some sort of order among sense impressions, this order being produced by the creation of general concepts, relations between these concepts, and by relations between the concepts and sense experience, these relations being determined in any possible manner. It is in this sense that the world of our sense experiences is comprehensible. The fact that it is comprehensible is a miracle.

In my opinion, nothing can be said concerning the manner in which the concepts are to be made and connected, and how we are to coordinate them to the experiences. In guiding us in the creation of such an order of sense experiences, success in the result is alone the determining factor. All that is necessary is *the statement* of a set of rules, since without such rules the acquisition of knowledge in the desired sense would be impossible. One may compare these rules with the rules of a game in which, while the rules themselves are arbitrary, it is their rigidity alone which makes the game possible. However, the fixation will never be final. It will have validity only for a special field of application (i.e., there are no final categories in the sense of Kant).

The connection of the elementary concepts of every day thinking with complexes of sense experiences can only be comprehended intuitively and it is unadaptable to scientifically logical fixation. The totality of these connections—none of which is expressible in notional terms—is the only thing which differentiates the great building which is science from a logical but empty scheme of concepts. By means of these connections, the purely notional theorems of science become statements about complexes of sense experiences.

We shall call "primary concepts" such concepts as are directly and intuitively connected with typical complexes of sense experiences. All other notions are—from the physical point of view—possessed of meaning, only in so far as they are connected, by theorems, with the primary notions. These theorems are partially definitions of the concepts (and of the statements derived logically from them) and partially theorems not derivable from the definitions, which express at least indirect relations between the "primary concepts," and in this way between sense experiences. Theorems of the latter kind are "statements about reality" or laws of nature, i.e. theorems which have to show their usefulness when applied to sense experiences comprehended by primary concepts. The questions as to which of the theorems shall be considered as definitions and which as natural laws will depend largely upon the chosen representation. It really becomes absolutely necessary to make this differentiation only when one examines the degree to which the whole system of concepts considered is not empty from the physical point of view.

STRATIFICATION OF THE SCIENTIFIC SYSTEM

The aim of science is, on the one hand, a comprehension, as *complete* as possible, of the connection between the sense experiences in their totality, and, on the other hand, the accomplishment of this aim *by the use of a minimum of primary concepts and relations.* (Seeking, as far as possible, logical unity in the world picture, i.e. paucity in logical elements.)

Science concerns the totality of the primary concepts, i.e. concepts directly connected with sense experiences, and theorems connecting them. In its first stage of development, science does not contain anything else. Our everyday thinking is satisfied on the whole with this level. Such a state of affairs cannot, however, satisfy a spirit which is really scientifically minded; because, the totality of concepts and relations obtained in this manner is utterly lacking in logical unity. In order to supplement this deficiency, one invents a system poorer in concepts and relations, a system retaining the primary concepts and relations of the "first layer" as logically derived concepts and relations. This new "secondary system" pays for its higher logical unity by having, as its own elementary concepts (concepts of the second layer), only those which are no longer directly connected with complexes of sense experiences. Further striving for logical unity brings us to a tertiary system, still poorer in concepts and relations, for the deduction of the concepts and relations of the secondary (and so indirectly of the primary) layer. Thus the story goes on until we have arrived at a system of the greatest conceivable unity, and of the greatest poverty of concepts of the logical foundations, which are still compatible with the observation made by our senses. We do not know whether or not this ambition will ever result in a definite system. If one is asked for his opinion, he is inclined to answer no. While wrestling with the problems, however, one will never give up the hope that this greatest of all aims can really be attained to a very high degree.

An adherent to the theory of abstraction or induction might call our layers "degrees of abstraction;" but, I do not consider it justifiable to veil the logical independence of the concept from the sense experiences. The relation is not analogous to that of soup to beef but rather of wardrobe number to overcoat.

The layers are furthermore not clearly separated. It is not even absolutely clear which concepts belong to the primary layer. As a matter of fact, we are dealing with freely formed concepts, which, with a certainty sufficient for practical use, are intuitively connected with complexes of sense experiences in such a manner that, in any given case of experience, there is no uncertainty as to the applicability or non-applicability of the statement. The essential thing is the aim to represent the multitude of concepts and theorems, close to experience, as theorems, logically deduced and belonging to a basis, as narrow as possible, of fundamental concepts and fundamental relations which themselves can be chosen freely (axioms). The liberty of choice, however, is of a special kind; it is not in any way similar to the liberty of a writer of fiction. Rather, it is similar to that of a man engaged in solving a well designed word puzzle. He may, it is true, propose any word as the solution; but, there is only *one* word which really solves the puzzle in all its forms. It is an outcome of faith that nature—as she is perceptible to our five senses—takes the character of such a well formulated puzzle. The successes reaped up to now by science do, it is true, give a certain encouragement for this faith.

The multitude of layers discussed above corresponds to the several stages of progress which have resulted from the struggle for unity in the course of development. As regards the final aim, intermediary layers are only of temporary nature. They must eventually disappear as irrelevant. We have to deal, however, with the science of today, in which these strata represent problematic partial successes which support one another but which also threaten one another, because today's systems of concepts contain deep seated incongruities which we shall meet later on.

PART ONE

THE CLASSICAL ERA

CHAPTER 2

MOTION AND ENERGY

2.1

A Theory that the Earth Moves Around the Sun

Nicholas Copernicus

THAT THE UNIVERSE IS SPHERICAL

First of all we assert that the universe is spherical; partly because this form, being a complete whole, needing no joints, is the most perfect of all; partly because it constitutes the most spacious form, which is thus best suited to contain and retain all things; or also because all discrete parts of the world, I mean the sun, the moon and the planets, appear as spheres; or because all things tend to assume the spherical shape, a fact which appears in a drop of water and in other fluid bodies when they seek of their own accord to limit themselves. Therefore no one will doubt that this form is natural for the heavenly bodies.

THAT THE EARTH IS LIKEWISE SPHERICAL

That the earth is likewise spherical is beyond doubt, because it presses from all sides to its center. Although a perfect sphere is not immediately recognized because of the great height of the mountains and the depression of the valleys, yet this in no wise invalidates the general spherical form of the earth. This becomes clear in the following manner: To people who travel from any place to the North, the north pole of the daily revolution rises gradually, while the south pole sinks a like amount. Most of the stars in the neighborhood of the Great Bear appear not to set, and in the South some stars appear no longer to rise. Thus Italy does not see Canopus, which is visible to the Egyptians. And Italy sees the outermost star of the River, which is unknown to us of a colder zone. On the other hand, to people who travel toward the South, these stars rise higher in the heavens, while those stars which are higher to us become lower. Therefore,

SOURCE. From *Concerning the Revolutions of the Heavenly Bodies,* 1543.

it is plain that the earth is included between the poles and is spherical. Let us add that the inhabitants of the East do not see the solar and lunar eclipses that occur in the evening, and people who live in the West do not see eclipses that occur in the morning, while those living in between see the former later, and the latter earlier.

That even the water has the same shape is observed on ships, in that the land which can not be seen from the ship can be spied from the tip of the mast. And, conversely, when a light is put on the tip of the mast, it appears to observers on land gradually to drop as the ship recedes until the light disappears, seeming to sink in the water. It is clear that the water, too, in accordance with its fluid nature, is drawn downwards, just as is the earth, and its level at the shore is no higher than its convexity allows. The land therefore projects everywhere only as far above the ocean as the land accidentally happens to be higher.

.   .   .   .   .

WHETHER THE EARTH HAS A CIRCULAR MOTION, AND CONCERNING THE LOCATION OF THE EARTH

Since it has already been proved that the earth has the shape of a sphere, I insist that we must investigate whether from its form can be deduced a motion, and what place the earth occupies in the universe. Without this knowledge no certain computation can be made for the phenomena occurring in the heavens. To be sure, the great majority of writers agree that the earth is at rest in the center of the universe, so that they consider it unbelievable and even ridiculous to suppose the contrary. Yet, when one weighs the matter carefully, he will see that this question is not yet disposed of, and for that reason is by no means to be considered unimportant. Every change of position which is observed is due either to the motion of the observed object or of the observer, or to motions, naturally in different directions, of both; for when the observed object and the observer move in the same manner and in the same direction, then no motion is observed. Now the earth is

the place from which we observe the revolution of the heavens and where it is displayed to our eyes. Therefore, if the earth should possess any motion, the latter would be noticeable in everything that is situated outside of it, but in the opposite direction, just as if everything were traveling past the earth. And of this nature is, above all, the daily revolution. For this motion seems to embrace the whole world, in fact, everything that is outside of the earth, with the single exception of the earth itself. But if one should admit that the heavens possess none of this motion, but that the earth rotates from west to east; and if one should consider this seriously with respect to the seeming rising and setting of the sun, of the moon and the stars; then one would find that it is actually true. Since the heavens which contain and retain all things are the common home of all things, it is not at once comprehensible why a motion is not rather ascribed to the thing contained than to the containing, to the located rather than to the locating. This opinion was actually held by the Pythagoreans Heraklid and Ekphantus and the Syracusean Nicetas (as told by Cicero), in that they assumed the earth to be rotating in the center of the universe. They were indeed of the opinion that the stars set due to the intervening of the earth, and rose due to its receding.

.   .   .   .   .

REFUTATION OF THE ARGUMENTS, AND THEIR INSUFFICIENCY

It is claimed that the earth is at rest in the center of the universe and that this is undoubtedly true. But one who believes that the earth rotates will also certainly be of the opinion that this motion is natural and not violent. Whatever is in accordance with nature produces effects which are the opposite of what happens through violence. Things upon which violence or an external force is exerted must become annihilated and cannot long exist. But whatever happens in the course of nature remains in good condition and in its best arrangement. Without cause, therefore, Ptolemy feared that the earth and all earthly things if set in rotation would be dissolved by the action of nature, for the functioning of nature is something entirely different from artifice, or from that which could be contrived by the human mind. But why did he not fear the same, and indeed in much higher degree, for the universe, whose motion would have to be as much more rapid as the heavens are larger than the earth? Or have the heavens become infinite just because they have been removed from the center by the inexpressible force of the motion; while otherwise, if they were at rest, they would collapse? Certainly if this argument were true the extent of the heavens would become infinite. For the more they were driven aloft by the outward impulse of the motion, the more rapid would the motion become because of the ever increasing circle which it would have to describe in the space of 24 hours; and, conversely, if the motion increased, the immensity of the heavens would also increase. Thus velocity would augment size into infinity, and size, velocity. But according to the physical law that the infinite can neither be traversed, nor can it for any reason have motion, the heavens would, however, of necessity be at rest.

But it is said that outside of the heavens there is no body, nor place, nor empty space, in fact, that nothing at all exists, and that, therefore, there is no space in which the heavens could expand; then it is really strange that something could be enclosed by nothing. If, however, the heavens were infinite and were bounded only by their inner concavity, then we have, perhaps, even better confirmation that there is nothing outside of the heavens, because everything, whatever its size, is within them; but then the heavens would remain motionless. The most important argument, on which depends the proof of the finiteness of the universe, is motion. Now, whether the world is finite or infinite, we will leave to the quarrels of the natural philosophers; for us remains the certainty that the earth, contained between poles, is bounded by a spherical surface. Why should we hesitate to grant it a motion, natural and corresponding to its form; rather than assume that the whole world, whose boundary is not known and cannot be known, moves? And why are we not willing to acknowledge that the *appearance* of a daily revolution belongs to the heavens, its *actuality* to the earth? The relation is similar to that of which Virgil's *Aeneas* says: "We sail out of the harbor, and the countries and cities recede." For when a ship is sailing along quietly, everything which is outside of it will appear to those on board to have a motion corresponding to the movement of the ship, and the voyagers are of the erroneous opinion that they with all that they have with them are at rest. This can without doubt also apply to the motion of the earth, and it may appear as if the whole universe were revolving.

.   .   .   .   .

CONCERNING THE CENTER OF THE UNIVERSE

. . . Since nothing stands in the way of the movability of the earth, I believe we must now investigate whether it also has several motions, so that it can be considered one of the planets. That it is not the center of all the revolutions is proved by the irregular motions of the planets, and their varying distances from the earth, which cannot be explained as concentric circles with the earth at the center. Therefore, since there are several central points, no one will without cause be uncertain whether the center of the universe is the center of gravity of the earth or some other central point. I, at least, am of the opinion that gravity is nothing else than a natural force planted by the divine providence of the Master of the World into its parts, by means of which they, assuming a spherical shape, form a unity and a whole. And it is to be assumed that the impulse is also inherent in the sun and the moon and the other planets, and that by the operation of this force they remain in the spherical shape in which they appear; while they, nevertheless, complete their revolutions in diverse ways. If then the earth, too, possesses other motions besides that around its center, then they must be of such a character as to become apparent in many ways and in appropriate manners; and among such possible effects we recognize the yearly revolution.

Kepler's Concept of Gravity
Edward Rosen

Pre-Copernican thought about gravity had in the main been dominated by Aristotle's view that "the earth and the universe happen to have the same center; a heavy body moves also toward the center of the earth, but it does so only incidentally, because the earth has its center at the center of the universe."

When Copernicus moved the earth away from the center of the universe, he had to abandon Aristotle's conception that heavy bodies fell toward the center of the universe, because that center was no longer identical with the center of the earth. To emphasize his departure from Aristotle's theory of gravity, Copernicus declared:

"For my part, I think that gravity is nothing but a certain natural striving with which parts have been endowed ... so that by assembling in the form of a sphere they may join together in their unity and wholeness. This tendency may be believed to be present also in the sun, the moon, and the other bright planets, so that it makes them keep that roundness which they display."

The constituent parts of each planet gathered together to form one spherical body, according to Copernicus, but none of those spherical bodies affected any other. In this respect Kepler's thinking went far beyond Copernicus'. In the New Astronomy Kepler defined gravity as a "mutual corporeal tendency of kindred bodies to unite or join together." Whereas in Copernicus' conception parts of a whole united, in Kepler's theory kindred bodies unite. And for Kepler, "the

moon is a body akin to the earth." Hence, if the moon and the earth were not restrained "each in its own orbit, the earth would move up toward the moon ... and the moon would come down toward the earth ... and they would be joined together."

In the New Astronomy Kepler restricted mutual gravitational attraction to kindred bodies. But in his Note 66 on the Dream he did not repeat that restriction in his definition of gravity. How much did his removal of that restriction contribute to the universalization of gravitational attraction? By pushing aside the kindred bodies of the Introduction to his New Astronomy, and defining gravity more generally "as a force of mutual attraction," ... Kepler took a long stride in the direction of Newton's universal gravitation. What was still needed was the concept that all physical bodies are akin; in other words, Aristotle's distinction between heaven and earth had to be swept away completely before Newtonian gravitation could be proclaimed. This proclamation may be regarded as a recognition of what was implied in Copernicus' classification of the earth as a planet. In the long process of eliciting that implication, Kepler played a notable part, particularly by his more general definition of gravity ... and by his statement ... on the Dream that "the causes of the ocean tides seem to be the bodies of the sun and moon attracting the ocean waters by a certain force similar to magnetism." But in his universe the stars remain outside the planetary realm of mutual gravitational attractions, so that Kepler's

SOURCE. From Kepler's Somniun, translated with a commentary by Edward Rosen (Madison, Wisc.: University of Wisconsin Press, 1967) Appendix H; copyright by the Regents of the University of Wisconsin.

concept of gravity is not universal.

Nor did the concept of gravity enter into Kepler's explanation of the motion of the planets. Instead, as he wrote early in July, 1600:

In the body of the sun (whether it is at rest or itself also in motion), there is present a force which spreads out from the sun, as the force's own dwelling-place, to all the planets and whirls them around the sun, slowly or swiftly, in proportion to the distance of each [planet from the sun]. To this general category there is only one exception, the moon, which revolves, not around the sun, like the other five [planets], but around the earth. And yet we cannot with good reason exempt even the moon from the moving force which is common to all the rest of the universe. Let us therefore admit that we must seek primarily in the body of the sun for the dwelling-place of the force which is the origin of the moon's monthly motion. On the other hand, since all the other planets are affected by the body around which they revolve, namely, the sun, as the source of the force that moves them, so, too, the moon will be acted upon to some extent by the body around which it revolves. But it revolves around the earth. Therefore, in the earth there is present a force which moves the moon. But, [as was said] previously, the primary origin of that force was in the sun. Therefore we have no alternative but to say that a sort of ray of moving force emerges from the sun and continues straight ahead through the body of the earth, where it nests, as it were. Through this continuation a kind of secondary force, as an offshoot of that solar force, is created and persists. It spreads out again spherically from the earth, as its new dwelling-place, in order to make the moon revolve around the earth.

In 1600, then, Kepler believed that a solar force drove all the planets around the sun. Among the planets so driven was the earth. As the solar force passed through the earth, it left behind as an offshoot a secondary force which drove the moon around the earth. This earth-force originated in the sun, not in the earth, and was not independent of the sun-force, which was unique of its kind. That unique sun-force drove the planets around. It did not attract them. Nor did the planets attract the sun. There was no mutual

attraction between sun and planets. But since Kepler conceived of gravity as a mutual attraction, gravity had nothing to do with the motion of Kepler's planets.

Although Kepler's primary sun-force and secondary earth-force had nothing whatever to do with gravity, he did profoundly modify the traditional doctrine of gravity. This had interpreted the falling of a heavy body toward the earth as a "striving" on the part of the heavy body. For this striving Kepler substituted a pull, or gravitational attraction, exerted by kindred bodies on each other. The contrast between the traditional doctrine and his own conception was expressed by Kepler in the Introduction to his *New Astronomy* as follows: "Rather than the stone seeking (*petit*) the earth, the earth attracts (*trahat*) the stone." Kepler's statement that the earth attracts a stone is no casual utterance. ... He defines gravity as a force of mutual attraction. This attraction is no mere logical consequence of his definition; on the contrary, he presents it as a physical property of the earth, the other planets, and the bodies on them.

... Kepler's attention was directed to the distance between bodies as affecting the strength of their mutual gravitational attraction. In his *New Astronomy*, however, he had been concerned with a body's size as governing the strength of its gravitational attraction: "Suppose that somewhere in the universe two stones were put near each other, yet outside the sphere of the force exerted by any third kindred body. Like two magnetic bodies, those stones would come together at an intermediate place. As each one approached the other, the distance traversed would be in proportion to the other's size." Likewise, in the case of the moon and the earth, "on the assumption that the substance of both is of one and the same density, the earth would rise toward the moon one of the fifty-four parts of the interval between them, and the moon would drop down toward the earth about fifty-three parts of the interval."

The spectacular discovery announced to the world in the *New Astronomy*—that a planet's orbit is elliptical—was first communicated to Fabricius in a private letter in which Kepler explained:

[Gravity] is exactly the same in a big body and a small body; it is divided according to the size of the bodies, and it receives the same dimensions as the bodies. Suppose that a stone of a magnitude having some perceptible ratio to the size of the earth were placed behind the earth. Let it be the case that both are exempt from all other motions. Then I say that not only will the stone move toward the earth, but also the earth will move toward the stone. They will divide the intervening space in the inverse ratio of their weights. Then C being the place where they will meet, as A [the earth] is to B [the stone] in size, so [the distance] BC will be to [the distance] CA, in exactly the same ratio as is utilized in a balance with unequal arms.

In a balance with unequal arms, two unequal weights are in equilibrium when they are suspended at such distances from the fulcrum that the product of weight times distance is equal on both sides; the smaller weight will hang farther away from the fulcrum. Similarly, in Kepler's theory of gravity, a big body will pull a small body a greater distance toward their meeting-place than the small body will pull the big body. He asserts that gravity is exactly (*numero*) the same in a big body and in a small body, qualitatively, not quantitatively. Similarly he observes the same phenomena at different times "around exactly (*numero*) the same spot."

Quantitatively, Kepler measured the force of gravitation as the product of a weight times a distance. He believed that gravitational force, like magnetic force and his solar force, was inversely proportional to the distance, diminishing directly with the simple distance, while light was attenuated with the square of the distance. How much did his speculations contribute to Newton's measurement of the gravitational attraction between any two bodies as the product of their masses divided by the square of the distance between them?

2.3   *Galileo and the Leaning Tower*

A. Sutcliffe and A. P. D. Sutcliffe

Until about the fifteenth century, scholars, with a few exceptions, had accepted without question the teachings of the ancient writers. But in the fifteenth and sixteenth centuries important discoveries were made and changes of many kinds took place. New lands, such as America, were discovered; religion went through a great upheaval with the Reformation; printing was invented; and a few scholars who had the curiosity to examine Nature often obtained surprising results.

About the year 1500 a Polish philosopher named Copernicus startled the intellectual world by his views that the sun was the center of the universe and that the earth moved round it. This view was contrary to the teach-ings of the ancients. Such a new belief was not very popular nor was it universally accepted, and most of the universities and schools continued to teach the traditional science as laid down by the ancients, and especially by Aristotle, the Greek who lived 350 years before Christ.

Galileo was born in the year 1564. As a young man he studied medicine at first, but early in his university life he decided to change to the study of mathematics. In this subject he showed so much originality that before long his approach to the subject had become quite different from the customary one of merely reading the works of Aristotle and other ancient writers and then discussing them. Galileo could satisfy his curiosity only

SOURCE. From *Stories from Science*; (Cambridge University Press, 1962) Vol. 2, Chapter 27.

by making experiments. A few men before him had also studied science by the experimental method and had met with much opposition from the scholars of their day. Galileo did not escape criticism himself, as the following traditional story relates.

In the year 1590 Galileo, who was then a young man of twenty-five and a lecturer in mathematics at the University of Pisa in Tuscany, Italy, decided to make a public experiment on the speed at which objects fall in air.

Knowledge about the force of gravity became most important after the introduction of artillery into warfare . . . , for then problems concerning the flight of cannon balls in air had to be considered. It was obvious to some scholars that there were two forces acting on the moving cannon ball. There was the force produced by the explosion which shot the ball high into the air, and the force of gravity, which, pulling earthwards, brought the ball back to the ground.

The manner in which an object falls in the air had been the subject of study for centuries. Aristotle had written that a heavy body falls to the ground, from a height, much faster than a lighter one, stating that an object whose weight is a hundred times greater than that of another will fall a hundred times faster than it. Galileo questioned this statement and determined to test its truth by actually dropping a heavy and a light ball from a great height at the same time.

He could have found no better place to try this experiment than Pisa, because in that town there is the famous "leaning tower." The building was begun in the twelfth century as the bell tower, or *campanile*, of the Cathedral and is about one hundred and eighty feet high with seven tiers or floors, and a belfry. It leans over to what seems a frightening degree, being at the top about fourteen feet out of the perpendicular. For centuries it was supposed that the tower had been purposely built that way; now it is thought that the foundations were laid on wooden piles driven into the boggy ground and that when the tower had been built to

a height of about thirty feet it began to settle on one side. Despite the slant it was then decided to finish building it up to its present height. People looking out over the seventh balcony can see straight down to the ground more than a hundred feet below them.

One day in 1590, so the traditional story relates, Galileo climbed the long spiral staircase of the tower up to the seventh gallery, carrying with him two metal balls, one weighing ten pounds and the other weighing only one pound. (Some writers give the weight of one ball as a hundred pounds and that of the other as one pound; others merely state that one ball weighed ten times as much as the other.)

He leaned over the gallery and saw the crowd which had gathered to watch this public experiment. It included members of the University of Pisa—the professors, philosophers and students. All these people knew that Galileo's belief was opposed to that which had been accepted for hundreds of years, and according to one version of the story there were many angry mutterings as the young upstart slowly mounted the tower to try to disprove the beliefs of his elders.

He balanced the two balls carefully on the overhanging edge of the parapet of the gallery and allowed both of them over to begin their fall at the same time. The crowd saw that they kept together in falling through the air and heard a single crash as both hit the ground at exactly the same time. The people were amazed for, in accordance with the long accepted belief, they had expected that the heavier ball would fall much faster than the other and thus hit the ground before it.

Some of the versions of this story give so many details that it is interesting to turn to the first recorded one, which was written in 1654. The main details given there are simply that Galileo demonstrated by repeated experiments from the height of the *campanile*—and in the presence of teachers, philosophers and the whole assembly of students—that heavy bodies falling in air all moved at the same speed; and that these experiments dismayed the philosophers.

This story of Galileo and the Leaning Tower is one of the most familiar ones in the popular histories of science but there are reasons for regarding it as fictitious.

No mention of the demonstration is given in the writings of any one living at the time when it is supposed to have taken place and not even Galileo himself ever once alluded to it in any of his many books. Yet had it really occurred it would have been such a striking event that some one person then alive would surely have mentioned it. The first account of it is found in a biography of Galileo, written by Viviani, a great admirer of his, which appeared in print sixty-four years after the reputed date of the experiment.

There are many instances in the history of science where an admirer has given his hero the credit of doing some outstanding thing which in fact was done by some other person. Viviani might have done so in this case, for it has been well established that others before Galileo had attacked Aristotle's statement that bodies fall at a speed proportional to their weight. It is also well established that a very similar experiment to the one reputedly done by Galileo was made before 1590 by a certain Simon Stevin of Bruges.

Stevin was a brilliant military engineer who became quartermaster general to the army of Holland; he is also known for his mathematical skill, for he was largely responsible for introducing the decimal system into mathematics.

Stevin was assisted in his famous experiment by his friend De Groot. Two balls of lead, one ten times heavier than the other, were dropped together out of an upstairs window on to a plank. The lighter ball did not take ten times longer in falling than the heavier one, as Aristotle and others had taught; on the contrary, the two balls hit the board underneath the window "so simultaneously that the two sounds seemed to be one and the same rap."

This experiment was done in 1587 but there seems to be no evidence that Galileo knew about it. Viviani may have heard of it and then had the idea of transferring the credit for doing it first to Galileo; and the fact that Galileo lived near such an ideal setting as the Leaning Tower of Pisa could strongly have tempted him to do so.

It appears, therefore, that Galileo was not the first even to think of this experiment. Nevertheless, if he did not actually do the experiment he most certainly taught the result that would have been obtained from it as the following passage from one of his books shows: "I can assure you that a cannon ball weighing one or two hundred pounds will not reach the ground by as much as a span ahead of a musket ball provided both are dropped from a height of two hundred cubits."

This passage may well have caused Viviani to believe that Galileo had actually dropped the two balls from the Leaning Tower, which was of approximately such a height.

2.4   *Galileo's Projectile Motion*

                  G. Holton and D. H. D. Roller

... To this point we have been solely concerned with the motion of objects as characterized by their speed; we have not given much consideration to the *direction* of motion, or to changes in direction of motion.

Turning now to the more general problem of projectile motion, we leave the relatively simple case of bodies moving in a *straight line* only and expand our methods to deal with projectiles moving along curved paths. Our

SOURCE. From *Foundations of Modern Physical Science* (Reading, Mass.: Addison-Wesley, 1958) Chapter 3.

understanding of this field will hinge largely on a far-reaching idea: the observed motion of a projectile may be thought of as the result of two *separate* motions, combined and occurring *simultaneously*; one component of motion is in a horizontal direction and without acceleration, whereas the other is in a vertical direction and has a constant acceleration downward in accordance with the laws of free fall. Furthermore, these two components do not interfere with each other; each component may be studied as if the other were not present. Thus the whole motion of the projectile at every moment is simply the result of the two individual actions.

This principle of the independency of the horizontal and vertical components of projectile motion was set forth by Galileo in his *Dialogue on the great world systems* (1632). Although in this work he was principally concerned with astronomy, Galileo already knew that terrestrial mechanics offered the clue to a better understanding of planetary motions. Like the *Two new sciences*, this earlier work is cast in the form of a discussion among the same three characters, and also uses the Socratic method of the Platonic dialogues. Indeed, the portion of interest to us here begins with Salviati reiterating one of Socrates' most famous phrases, as he tells the Aristotelian Simplicio that he, Simplicio, knows far more about mechanics than he is aware:[1]

Salviati: ... Yet I am so good a midwife of minds that I will make you confess the same whether you will or no. But Sagredus stands very quiet, and yet, if I mistake not, I saw him make some move as if to speak.

Sagredo: I had intended to speak a fleeting something; but my curiosity aroused by your promising that you would force Simplicius to uncover the knowledge which he conceals from us has made me depose all other thoughts. Therefore I pray you to make good your vaunt.

Salviati: Provided that Simplicius consents to

reply to what I shall ask him, I will not fail to do it.

Simplicio: I will answer what I know, assured that I shall not be much put to it, for, of those things which I hold to be false, I think nothing can be known, since Science concerns truths, not falsehoods.

Salviati: I do not desire that you should say that you know anything, save that which you most assuredly know. Therefore, tell me; if you had here a flat surface as polished as a mirror and of a substance as hard as steel that was not horizontal but somewhat inclining, and you put upon it a perfectly spherical ball, say, of bronze, what do you think it would do when released? Do you not believe (as for my part I do) that it would lie still?

Simplicio: If the surface were inclining?

Salviati: Yes, as I have already stated.

Simplicio: I cannot conceive how it should lie still. I am confident that it would move towards the declivity with much propenseness.

Salviati: Take good heed what you say, Simplicius, for I am confident that it would lie still in whatever place you should lay it.

Simplicio: So long as you make use of such suppositions, Salviatus, I shall cease to wonder if you conclude most absurd conclusions.

Salviati: Are you assured, then, that it would freely move towards the declivity?

Simplicio: Who doubts it?

Salviati: And this you verily believe, not because I told you so (for I endeavored to persuade you to think the contrary), but of yourself, and upon your natural judgment?

Simplicio: Now I see your game; you did not say this really believing it, but to try me, and to wrest words out of my mouth with which to condemn me.

Salviati: You are right. And how long and with velocity would that ball move? But take notice that I gave as the example a ball exactly round, and a plane exquisitely polished, so that all external and accidental impediments might be taken away. Also I would have you remove all obstructions caused by the air's resistance and any other causal obstacles, if any other there can be.

Simplicio: I understand your meaning very well and answer that the ball would continue to move

[1]These extracts from Galileo's *Dialogue on the great world systems* are taken from the translation of T. Salisbury, edited and corrected by Giorgio de Santillana (Chicago, Ill.: University of Chicago Press, 1953), reprinted by permission of the University of Chicago Press.

in infinitum if the inclination of the plane should last so long, accelerating continually. Such is the nature of ponderous bodies that they acquire strength in going, and, the greater the declivity, the greater the velocity will be.

Simplicio is next led to express his belief that if he observed the ball rolling *up* the inclined plane he would know that it had been pushed or thrown, since it is moving contrary to its natural tendencies. Then Salviati turns to the intermediate case:

Salviati: It seems, then, that hitherto you have well explained to me the accidents of a body on two different planes. Now tell me, what would befall the same body upon a surface that had neither acclivity nor declivity?

Simplicio: Here you must give me a little time to consider my answer. There being no declivity, there can be no natural inclination to motion; and there being no acclivity, there can be no resistance to being moved. There would then arise an indifference between propulsion and resistance; therefore, I think it ought naturally stand still. But I had forgot myself; it was not long ago that Sagredus gave me to understand that it would do so.

Salviati: So I think, provided one did lay it down gently; but, if it had an impetus directing it towards any part, what would follow?

Simplicio: That it should move towards that part.

Salviati: But with what kind of motion? Continually accelerated, as in declining planes; or successively retarded, as in those ascending?

Simplicio: I cannot tell how to discover any cause of acceleration or retardation, there being no declivity or acclivity.

Salviati: Well, if there be no cause of retardation, even less should there be any cause of rest. How long therefore would you have the body move?

Simplicio: As long as that surface, neither inclined nor declined, shall last.

Salviati: Therefore if such a space were interminate, the motion upon it would likewise have no termination, that is, would be perpetual.

Simplicio: I think so, if the body is of a durable matter.

Salviati: That has been already supposed when it was said that all external and accidental impediments were removed, and the brittleness of the body in this case is one of those accidental impediments. Tell me now, what do you think is the cause that that same ball moves spontaneously upon the inclining plane, and does not, except with violence, upon the plane sloping upwards?

Simplicio: Because the tendency of heavy bodies is to move towards the center of the Earth and only by violence upwards towards the circumference. [This is the kernel of the Scholastic viewpoint on falling bodies. Salviati does not refute it, but turns it to Galileo's purposes.]

Salviati: Therefore a surface which should be neither declining nor ascending ought in all its parts to be equally distant from the center. But is there any such surface in the world?

Simplicio: There is no want of it, such is our terrestrial globe, for example, if it were not rough and mountainous. But you have that of the water, at such time as it is calm and still.

Here is the genesis of one of the fundamental principles of the new mechanics: if all "accidental" interferences with an object's motion are removed, the motion will endure. The "accidents" are eliminated in this thought experiment by: (1) proposing the use of a perfectly round, perfectly hard ball on a perfectly smooth surface, and (2) by imagining the surface to be a globe whose surface is everywhere equidistant from the center of the earth, so that the ball's "natural tendency" to go downward is balanced by the upward thrust of the surface.

Note carefully the drastic change from the Scholastic view: instead of asking "What makes the ball move?" Galileo asks "What might change its motion?"

Having turned the conversation to smooth water, Galileo brings in the motion of a stone dropping from the mast of a moving ship. Since the stone is moving horizontally with the ship before it is dropped, it should continue to move horizontally while it falls.

Sagredo: If it be true that the impetus with which the ship moves remains indelibly impressed in the stone after it is let fall from the mast; and if it be further true that this motion brings no impediment or retardment to the motion directly downwards natural to the stone, then there ought to ensue an effect of a very wonderful nature. Suppose a ship stands still, and the time of the falling of a stone from the mast's round top to the deck is two beats of the pulse. Then afterwards have the ship under sail and let the same stone depart from the same place. According to what has been premised, it shall still take up the time of two pulses in its fall, in which time the ship will have gone, say, twenty yards. The true motion of the stone then will be a transverse line [i.e., a curved line in the vertical plane], considerably longer than the first straight and perpendicular line, the height of the mast, and yet nevertheless the stone will have passed it in the same time. Increase the ship's velocity as much as you will, the falling stone shall describe its transverse lines still longer and longer and yet shall pass them all in those selfsame two pulses. In this same fashion, if a cannon were leveled on the top of a tower, and fired point-blank, that is, horizontally, and whether the charge were small or large with the ball falling sometimes a thousand yards distant, sometimes four thousand, sometimes ten, etc., all these shots shall come to ground in times equal to each other. And every one equal to the time that the ball would take to pass from the mouth of the piece to the ground, if, without other impulse, it falls simply downwards in a perpendicular line. Now it seems a very admirable thing that, in the same short time of its falling perpendicularly down to the ground from the height of, say, a hundred yards, equal balls, fired violently out of the piece, should be able to pass four hundred, a thousand, even ten thousand yards. All the balls in all the shots made horizontally remain in the air an equal time.

Salviati: The consideration is very elegant for its novelty and, if the effect be true, very admirable. Of its truth I make no question, and, were it not for the accidental impediment of the air, I verily believe that, if at the time of the ball's going out of the piece another were let fall from the same height directly downwards, they would both come to the ground at the same instant, though one should have traveled ten thousand yards in its range, and another only a hundred, presupposing the surface of the Earth to be level. As for the impediment which might come from the air, it would consist in retarding the extreme swift motion of the shot.

2.5

The Crime of Galileo
William Bixby

Galileo's early experiments on the motion of falling bodies had disproved Aristotle; his later astronomical discoveries were interpreted as casting doubt on the Bible. Angry churchmen pointed to the contradiction between Galileo's belief that the earth moved about the sun and the biblical assertion that Joshua had commanded the sun to stand still. Stubbornly, these dogmatists argued that if the sun was fixed at the center of the universe—as Copernicus and Galileo insisted—then how could Joshua have observed the sun to be moving? Their interpretation of the Bible was literal and unyielding.

In Pisa and in Florence, the rumor spread that Galileo was a heretic who questioned biblical truth. The sarcasm with which he replied to his critics did nothing to win new allies, and the ranks of his enemies grew. Soon university officials at Pisa ordered that no theory or discovery of Galileo's was ever to be mentioned or taught—to which Galileo responded once again contemptuously. He declared that the ignorance of his enemies had been the best master he had ever had, since their blindness had forced him to make many experiments to demonstrate the validity of his discoveries.

Galileo was waging a fight over the right of a scientist to teach and defend his beliefs. Thus, even though he recognized the undercurrent rising against him, he refused to suppress his discoveries or temper his statements. He had proof, he insisted, that Aristotle's concept of a changeless heaven was false. Further, he said he felt no compunctions about modifying or revising outmoded theories. Nature, he said, "in order to aid our understanding of her great works, has given us two thousand more years of observations, and sight twenty times as acute as that which she gave Aristotle." This

SOURCE. From *The Universe of Galileo and Newton* (New York: American Heritage Publishing Co., 1964).

explanation, despite its logic, would not satisfy his enemies, and Galileo knew it. He was always aware that his work would provoke criticism, even condemnation. In a letter to a friend, following an important discovery, Galileo expressed certainty that the Aristotelians would immediately "put forth some grand effort to maintain the immutability of the heavens."

Just what this "grand effort" would be, Galileo did not of course know. He received a hint of it early in 1613, after a friend of his, a monk named Benedetto Castelli, had dined with members of the Tuscan court. Mealtime conversation that evening had focused on Galileo's discovery of the moons of Jupiter. One of the guests, a distinguished professor of physics, told the Grand Duchess Christina that while Galileo's discoveries might be true, the conclusion drawn from them that the earth moved about the sun was contrary to the Bible.

To Castelli, who overheard the scholar's remark, hiding behind biblical authority to denounce a scientific theory seemed contrary to the rules of academic fair play. The monk felt obliged to defend Galileo. Apparently he made a strong impression on the Grand Duchess, who was alarmed at the widening breach between science and religion.

After the dinner, Castelli wrote immediately to Galileo to inform him of the evening's discussion. Both men knew that if university professors were to convince the Church that Galileo's beliefs were heretical, his career would be ruined.

Until this incident, Galileo did not allow his religious faith and his study of science to interfere with each other. Though a devout Catholic, he was able to separate the two spheres of his life. To him each step out of the darkness of ignorance into the light of truth was further cause for belief in divine creation and control. But after Castelli's warning, Galileo recognized the need to set forth his views on the relationship between science and religion. On December 13, 1613, he wrote to Castelli saying:

... I think it would be the better part of wisdom not to allow anyone to apply passages of Scripture in such a way as to force them to support, as true, conclusions concerning nature the contrary of which may afterward be revealed by the evidence of our senses or by necessary demonstration. Who will set bounds to man's understanding? Who can assure us that everything that can be known in the world is known already?

Castelli allowed the letter to circulate. When its contents became known there was a noticeable easing of tensions. Soon, even the Grand Duchess Christina's doubts had been removed. For all the surface calmness, however, the seeds of opposition were finding fertile soil deep within the Church. And nowhere was the opposition so actively at work as in Florence.

The first overt clerical attack took place in December, 1614. Standing in the pulpit of the Dominican Church of Santa Maria Novella in Florence, Father Thomas Caccini delivered a sermon that denounced mathematics as inconsistent with the Bible and detrimental to the State. And then, to drive home his venomed attack, the angry monk declaimed this biblical passage:

Then spake Joshua to the Lord in the day when the Lord delivered up the Amorites before the children of Israel, and he said in the sight of Israel, Sun, stand thou still upon Gibeon. . . . So the sun stood still in the midst of heaven. . . .

Word of Caccini's sermon, which was an indirect censure of Galileo, spread quickly through Florence. Many people were shocked, and Galileo's friends grew fearful and uneasy. It had begun to seem that Galileo's beliefs really did contradict the Bible. Not so, said the scientist, and then explained his reasoning:

... I do not think it necessary to believe that the same God who gave us our senses, our speech, our intellect, would have us put aside the use of these, to teach us instead such things as with their help we could find out for ourselves, particularly in the case of these sciences, of which there is not the smallest mention in the Scriptures. . . .

Many intelligent churchmen sided with Galileo, and a high dignitary in the Dominican order, Father Maraffi, even apologized to

Galileo for Caccini's outburst. But the damage had been done, and Florence was in an uproar. In Rome, however, many of the strictest clerics seemed unconcerned—and were still reading Galileo's works with interest. A number believed, in fact, that his writings were destined to be recognized as established truths.

Then Father Caccini unleashed the ultimate weapon. He went to Rome and asked to be received in secret by the Inquisition. He told the tribunal that Galileo was a scoffer at saints and miracles, even at God Himself. He declared that Galileo corresponded not only with suspected heretics but actually with German Protestants—an obvious slur against Kepler. Caccini was a master at mixing innuendo with derision, and in the judges of the Inquisition he had a ready audience. When he had finished attesting to the godlessness of Galileo, the Inquisitors were aroused. It was not long before another rumor was being whispered among Florentines—to the effect that an undercover investigation of Galileo had begun.

So vigorous was the reaction in Florence that Galileo decided to go to Rome to explain his position and outline his beliefs to the heads of the Church. It would not have been in character for him either to capitulate to the pressures placed on him or to compromise. In June, 1615, six months before leaving for Rome, Galileo expanded the explanatory letter he had written to Castelli and sent it as a formal memorandum to the Grand Duchess Christina.

In this paper he set down his position as a scientist and as a true Catholic. He tried to emphasize that the Holy Spirit intended that the Bible should teach man how to go to heaven, not how the heavens go. He hoped that such an argument, when passed on to Rome, would help his cause.

A Roman prelate, Robert Cardinal Bellarmine, who served as a member of the Holy Office (the Inquisition), summed up the view of the Church in the matter of how the heavens went in this extract from a letter written to a Carmelite monk Paolo Antonio Foscarini:

. . . the words "The sun also ariseth, and the sun goeth down, and hasteth to his place where he arose" were written by Solomon, who not only spoke by divine inspiration, but was a man wise above all others and learned in the human sciences and in the knowledge of all created things, which wisdom he had from God. . . .

Implied strongly in the text of this statement was Bellarmine's assertion that anyone who believed the theories of Copernicus was questioning the wisdom of Solomon and hence the Bible. Galileo was obviously in this category.

Although Bellarmine had been present when Galileo had been denounced by Caccini, the Cardinal's evaluation of Galileo's argument was not reported to the Inquisition. Bellarmine knew that the scientist was being investigated, but at this point he did not press for a ruling against him. Nor was he particularly eager to have Galileo come to Rome to answer his critics. Probably Bellarmine knew that the Inquisition did not yet have a solid case against Galileo. He must also have realized that despite frequent illnesses, Galileo was a persuasive and dynamic man whose presence in Rome might score heavily against Church opposition.

But in December, 1615, Galileo did go to Rome, and soon his efforts to win clerical approval seemed to be meeting success. Even Caccini, the man who had maligned him, was told to apologize. Galileo's eloquence won him many admirers. As one witness to his forceful debating wrote to a certain Cardinal d'Este in Florence:

Your Reverence would be delighted with Galileo if you heard him holding forth, as he often does, in the midst of fifteen or twenty, all violently attacking him, sometimes in one house and sometimes in another. But he is armed after such fashion that he laughs all of them to scorn. . . .

Certainly Galileo was well armed against verbal attack, but against the forces that were already conniving against him, there was no defense. At last he was summoned to appear before Cardinal Bellarmine. To Galileo's surprise the Cardinal told him that the Copernican theory had been determined to be "in error." Therefore, Bellarmine advised, Galileo should abandon the idea that the sun was the center around which the earth and the other planets revolved.

The meeting apparently took place in February, 1616, although the only official record of it is a note in the Inquisition files. The note, which was unsigned and lacked a notary's seal, states that after the Cardinal had spoken, the Commissary General of the Inquisition stepped forward, and Galileo was "commanded and enjoined, in the name of His Holiness the Pope and the whole Congregation of the Holy Office, to relinquish altogether the said opinion . . . nor further to hold, teach, or defend it in any way whatsoever, verbally or in writing; otherwise proceedings would be taken against him by the Holy Office; which injunction the said Galileo acquiesced in and promised to obey."

This injunction by the Commissary General has been a source of unending puzzlement to historians. There is no evidence that Galileo "acquiesced" to anything—or that he knew he was forbidden to "hold, teach, or defend" what he felt certain was true. Some historians suggest that key phrases of the injunction may have been added later, which would explain why the document did not appear to have been authenticated.

Galileo's subsequent actions raise doubt that he ever realized the seriousness of his situation. He never imagined that he was liable to immediate arrest if he ever showed "in any way whatsoever" his Copernican leanings. Even so, he left Bellarmine's palace shaken and somewhat depressed. He had lost the battle to convince his Church that truth— even scientific truth—need not be considered contrary to belief in God.

By this time his problems were no longer theoretical, but actual. Rome was blackened by rumors that Galileo had at last been brought down, that he had been forced to recant. Hearing this whispered gossip, Galileo requested a statement from Cardinal Bellarmine describing what had occurred at their meeting. The Cardinal responded willingly. He certified that Galileo had not been forced to recant or to do penance for his erroneous beliefs but had been advised that the Copernican theory was contrary to the Bible and therefore was not to be held or defended. Apparently Bellarmine was unaware of the intervention of the Commissary General—or perhaps pretened to be.

Armed with the Cardinal's certificate, Galileo hoped once more to suppress his enemies, and for a time it appeared that he had done so. He lingered four more months in Rome, and before his departure he was received by Pope Paul V. The Pope, and many high churchmen, bore Galileo no ill will. How could they? for many had not read his books thoroughly nor thought out his logical and upsetting conclusions. Despite the pardon from the Church, Galileo still feared persecution. He expressed this concern in an audience with Paul V and was assured that neither the Pope nor the Holy Office would give heed to rumours spread against him.

Outwardly, at any rate, it seemed that Galileo had triumphed once again. Despite the rap administered by the Inquisition, the Church appeared to have forgiven him—and to have granted him permission to continue on his way as long as he said nothing that would collide with accepted doctrines. Though this may seem a suffocating kind of freedom, Galileo was comforted by it. It offered a way for him to avert persecution, and it gave him the impetus to make plans to go home. A Roman churchman wrote the Grand Duke of Tuscany that Galileo would leave Rome with "the best reputation" and that none believed the rumors of his enemies. Yet another witness to Galileo's activities in Rome, the Tuscan ambassador, did not feel so positive of the scientist's future. He wrote also to the Grand Duke, saying: "Galileo seems disposed to take on the monks and to contend with personages whom you cannot attack without ruin to yourself. It may any moment be heard in Florence that he has stumbled into some bad precipices . . ." His friends in Florence became alarmed. The secretary of state to the Grand Duke wrote to Galileo, hoping to speed his return:

You have had enough of monkish persecutions and ought to know by this time what the flavor of them is. His Highness [the Grand Duke] fears that your longer stay in Rome may involve you in fresh difficulties and would therefore be glad if ... you would not tease the sleeping dog any more and would return here as soon as possible. There are rumors flying about which we do not like, and the monks are all powerful ...

Coming from a trusted friend, this counsel would have constituted words of caution, but as written by the Grand Duke's deputy, it was an order. On June 30, 1616, Galileo left Rome. He carried with him the signed statement of Cardinal Bellarmine, but the germ of his eventual downfall lay in that unsigned note buried in the Inquisition files.

Galileo had not directed his energies solely toward debate during the time he was in Rome. His questing mind had grappled with an increasing number of scientific problems. He had rarely been at rest. A problem solved always led to another one awaiting solution. Ideas passed through his inquiring mind to which he had to give voice. He wrote of the tides, and he devised a means for determining longitude. But his health suffered—it had never been very good— and finally broke down. For years following his return to Florence he was bedridden. Still he continued to work and write, and to correspond. But he lived in virtual retirement at the villa of Bellosguardo near Florence, and from there, for eight years, he maintained an almost unbroken silence.

In 1621 Pope Paul V died. The next pope, Gregory XV, lived only two years. Then to Galileo's delight, Maffeo Cardinal Barberini, a friend and admirer, was elected to the papacy, choosing the name Urban VIII. Galileo was determined now to go back to Rome. He would press for the withdrawal of the 1616 prohibition against his publicly championing the Copernican theory.

In planning a return to his former battleground, Galileo disregarded the advice of his longtime friend Giovan Francesco Sagredo, whose name Galileo eventually used in his *Dialogue*. Urging him to be content with what he already had, Sagredo wrote: "Philoso-phize comfortably in your bed, and leave the stars alone. Let fools be fools, let the ignorant plume themselves on their ignorance. Why should you court martyrdom for the sake of winning from their folly?"

For Galileo, no less of a wrangler than before, felt compelled to fight. In 1624, he traveled to Rome to talk with the Pope. To his dismay he found that as a cardinal Barberini had been able to hold private views favorable to Galileo and his scientific endeavors, but as pope he must steer a different course. Urban VIII welcomed his old friend from Florence, feted him, assured him of his warmest regard, and even conferred a pension on the learned scientist. But he would not withdraw the admonition. Instead he suggested that Galileo put his literary skills at the service of the authorities and enumerate all the pros and cons relative to new scientific theories. One stipulation was that Galileo should conclude his dissertation by stressing that the problem had not been solved, that God might have solved it by means beyond human consideration, and that Galileo must not constrain divine omnipotence within the limits of his theories. The Pope even dictated this conclusion for the proposed work, but Galileo would not agree to undertake the project.

For years Galileo had dreamed of giving the Copernican theory to the world, of explaining it so persuasively that no one could fail to see its truth. He had been admonished by the Church to which he remained loyal and, in spiritual matters, subservient. But he could not stop searching for a way to satisfy both his love of scientific truth and his love for his Church.

Finally, a way had become apparent. During the years of his semiretirement in Florence, he had worked on a book called *The Assayer*. It dealt with the nature of comets but was in reality a reply to attacks made on his physical discoveries. His views in it were erroneous, for he believed that comets, like rainbows, were atmospheric peculiarities that reflected sunlight. However, the significance of *The Assayer*, as far as Galileo's immediate future was concerned, was in its skilful writing rather than in its science. For the book managed to discomfit his enemies without violating Church doctrine; and though it stung the opposition, the book received the applause of the Pope himself, who had been cool to Galileo's pursuit of scientific discovery.

Galileo's method, which proved a most successful tactic, was to present his own theories simply as plausible ideas without stressing his firm belief that they were true. And as he deftly skirted the question of Copernicanism, he could not be accused of lending support to theories that were in conflict with the dogma of the Church.

The Assayer was printed in Rome in October, 1623, and its success had been one of the reasons Galileo had ventured to Rome to plead before the new pope. Its success also encouraged Galileo to turn to a long-cherished project, *Dialogue Concerning the Great World Systems*, after his return from Rome. He was sixty when he began it, and for five years thereafter, despite recurrent illness, he poured his best reasoning and most skilful writing into the volume. It is the book that introduces the three characters Salviati (champion of Galilean physics), Simplicio (an Aristotelian dogmatist), and Sagredo, the object of their persuasive speeches.

Galileo finished the manuscript at the end of 1629. By that time several changes had taken place in the Church that made Galileo hope he could obtain speedy permission to publish. His old friend Benedetto Castelli had been made mathematician to the Pope. Another friend, Niccolo Riccardi, had been named chief censor to the Vatican and would pass on the submitted manuscript. In addition, word had reached Galileo of Pope Urban's statement that if he had been pope in 1616, the prohibition against Galileo would not have been made. Had the Pope changed his mind? Galileo could not suppress his optimism.

With great expectations Galileo set out once more for Rome. Castelli, Riccardi, and Pope Urban each received him warmly. His manuscript was read, and after minor alterations had been made, it was returned to him

with the censor's general approval. Galileo had also written a preface and an epilogue designed to forestall all possible criticism and even incorporating conclusions dictated by the Pope. But at this point Riccardi began to realize that the text might not be so harmless as he had imagined and that by allowing it to be published he was becoming a kind of accomplice.

Galileo was allowed to return to Florence with the text, but the chief censor withheld the preface and the conclusion, thus preventing the book's publication. With the exercise of some pressure by Galileo, and with the help of the Tuscan ambassador to Rome, Galileo was given permission to submit the text to ecclesiastical censors in Florence and to have it approved there. With further prodding from the Tuscan ambassador, Galileo managed to retrieve the preface and the conclusion from Rome. So at last, after more than a year's delay, the book was published in February, 1632.

With its circulation throughout Europe, a virtual thunder of applause reached the old scientist. Congratulations poured in from every important center of learning—from kings as well as scientists, and from laymen who had read the book and had seen for the first time the truth contained in Galileo's new science. But the applause was not unanimous. One of Galileo's most embittered enemies, Father Christopher Scheiner, a Jesuit astronomer from Ingolstadt, Germany, wrote in protest to an influential friend, "In these dialogues the author has made null all my mathematical researches. . . . I am preparing to defend myself and the truth." But his preparations were, for a time, kept secret.

In Rome, Father Riccardi read the *Dialogue* and could not contain his growing uneasiness. Reportedly he said, "The Jesuits will persecute this with the utmost bitterness."

Father Riccardi prophesied correctly. For though the *Dialogue* appeared an impartial and noncommittal discussion, it presented an inescapable argument for the Copernican system and thus dealt a death blow to traditional theory. Father Scheiner and his colleagues succeeded in proving to the censors that they had been duped. The censors were persuaded that by allowing the *Dialogue* to be published, they had inadvertently allowed irreparable damage to be inflicted on the established system of teaching and on the authorities who backed it.

Shortly after its publication, orders came from Rome to stop printing the book. And in another order, the printer was commanded to remit to the Vatican all copies of the book remaining in stock. The printer wrote that he could not comply with the order as his stock was completely sold out.

Friends of Galileo pleaded with Pope Urban to intercede, but they were quickly informed that the Pope had grown furious with the scientist. Urban had come to believe that the Simplicio of the *Dialogue*, the Aristotelian, was a caricature of himself. Angrily he described Galileo as a man "who did not fear to make game of me." The scientist's friends did not doubt that the papal wrath would find an outlet.

On September 23, 1632, a message from Rome arrived at the residence of the Florentine Inquisitor. It directed him to "inform Galileo in the name of the Holy Office that he is to appear as soon as possible in the course of the month of October, at Rome, before the Commissary General of the Holy Office...."

Remembering all too clearly the fate of other men at the hands of the Inquisition, Galileo delayed as long as possible. Doctors attested to his advancing age and poor health. Friends and officials sought in vain to have the order withdrawn. In despair Galileo wrote to Francesco Cardinal Barberini, who was Urban's nephew:

This vexes me so much that it makes me curse the time devoted to these studies in which I strove and hoped to deviate somewhat from the beaten path generally pursued by learned men. I not only repent having given the world a portion of my writings, but feel inclined to suppress those still in hand, and to give them all to the flames. ...

But all protestations failed. On January 20, 1633, the ailing scientist wearily set out to face the Inquisition. Once in Rome he waited several months for the tribunal to request his appearance. During this time he was kept in close confinement and interrogated now and then, but he was not mistreated. Surely there was no basis for either punishment or torture; despite the conviction of the Jesuits and other clerics that Galileo was a heretic, proof for this conclusion was not forthcoming.

His *Dialogue* had received the censor's approval; all the requested corrections had been made; the book had been granted the seal of approval required for the publication of any book in a Catholic country. Thus there seemed to be no basis upon which to substantiate charges against him. But one day someone riffling through the Inquisition files discovered the unsigned note setting forth the injustice of 1616. In it was the statement that Galileo had agreed no longer to "hold, teach, or defend it [the Copernican theory] in any way whatsoever, verbally or in writing. ..."

Galileo was sure that the words "in any way whatsoever, verbally or in writing" had not appeared in the original admonition of Cardinal Bellarmine or in the prelate's later explanatory note. He was certain too that though he had agreed not to hold or defend the controversial theory—which the *Dialogue* gracefully avoided—he had not been asked to refrain from teaching it.

Had his memory failed him? Was this an error, a forgery? Galileo would never know, for with these words as evidence, the Inquisition now had the means to deliver the final blow against him. On June 21, 1633, Galileo was convicted of disregarding the prohibition of 1616. The following day he knelt before his inquisitors and read slowly in a halting voice from a prepared apology:

I, Galileo ... kneeling before you, most Eminent and Reverend Lord Cardinals Inquisitors-General against heretical depravity throughout the whole Christian Republic, having before my eyes and touching with my hands the Holy Gospels, swear that I have always believed, do believe, and by God's help will in the future believe, all that is held, preached, and taught by the Holy Catholic and Apostolic Church. But whereas—after an injunction had been judicially intimated to me by the Holy Office to the effect that I must altogether abandon the false opinion that the sun is the center of the world and immovable, and that the earth is not the center of the world, and moves, and that I must not hold, defend, or teach in any way whatsoever, verbally or in writing, the said false doctrine. . . .[1]

His voice droned on and on as he enumerated the sins he had committed and of which the tribunal had found him guilty. Upon completing his recantation—the true purpose of the trial—Galileo was not imprisoned, as many people believe. He was allowed to return to Florence and remained there under house arrest. His movements were restricted, and he was continually watched. The father of modern science had been humbled at last by his enemies—but not wholly subdued.

He was bedridden much of the time now, felled by bouts with asthma and painful ruptures, but he still fought back. In 1638, he completed the *Discourses*, a work which paved the way for those who later developed the science of mechanics. Not only did the treatise deal with theories of motion in falling bodies, but it discussed light waves, combustion, and the vibration of instruments to produce musical sound. The volume was printed in Protestant Holland, safely outside the jurisdiction of the Church of Rome. The *Discourses* as dictated to the Duke of Noailles, a French nobleman whose major virtue, from Galileo's point of view, was his distance from the Italian disputes which had so vexed Galileo.

Shortly before the *Discourses* was completed, inflammations in both eyes had caused Galileo to become blind. Nearly helpless now, he was cared for by his daughter and his friends; even the Grand Duke came to cheer him with conversation and a bottle of his favorite wine. But his spirit and intellect did not fail, and so he was able to continue his work almost to his last days by dictating his thoughts and theories to two of his loyal disciples. Finally a slow fever overcame him. In 1642, when he was nearly seventy-eight, he died. Still the Roman Church would not relax its judgment, and Galileo was buried in an unmarked grave.

His adherents mourned his death, certain that his years of work and wrangling had perished with him, that accepted Aristotelian and Ptolemaic doctrines would continue to be taught and believed. They could not have foreseen that the year Galileo died in Italy, Isaac Newton would be born in England. Whereas Galileo had discovered how things moved, Newton would discover why. And in so doing, he would advance Galileo's theories and refine them to the point of ultimate acceptance.

As Newton gained recognition for his discoveries, he would in effect be building a permanent monument to research and scientific discovery—and of course to his great predecessor, Galileo.

2.6   *Galileo's Sentence*[2]

The Inquisition

Whereas you, Galileo, son of the late Vincenzio Galilei, of Florence, aged 70 years, were denounced in 1615, to this Holy Office, for holding as true a false doctrine taught by many, namely, that the sun is immovable in the center of the world, and that the earth moves, and also with a diurnal motion; also, for having pupils whom you instructed in the same opinions; also, for maintaining a correspondence on the same with some German mathematicians; also for publishing certain letters on the sunspots, in which you

[1]The complete text is given in the following section. [Ed.]
[2]Pronounced in 1633.

developed the same doctrine as true; also for answering the objections which were continually produced from the Holy Scriptures, by glozing the said Scriptures according to your own meaning; and whereas thereupon was produced the copy of a writing, in form of a letter, professedly written by you to a person formerly your pupil, in which, following the hypothesis of Copernicus, you include several propositions contrary to the true sense and authority of the Holy Scriptures; therefore (this Holy Tribunal being desirous of providing against the disorder and mischief which were thence proceeding and increasing to the detriment of the Holy Faith) by the desire of his Holiness and of the Most Eminent Lords, Cardinals of this supreme and universal Inquisition, the two propositions of the stability of the sun, and the motion of the earth, were qualified by the Theological Qualifiers as follows:

1. The proposition that the sun is in the center of the world and immovable from its place is absurd, philosophically false, and formally heretical; because it is expressly contrary to the Holy Scriptures.

2. The proposition that the earth is not the center of the world, nor immovable, but that it moves, and also with a diurnal action, is also absurd, philosophically false, and, theologically considered, at least erroneous in faith.

But whereas, being pleased at that time to deal mildly with you, it was decreed in the Holy Congregation, held before his Holiness on the twenty-fifth day of February, 1616, that his Eminence the Lord Cardinal Bellarmine should enjoin you to give up altogether the said false doctrine; and if you should refuse, that you should be ordered by the Commissary of the Holy Office to relinquish it, not to teach it to others, nor to defend it; and in default of acquiescence, that you should be imprisoned; and whereas in execution of this decree, on the following day, at the Palace, in the presence of his Eminence the said Lord Cardinal Bellarmine, after you had been mildly admonished by the said Lord Cardinal, you were commanded by the Commissary of the Holy Office, before a notary and witnesses, to relinquish altogether the said false opinion, and, in future, neither to defend nor teach it in any manner, neither verbally nor in writing, and upon your promising obedience you were dismissed.

And, in order that so pernicious a doctrine might be altogether rooted out, not insinuate itself further to the heavy detriment of the Catholic truth, a decree emanated from the Holy Congregation of the Index prohibiting the books which treat of this doctrine, declaring it false, and altogether contrary to the Holy and Divine Scripture.

And whereas a book has since appeared published at Florence last year, the title of which showed that you were the author, which title is *The Dialogue of Galileo Galilei*, on the two principal Systems of the World—the Ptolemaic and Copernican; and whereas the Holy Congregation has heard that, in consequence of printing the said book, the false opinion of the earth's motion and stability of the sun is daily gaining ground, the said book has been taken into careful consideration, and in it has been detected a glaring violation of the said order, which had been intimated to you; inasmuch as in this book you have defended the said opinion, already, and in your presence, condemned; although, in the same book, you labor with many circumlocutions to induce the belief that it is left undecided and merely probable; which is equally a very grave error, since an opinion can in no way be probable which has been already declared and finally determined contrary to the Divine Scripture.

Therefore, by Our order, you have been cited to this Holy Office, where, on your examination upon oath, you have acknowledged the said book as written and printed by you. You also confessed that you began to write the said book ten or twelve years ago, after the order aforesaid had been given. Also, that you had demanded licence to publish it, without signifying to those who granted you this permission that you had been commanded not to hold, defend, or teach, the said doctrine in any manner. You also confessed that the reader might think the arguments adduced on the false side to be so worked as more effectually to compel conviction that to be easily refutable, alleging, in excuse, that you had thus run into an error, foreign (as you say) to your intention, from writing in the form of a dialogue, and in consequence of the natural complacency which everyone feels with regard to his own subtleties, and in showing himself more skilful than the generality of mankind in contriving, even in favor of false propositions, ingenious and plausible arguments.

And, upon a convenient time being given you for making your defense, you produced a certificate in the handwriting of his Eminence the Lord Cardinal Bellarmine, procured, as you said, by yourself, that you might defend yourself against the calumnies of your enemies, who reported that you had adjured your opinions, and had been punished by the Holy Office; in which certificate it is declared that you had not abjured nor had been punished, but merely that the declaration made by his Holiness, and promulgated by the Holy Congregation of the Index, had been announced to you, which declares that the opinion of the motion of the earth and stability of the sun is contrary to the Holy Scriptures, and, therefore, cannot be held or defended. Wherefore, since no mention is there made of two articles of the order, to wit, the order "not to teach" and "in any manner," you argued that we ought to believe that, in the lapse of fourteen or sixteen years, they had escaped your memory, and that this was also the reason why you were silent as to the order when you sought permission to publish your book, and that this is said by you, not to excuse your error, but that it may be attributed to vain-glorious ambition rather than to malice. But this very certificate, produced on your behalf, has greatly aggravated your offence, since it is therein declared that the said opinion is contrary to the Holy Scriptures, and yet you have dared to treat of it, and to argue that it is probable. Nor is there any extenuation in the licence artfully and cunningly extorted by you, since you did not intimate the command imposed upon you. But whereas it appeared to Us that you had not disclosed the whole truth with regard to your intention, We thought it necessary to proceed to the rigorous examination of you, in which (without any prejudice to what you had confessed, and which is above detailed against you, with regard to your said intention) you answered like a good Catholic.

Therefore, having seen and maturely considered the merits of your cause, with your said confessions and excuses, and everything else which ought to be seen and considered, we have come to the underwritten final sentence against you:

Invoking, therefore, the most holy name of our Lord Jesus Christ, and of His Most Glorious Virgin Mother, Mary, We pronounce this Our final sentence, which, sitting in council and judgment with the Reverend Masters of Sacred Theology and Doctors of both Laws, Our Assessors, We put forth in this writing in regard to the matters and controversies between the Magnificent Carlo Sincereo, Doctor of both Laws, Fiscal Proctor of the Holy Office, of the one part, and you, Galileo Galilei, defendant, tried and confessed as above, of the other part, We pronounce, judge, and declare, that you, the said Galileo, by reason of these things which have been detailed in the course of this writing, and which, as above, you have confessed, have rendered yourself vehemently suspected by this Holy Office of heresy, that is of having believed and held the doctrine (which is false and contrary to the Holy and Divine Scriptures), that the sun is the center of the world, and that it does not move from east to west, and that the earth does move, and is not the centre of the world; also, that an opinion can be held and supported and probable, after it has been declared and finally decreed contrary to the Holy Scripture, and, consequently, that you have incurred all the censures and penalties enjoined and promulgated in the sacred canons and other general and particular constitutions against delinquents of this description. From which it is Our pleasure that you be absolved, provided that with a sincere heart and unfeigned faith, in Our presence, you abjure, curse, and detest, the said errors and heresies, and every other error and heresy, contrary to the Catholic and Apostolic Church of Rome, in the form now shown to you.

But that your grievous and pernicious error and transgression may not go altogether unpunished, and that you may be made more cautious in future, and may be a warning to others to abstain from delinquencies of this sort, We decree that the book *Dialogues of Galileo Galilei* be prohibited by a public edict, and We condemn you to the formal prison of this Holy Office for a period determinable at Our Pleasure; and by way of salutary penance, We order you during the next three years to recite, once a week, the seven penitential psalms, reserving to Ourselves the power of moderating, commuting, or taking off, the whole or part of the said punishment or penance.

THE FORMULA OF ABJURATION

I, Galileo Galilei, son of the late Vincenzio Galilei of Florence, aged seventy years, being brought personally to judgment, and kneeling before you, Most Eminent and Most Reverend Lords Cardinals, General Inquisitors of the Universal Christian Republic against heretical depravity, having before my eyes the Holy Gospels which I touch with my own hands, swear that I have always believed, and, with the help of God, will in future believe, every article which the Holy Catholic and Apostolic Church of Rome holds, teaches, and preaches. But because I have been enjoined, by this Holy Office, altogether to abandon the false opinion which maintains that the sun is the centre and immovable, and forbidden to hold, defend, or teach, the said false doctrine in any manner; and because, after it had been signified to me that the said doctrine is repugnant to the Holy Scripture, I have written and printed a book, in which I treat of the same condemned doctrine, and adduce reasons with great force in support of the same, without giving any solution, and therefore have been judged grievously suspected of heresy; that is to say, that I held and believed that the sun is the centre of the world and immovable, and that the earth is not the centre and movable, I am willing to remove from the minds of your Eminences, and of every Catholic Christian, his vehement suspicion rightly entertained toward me, therefore, with a sincere heart and unfeigned faith, I abjure, curse, and detest the said errors and heresies, and generally every other error and sect contrary to the said Holy Church; and I swear that I will never more in future say, or assert anything, verbally or in writing, which may give rise to a similar suspicion of me; but that if I shall know any heretic, or anyone suspected of heresy, I will denounce him to this Holy Office, or to the Inquisitor and Ordinary of the place in which I may be. I swear, moreover, and promise that I will fulfil and observe fully all the penances which have been or shall be laid on me by this Holy Office. But if it shall happen that I violate any of my said promises, oaths, and protestations (which God avert!), I subject myself to all the pains and punishments which have been decreed and promulgated by the sacred canons and other general and particular constitutions against delinquents of this description. So, may God help me, and His Holy Gospels, which I touch with my own hands, I, the above-named Galileo Galilei, have abjured, sworn, promised, and bound myself as above; and, in witness thereof, with my own hand have subscribed this present writing of my abjuration, which I have recited word for word.

At Rome, in the Convent of Minerva, June 22, 1633, I, Galileo Galilei, have abjured as above with my own hand.

2.7   *Planetary Motion*

                                     Norman Feather

In our catalogue of the basic facts of observational astronomy as known to the ancients, we [note the recognition of the fact that] except for a very few "wanderers," the stars keep their relative positions constant night after night. The "wanderers" referred to in that statement are the planets. There is no written account of the astronomical knowledge of early man in any part of the world which shows him lacking in appreciation of the peculiar characteristics of "the five stars called 'planets'." The emperor Yao, whose accession the traditional chronology of the Chinese places in the year 2317 BC, and in whose reign the length of the tropical year was officially accepted as $365\frac{1}{4}$ solar days recalled his astronomers to the sterner disciplines of earlier times and ordered them to be diligent in observing the motions of the sun and the moon, of the planets and the stars. Again, the first known mention of Babylon is in a dated tablet of the reign of Sargon of Akkad (about 3800 BC), yet records of the same reign imply an expert and detailed knowledge of the heavens.

In Egypt, the priest-astronomers of a past age achieved a bold synthesis in classing together the five planets with the sun and the moon: all are indeed wanderers against the background of the stars.

In that way, it has been said, the names of the days of the week had their origin. Saturn, Jupiter, Mars, Sun, Venus, Mercury, Moon: this was the supposed order of decreasing distance from the solid ground of Earth. If we go through this list cyclically, taking one and missing two until all the names have been used, we recognize, either in English or in French, the proper names of the seven days. In Egypt and the East this seven-day "week" was already an independent calendar unit more than three thousand years before the birth of Christ.

We have referred to the five planets of the ancient astronomers. Today, of course, it is common knowledge that there are others, but it should not be forgotten that the first of these, Uranus, was not discovered until 1781, not until William Herschel had greatly improved the optical power of the Newtonian telescope. There are only five "naked-eye" planets, and the antiquity of knowledge concerning them is certainly in large measure explained by a curious fact. Following the classification introduced by Hipparchus in the second century BC, the brightness of a star is reckoned in "magnitudes." Hipparchus placed six such magnitudes within the compass of naked-eye visiblity, and the curious fact is that in favorable circumstances each of the five planets of the ancients exceeds in brightness a standard first-magnitude star. At its brightest—for this brightness varies, as we shall see—each is a conspicuous object in the night sky. It is surely curious that, over the much wider range from the second to the sixth magnitude, inclusive, this small class of wanderers has no single representative. (Modern, more exact, definition of star magnitudes may place Uranus, at its brightest, in the sixth magnitude, and acute observers have claimed to have seen it with the naked eye, but the statement which we have made is not seriously misleading.) The upshot is the situation that we have noted: five "wandering stars" were recognized as soon as men turned intelligently to the contemplation of the heavens, beyond those five their number received no addition for five thousand years.

This part of the story ends, as we shall discover, in 1609. In 1609 the complexities

SOURCE. From *An Introduction to the Physics of Mass, Length and Time* (Edinburgh: Edinburgh Univ. Press, 1959) Chapter 7; copyright by Edinburgh University Press.

of the apparent motions of the "naked-eye" planets were finally elucidated by Kepler, and in that year, also, by a strange coincidence, Galileo (1564–1642), happening to be in Venice, heard that, in Holland, Lippershey, Metius and others had independently discovered the properties of certain combinations of two lenses, and in so doing had invented the first practical telescope. With the invention of the telescope naked-eye astronomy virtually came to an end, but Kepler's mathematical work is a fitting commentary on the unremitting labors, and a tribute to the skill, of the last and in some ways the greatest of the long succession of observers who measured the sky without optical aid, Tycho Brahe (1546–1601), a Danish nobleman. For a short time Kepler was assistant to Brahe, and without his master's observations his own attempts at systematization would have been of no avail.

It has been stated that the ancient Egyptians arranged the planets, with the sun and moon, in order of supposed distance from the earth: the moon was the nearest, then Mercury and Venus, next the sun, and beyond that Mars, Jupiter and Saturn. We have to see on what evidence this order was decided. We have described the apparent motion of the sun against the background of the stars: steadily throughout the year it appears to move from west to east through the constellations of the zodiac, completing the cycle in a little more than 365 days. The apparent motion of the moon is in many ways similar, though it is much more rapid, a period of 27.3 days sufficing for a full revolution. In general the planets, too, move from west to east in relation to the stars, but once each year there is for each a phase of retrogression, when the apparent motion is in the reverse direction. It is the nature of this retrogressive phase which provides a ready-made basis of classification—Mars, Jupiter and Saturn in the one class, and Mercury and Venus in the other. For the three first-named the rate of retrogression is greatest at the time when the planet crosses the southern meridian at midnight, for the other two it is greatest when it crosses the

meridian at noon (in the nature of things this last conclusion is based on an interpolation of observations). Let us, then, consider the question of distance in relation to these two classes separately.

First, in relation to the class of three: the overall rate of progression through the zodiac provides a suitable ground of comparison. This is least for Saturn (roughly 12 degrees of arc per annum) and greatest for Mars (some 200 degrees per annum). We recognize here the facts upon which intuition operates and builds its case: Saturn is "obviously" the most distant, Mars the nearest of the three. Then the relative variations of brightness, to which we have already referred: these are most pronounced for Mars, least pronounced for Saturn. For all three planets, however, the main rhythm is an annual one, the time of greatest brightness coinciding roughly with the time of most rapid retrogressive motion. Intuition accepts these further differences as corroborative of an earlier conclusion.

With the second class of two the situation is different, and the evidence more confusing. The overall rate of progression of Mercury and Venus through the zodiac is strictly the same: it is the sun's rate precisely: these planets appear to swing back and forth continuously, now leading now following the sun in its eastwards movement in relation to the stars. They are the evening and morning "stars": they never cross the southern meridian at midnight. If, intuitively, we seek for a periodic time, it is the period of one complete cycle with respect to the sun (rather than the stars) which we may take for comparison. For Venus this period is 584 days, for Mercury it is 116 days. Intuitively, though perhaps now no more slender grounds, we place Venus farther from us than Mercury. But we (or rather the ancient Egyptians) placed the sun in the mid-position between the two groups of planets, and we notice, with some misgiving, that Mercury appears to keep closer to the sun than Venus does. For Mercury the maximum "elongation" is about 23 degrees of arc, for Venus it is 46 degrees. But the order has been written down: in a simple order one cannot have it both ways. Moreover, if we were to consider the matter of brightness we should only be still further confused: with both Mercury and Venus, as the time of most rapid retrogressive motion is approached the brightness decreases: the maximum occurs at a point in the motion which, in terms of apparent velocity or position in relation to the sun, has no simple specification.

Finally, we have only the moon. As appropriate to the wanderer of most rapid motion, and possibly because we can make out its "features" with the naked eye, we place it closest to ourselves—then to explain its regularly varying aspects we call in the priest and the poet—for we are still, in imagination, in the land of the Pharaohs.

This, then, is the system of the ancient cosmology, but it was obvious to the early Greeks that it would not work. Plato (427–347 BC) pointed the problem of the kinematic description which would accurately "preserve the appearances of phenomena;" after him Eudoxus (c. 370 BC) made the first sustained attempt to meet the challenge. The earth is at the center, certainly, fixed and immobile, with the stars revolving in unison in the outer firmament. Taking their motions as the simplest and most perfect (or, as we should say, the most fundamental), and with the ancient cosmology as background recognizing that a single rigid rotating "crystal sphere" could be made their vehicle, Eudoxus attempted to describe the motions of the wanderers also in terms of sets of variously rotating spheres peculiar to each. Twenty-six spheres were required in all; three each for the sun and the moon, and four for each of the planets. According to Eudoxus, all these spheres were centered in the earth. For each wanderer one sphere had a motion of rotation essentially the same as that of the firmament of the stars. The others were contained within this, each rotating about an axis carried round by the sphere immediately outside it, the relative orientations of the axes, and the speeds of rotation, being suitably chosen. The wanderer—planet, sun or moon, as the case might be—was itself carried round in the equatorial plane of the innermost sphere of its own set.

We recognize in this description the earliest attempt at the systematic resolution of motions.

·   ·   ·   ·   ·

As we shall see, Eudoxus's system, and its derivatives, dominated astronomical thought for nineteen centuries, but we should note that it was not the only system to be proposed and to be seriously discussed by astronomers in the days when Greece was the center of the civilized world.

Eudoxus had adopted the ancient view that man is at the center of his universe: in the first place, therefore, we should note those other systems which differed from his in certain respects but were geocentric, nevertheless. Some authorities have believed that the Pythagoreans in the second half of the fifth century BC showed preference for a heliocentric system. If that is so, there is no clear record to establish their case. But it seems fairly certain that, in the following century, Hicetas of Syracuse, a contemporary of Eudoxus, and Ecphantus, taught that the earth rotates on its axis: at least the common motion of the stars could thus be explained. Then Heraclides of Pontus (c. 388–315 BC), a disciple of Plato's old age, not many years later, suggested that Mercury and Venus revolve round the sun (which, in the orthodox tradition, he still supposed to revolve round the earth).

These were the variants on orthodoxy; in relation to them all the system of Aristarchus of Samos (c. 310–230 BC) was certainly unorthodox, in the estimation of some it was heretical. We know little of Aristarchus's life other than that he was active in Alexandria in the period 280–264 BC. A mathematical school had been founded there, not long previously, by Euclid, and a school of astronomy, which was destined to retain its preeminence for close upon five hundred years, was already vigorous in its early fame. Aristarchus put forward the view that the sun is at the center of things and that the earth revolves round it, as the planets do. It should not be thought that this possibility was ignored as unworthy of serious consideration by other astronomers. Archimedes, as a young man, spent some time in Alexandria studying mathematics and astronomy (c. 265 BC), and on his return to Syracuse, in a small treatise addressed to the king's son, he referred to Aristarchus's system only to dismiss it as unsatisfactory. In Athens it became a subject of debate, sufficiently important, it would appear, to incur the judgment of impiety from Cleanthes the Stoic, head of that school of philosophy during the period 264–232 BC. Condemned by the greatest physicist of the day and by the most austere moralist, the heliocentric system might well have been forgotten, but such was the intellectual climate of the age that a century or so later it received favorable notice from the Babylonian astronomer Seleucus, and a new attempt was made to commend it to the philosophers. This again was of no avail: it was not accepted even by Hipparchus; thereafter science slowly hardened into dogmatism and nothing more was heard of this far-reaching notion for another sixteen hundred years.

It must be confessed that the objectors to Aristarchus's system had good reason for their doubts: once it was accepted that the earth revolved in an orbit round the sun, the common motion of the stars, the simplest and most perfect motion of all, was the more difficult to understand. For there was no experimental basis of knowledge of astronomical distances, and the idea that most stars might be a million times farther away than the sun had no place in the ancient cosmologies. It is to the further credit of Aristarchus, therefore, that, over and above his heliocentric system, he provided the first estimate of relative distances founded on a simple and unexceptionable hypothesis. In order to consider it, let us merely assume that the moon, spherical in shape as the other heavenly bodies are, shines by light reflected from the sun, then, whatever be the system of orbits and motions, when the moon is exactly at "first quarter" or "last quarter," that is with disc half-illuminated, the line

from the center of the sun to the center of the moon is at right angles to that from the center of the moon to the center of the earth. Suppose that, at that instant, the angular separation of sun and moon as seen from the earth is θ, and that R and r are their distances, respectively. Then, obviously, $r = R \cos \theta$, in our modern notation. We have, of course, assumed that light travels in straight lines, but this has been implicit in all our considerations, and we have neglected small effects due to the finite size of both the sun and the earth as seen from the moon. These imperfections in the argument need not detain us. In principle, at least, Aristarchus's method is valid: if we can measure θ we can calculate R/r. In practice the method is difficult to apply. The instant of half-illumination is not easy to determine, and great accuracy in the measurement of θ is necessary for success. The actual value of θ is about $89° 51'$ (R/r about 400); Aristarchus believed it to be about $87°$, and so concluded that the distance of the sun is some 20 times greater than that of the moon. Even this conclusion, so far from the truth, must be accounted a notable advance from speculation, a first step towards incontrovertible fact. Some forty years later Eratosthenes (276–196 BC), royal librarian at Alexandria, added another such fact to astronomical knowledge, calculating, with fair exactitude, the circumference of the earth from the lengths of the noontide shadows at Alexandria and Syene (Assuan).

But we must return to the history of our main concern. Within some thirty years of its formulation, the original system of Eudoxus had been further elaborated by Callippus of Cyzicus, and by Aristotle (384–322 BC). Here speculation entered once more, only to complicate the picture. Eudoxus had attempted merely to describe the motions of the wandering stars; Aristotle aspired to interpret them. Wishing to "explain" why the motions of the wanderers were independent one of another, he was compelled to postulate a series of "compensatory" spheres so arranged as to prevent movement being transmitted from the more distant wanderers to the less distant: in the end fifty-five spheres

were involved. It is noteworthy—even fortunate—that in this instance, at least, the
opinions of the great philosopher were not
accepted without question by those who
followed him. Quite apart from the hetero-
dox Aristarchus, whose ideas they discarded, Apollonius of Perga (born *c.* 262 BC), the
mathematician, and Hipparchus, famous
alike as mathematician and practical astronomer, restored the emphasis on description, eschewing speculation. Hipparchus, in
particular, regarded it as essential that the
astronomer should adopt the simplest hypothesis which it is possible to entertain in face
of ascertained fact. Moreover, being possessed, through his own labors (on the island
of Rhodes from 161 to 126 BC), of observations in greater variety and of greater
accuracy than any of his predecessors had
been, he was eminently well-placed to honor
this principle.

Yet the situation was not itself as simple
as it might appear. It is an ascertained fact
that the earth does not fly to pieces. The
question may be asked whether it is possible
to entertain the hypothesis of its axial rotation in face of this fact. Some of the ancients
held that it is not. Obviously, in such circumstances, the criterion of simplicity is not
entirely distinct from the criterion of plausi-
bility. For Hipparchus, at least, whatever his
reason, simplicity was not to be sought in
other than a fixed-earth system. But he replaced the spherical rotations of Eudoxus by
uniform circular motions—and he was strictly honest in his conclusions. He was satisfied
that the apparent motions of the sun and
the moon could be represented by this
method, if the centers of their circular orbits
were supposed displaced from the center of
the earth, but he was less satisfied in relation
to the five planets. This device of "eccentrics"
was certainly not adequate, nor was that of
concentric deferents and epicycles which
Apollonius had introduced entirely satisfactory, in respect of their more complicated
motions. (Apollonius had imagined the combination of two uniform circular motions,
the center of the circle of smaller radius, the
epicycle, being carried with uniform speed

round the deferent, or larger circle, which
was centered in the earth.) For the planets,
according to Hipparchus, the simplest description must involve epicycles moving
round eccentric deferents.

Another two and a half centuries passed
before this orthodox geocentric cosmology
was further systematized. Then in the period
AD 127 to 151, in Alexandria, Claudius
Ptolemy did for astronomy and trigonometry
what Euclid had done, in the same city more
than four hundred years earlier, for abstract
geometry. His *Syntaxis*, in thirteen books
(*Almagest*, its alternative designation, derives
from the fact that the work has come down
to us chiefly through its Arabic translations of
the ninth and tenth centuries), summarized
and codified the knowledge and accepted
belief in these two subjects as it stood at the
end of the classical era. And so Ptolemy's
great work remained, unquestioned, the
standard authority on each for fourteen
centuries. Modern trigonometry is indeed
securely built on the foundations which
Hipparchus and Ptolemy laid down, and if
modern cosmology bears little trace of the
geocentric system elaborated in such detail
by the successive contributions of these two
men, as a faithful description of the motions
of the wandering stars that system provided
all that was needed by the navigator (and the
astrologer) long after Columbus reached the
New World—and for a hundred years after
Copernicus at long last, had given the world
of learning another chance to reassess the
alternative hypothesis, that the motions
of the planets are sun-centered, not earth-
centered as the critics of Aristarchus had so
successfully maintained.

Nicolaus Copernicus (1473–1543) was not
a professional astronomer. He studied mathematics at the university of Cracow, canon
law at Bologna and medicine at Padua. Then,
in 1505, at the age of thirty-two, a young man
of immense learning, he returned to his
native Poland and spent the rest of his life in
ecclesiastical duties. From 1512 until his
death he was canon of the cathedral of
Frauenburg. As a student in Italy he had
sensed the excitement of the reawakening of

the spirit of critical enquiry in Europe: the doctrines of Aristotle and Ptolemy were at last subject to fresh scrutiny; his own attitude to them was plainly that of disbelief. He had become convinced that in cosmology a much simpler system could be devised, a system in which spherical bodies rotated as a consequence of their sphericity (so that the earth could not be fixed and immovable) and in which the sun, not the earth, was at the center of things. He had examined the writings of the ancients—such as were available to him—to discover whether similar views had previously been entertained. In the eighth book of the *Satyricon* of Martianus Capella, written in Carthage in the early years of the fifth century, he had found reference to the ascription of heliocentric motion to Mercury and Venus, which hypothesis we have attributed to Heraclides, though some have regarded the Egyptians as its originators. But it is reasonably certain that in his search he missed any record there might have been of Aristarchus and his fully heliocentric system: to all intents that had gone out through Babylon more than sixteen centuries earlier. So far as we know the only place where he might have found mention of it was in Archimedes's small treatise *Sandreckoner* (*Arenarius*), but, in spite of Eutocius's commentaries of the sixth century, the writings of the great man of Syracuse were still lost in general oblivion; they were not rescued for western civilization until they were printed at Basle in 1544 by Hervagius.

So Copernicus returned to Poland, and for seven years, as private physician to his uncle the bishop of Ermeland, he had leisure to mature his ideas. When he moved to Frauenburg their formulation was essentially complete. But they were not fully committed to paper for another eighteen years, and not published for thirty-one. During the intervening period of thirteen years (1530–1543), in the universities of central Europe, the new cosmology became more widely known, largely through oral transmission. At an early stage, it is true, and with papal approval, an official request for full publication had been made by Cardinal Schönberg, but this had produced no response. It remained for a young German, George Joachim (Rheticus), appointed professor of mathematics at Wittenberg in 1537, to resign his professorship two years later and devote four years of his life as Copernicus's disciple at Frauenburg, to bring the matter to issue. Ultimately, the first printed copy of *De revolutionibus orbium coelestium* came from the press only a few weeks before its author died, on May 24, 1543. The whole strange episode provides many parallels with that of Newton and Halley and the publication of the *Principia*.

.   .   .   .   .

Copernicus's system, though heliocentric, was essentially classical in structure. We have seen that he attached peculiar significance to the "perfection" of spherical form. He likewise accepted, even more explicitly than the ancients, the analogous "perfection" of circular orbits uniformly described. On this basis he could not hope to reconstruct the apparent motions of the sun, the moon and the planets without the device of eccentrics, though it was the sun not the earth which was now out of center in the planetary orbits— and he could not avoid the use of epicycles, though the planetary retrogressions were now reproduced without their aid. He had reduced the neo-Ptolemaic total of eighty independent motions to thirty-four; he had produced a system simpler than the classical system, indeed, but it was not one which carried instant conviction by its stark simplicity. That task was beyond him: if he was temperamentally unsuited to it, it is also fair to say that the observations at his disposal— for he relied essentially on the same observations as Ptolemy had used—were insufficiently detailed to sustain it. Fifty years later Tycho Brahe had made and recorded the necessary observations; not long after that Kepler had produced the cosmological system towards which the mathematical astronomers of two thousand years had been groping in vain.

As we have noted, Tycho Brahe was born three years after Copernicus's *De revolutionibus* was published. He was fortunate, as

Copernicus had been, in being able to follow academic studies in many centers of learning, Copenhagen, Leipzig, Wittenberg, Rostock and Augsburg, continuously over many years. He was reading Ptolemy in Latin at the age of fifteen, and two years later, when he should have been studying law, he was already spending his nights making observations on the heavens. When he returned to Denmark at the age of twenty-five he set up a modest observatory in his uncle's castle at Herritzvad. Before he was thirty he had secured the patronage of the king. Money was available without stint, and Brahe rapidly assembled at Uraniborg the instruments for an observatory conceived on a scale grander than anything which had been attempted since Hulagu Kahn had equipped the observatory at Maraga for the Persian astronomer Nasir-ud-din (1201–1274) three centuries earlier. For this purely "visual" observations—for it must be repeated that the telescope had yet to be invented—he had arranged to have constructed a quadrant of 19 ft. radius, and his other instruments were of commensurate size. Brahe made systematic observations in every field of naked-eye astronomy at Uraniborg for twenty-one years. He left Denmark in 1597, his former patron having died and the new king proving less well disposed towards his science and his person, and after two years moving from place to place (though still observing) he finally settled near Prague. The emperor Rudolph II had now befriended him, granting him the castle of Benatky for his residence and his observatory. There his instruments were re-assembled towards the end of 1599. But Tycho died on October 24, 1601, having achieved little in the country of his adoption—except the accession of Kepler as his assistant.

Tycho Brahe was essentially an observer, with a sound knowledge of the physics, and the chemistry, of his day. He was lucky in the time at which he lived. There can have been few occasions in history when a new star ("nova") of the brightness of that which appeared in November 1572 was available for observation, and hardly more on which a comet as spectacular as the great comet of 1577 provided unusual opportunity for measurement and speculation. Brahe seized these opportunities and profited by them: the ancient theory of the "perfection" of the heavens did not survive the conclusions which he drew from the observations which he made in these two "miraculous" years. In private conviction he was no Copernican: he believed that the earth was fixed and that the stars revolved daily in the firmament, but he accepted a heliocentric arrangement for the planets, believing them to revolve in orbits round the sun, as the sun circled the earth. Here, however, we are not concerned with his individual essay in cosmology; we need only record the fact that as a result of his patient observations he corrected the value of almost every accepted astronomical datum, leaving this vast material in perfect order for his successors. That is sufficient title to enduring fame.

The advantages of wealth and good family, which Copernicus and Brahe had enjoyed as young men, Johann Kepler did not possess. He was born of feckless and ill-educated parents (when he was nearly fifty he had to exert all his influence to secure the release of his mother, who, as an old woman, had been arrested as a witch). He survived smallpox in infancy, to be permanently maimed as a result. He had little formal education until he was placed in a seminary at the age of twelve. Growing to manhood, he for a time accepted the vocation chosen for him, and not until he was twenty-two did he finally abandon it. The intervening years, however, had shaped his future: he had obtained admission to the university of Tübingen as a foundationer, and, outside his classical and theological studies, he had become a pupil of the astronomer Maestlin. From Maestlin he had appropriated the view that astronomy was ripe for radical renovation, and it was the faint chance that he might be able himself to contribute to this end that had led him to renounce the ministry. This happened in 1594, when he was offered the professorship of astronomy in the Lutheran gymnasium at Gratz.

In 1596 Kepler published a fanciful dissertation in which he described a complicated figure wherein the five regular polyhedra, octahedron, icosahedron, dodecahedron, tetrahedron and cube, were built together in that order (from the center outwards), so scaled that the circumscribed sphere of the octahedron was the inscribed sphere of the icosahedron, the circumscribed sphere of the icosahedron was the inscribed sphere of the dodecahedron, and so on. In respect of this figure he claimed that the radii of the four "cementing" spheres and of the fifth (the circumscribed sphere of the cube) were in proportion as the mean distances of the five planets from the sun. This essay in geometrical mysticism brought him the gratifying reward of correspondence with Galileo and Tycho Brahe—and it did more for him than that, as we have already seen. In the following year Brahe started on his wanderings from Uraniborg in search of a new patron, and by the time that he had found one, and his observatory had been re-built at Benatky, Kepler also was in straitened circumstances under a new ruler. Soon after his accession in 1598 the archduke Ferdinand had issued a decree of banishment against Protestant professors, and, although Kepler was spared immediate exile as a result, he was glad to accept Brahe's offer of employment in 1599 as his personal assistant. After Brahe's death in 1601 the emperor appointed Kepler to succeed him. From that time, until the end of his life, at least he had no lack of observational material on which to work, and against which to test one speculative hypothesis after another.

For our present concern Kepler's contribution was complete by 1619, but long before then, his salary from the imperial treasury having fallen into great arrear, he had been forced into the first of a series of moves designed either to offer less gloomy prospects of financial return, or merely to escape the wars. So he transferred himself, his work and his family, from Prague to Linz, from Linz to Ulm, from Ulm to Sagan, and he died at Ratisbon (Regensburg) on November 15, 1630 appealing against the injustice of yet another move in prospect, to a professorship at Rostock, with which the duke of Friedland, his employer at the time, was preparing to "compensate" him for monies due. It is not the least of the wonders of his very considerable output of important original work that it was achieved amidst these vagaries and that it was continued until his death.

Brahe had been the first to make systematic observations on the planets throughout the whole year, and year after year, continually. He had observed the apparent motion of Mars in this way for thirteen years, that is over some seven circuits of the zodiac. Kepler was well advised (probably at the instigation of Brahe himself) to concentrate attention on these observations in his attempt to reduce them to simple formulation. If he were unable to describe the motion of Mars, concerning which the information was so nearly complete, obviously he could not hope to succeed with the other planets. In fact, it was nearly ten years before he reached his limited goal. Having postulated one shape after another for the orbit, and one law of velocity after another, at last he came upon assumptions which fitted the observations to his complete satisfaction. He tested his assumptions on the earth, and again they were satisfactory. This was confirmation indeed. He had now succeeded in explaining the apparent motions of Mars and the sun as seen from the earth on the assumption that the sun is at rest and that these two Copernican planets revolve around it in orbits having the same mathematical specification. He did not wait to confirm that the motions of the other planets could be similarly described. He assumed that that must be so— and time proved him to be right. In 1609 he published *De motibus stellae Martis*, wherein, along with much else that is quaint and much that is significant, appear the first and second laws essentially as we know them today:

1. Each planet moves in an ellipse with the sun at one focus.
2. For each planet the line from the sun to the planet sweeps out equal areas in equal times.

Kepler's third law was formulated in 1619. It appeared in *Harmonice mundi*, published at Augsburg and dedicated to James I of England. It may be stated in modern terms:

3. The squares of the periodic times of the planets are as the cubes of their mean distances from the sun.

Before he could formulate this empirical law, Kepler had to consolidate the observations of Brahe on the more distant planets Jupiter and Saturn, and those on Venus and Mercury, verifying the first and second laws for these four bodies as he went along. His final success provided the first real demonstration, intuitive rather than logical though it may be, that the system of Aristarchus and Copernicus is something more than the most convenient representation of the planetary motions: a system which is articulated in so simple a manner as the third law describes must certainly correspond with reality in some very intimate way. But the full kinematic simplicity of this law had to wait for the work of Newton to reveal it. Kepler's own presentation was still encumbered with the fantastic extravagance of his mysticism.

2.8 Newton's Principia
William Bixby

The work of Copernicus and Galileo had destroyed the whole structure of Aristotelian physics. But one of Aristotle's precepts about the physical universe remained unchallenged: the persistent notion that there was a difference between occurrences on earth and those elsewhere in the universe. Imperfect men lived on an imperfect earth, the Greek philosopher had declared; perfection was to be found in the heavens. Thus, Aristotle had reasoned, the laws governing earthly happenings could not possibly apply to the heavens.

But this theory too, which still appealed to thinkers of the seventeenth century, was soon to be exploded. In his quiet study at Cambridge, Newton was working carefully toward a scientific masterpiece that would overshadow the work of any previous scientist. His first objective, prompted by Halley, was to fashion one precise mathematical statement about gravity that would account not only for the motions of planets but for the fall of an apple from a tree and the course of a bullet from a rifle barrel.

Years before, Galileo had studied the path of a projectile. It neither fell abruptly nor soared endlessly into space; its route across the sky could be charted somewhere between the two dissimilar motions that Newton would one day study—the moon that fell constantly but never came to rest, and the apple that plunged directly earthward. Perhaps the projectile's route was a combination of both of these motions, Galileo concluded: the path of any projectile hurled upward at an angle—be it an arrow or a cannon ball—must surely be curved in the form of a parabola. He was correct, more or less. He would have been totally correct if he had seen that the problem was not a matter of a missile rising up from one point on a flat surface and then falling down toward another point on that surface; the problem actually involves a missile soaring up from a sphere (the earth) and then being drawn down toward the center of that sphere. If Galileo had charted the problem this way, he would have seen that the path was not a parabola, but part of an ellipse—like a section of the moon's orbit.

Newton began to think of a missile's trajectory in this new light. He knew that its path would initially be curved, and he perceived that it would ultimately be elliptical. If an apple could be hurled with sufficient

SOURCE. From *The Universe of Galileo and Newton* (New York: American Heritage Publishing Co., 1964).

force it might go into orbit around the earth.

Here then was the crucial relationship, the relationship between natural motion and gravity, that Newton had been seeking. But to measure the force of gravity that was constantly drawing an object toward or around the earth, he had to establish a trustworthy way of measuring the distances involved. When trying to gauge the forces acting on the falling apple, for example, should measurements be made merely from the apple's skin to the grass below, or from the core of the apple to the earth's center? Instinctively he answered that the proper figure was the distance between the center of the earth and the center of the apple. He did so on the same impulse that would prompt anyone desiring to balance a plate on one finger to place his finger at the center of the plate. Yet he was not satisfied with just an instinctive solution. Too many of these had been proved wrong. Applying his vast knowledge of mathematics, he set out to show that the combined gravitational effect on the apple of every cubic foot of earth—that beneath his feet and that on the opposite side of the earth—was exactly the same as though the entire bulk of the earth were concentrated at the center.

Completing this stupendous calculation, Newton knew that he finally had within his grasp a single statement that related heavenly motions to those on earth. And if he could express it accurately, it would overshadow any purely philosophical description of the universe, such as that of Aristotle's. Previously Newton had given astronomers useful information in relating the planets' elliptical orbits to the pull of gravity. Now he could offer to all other scientists who were concerned with force and motion on the earth a precise and irrefutable law about how gravity acts and why objects move as they do. Newton's law of universal gravitation states: "every particle in the universe is attracted to every other particle by a force that is directly related to the products of their masses and inversely related to the squares of the distances between them." Thus, for two boxes hanging two inches apart from a beam, the force pulling them towards each other would

be determined by dividing the product of their masses by the distance squared, that is, four inches.

Newton had found the formula that unraveled the mystery of the physical universe, and he knew it. Yet such was his nature that, finished with the problem and privately satisfied with its solution, he did not announce his revolutionary discovery. Instead he perfected his mathematical proof and arranged the material in a series of lectures which he delivered to his Cambridge classes. It is possible that he would have placed the text of his lectures in a file, once his classes were over, and turned his attention to other things. But he had promised to send Halley a report on the results of his work; so instead of merely storing his lectures, he dispatched copies of them to London.

At first, Halley was puzzled. He had left Newton at work on a problem in astronomy—or what might be called celestial mechanics. Yet here was a treatise relating motions on earth with celestial motions—motions everywhere, in fact. Though he did not grasp its entire significance at first, Halley had a suspicion that the paper contained a monumental discovery. He presented the paper to the Royal Society on December 10, 1684, and the minutes note the receipt of that "curious treatise." It was obvious that the Newton paper was of great importance, but it would have to receive further study.

Meanwhile Halley urged Newton to expand the paper and to prepare a book for publication. It is likely that without Halley's urging, Newton would never have undertaken to publish. Certainly he had no enthusiasm for the task; his past experience had drained that from him. But Halley's tact and good humor won him over. He set out to produce a small book that would set forth his proof of the law of gravitational force, show how it related to Kepler's laws of planetary motion, and also include some geometrical theorems that related to the subject.

He put aside other work and began to fulfill his promise to Halley. But as Newton developed his idea and began to illustrate it with examples, he realized the extent of what he must write, and the book grew larger and larger.

At his room in Cambridge, his very life was consumed by the book. Meals remained uneaten. Sleep was avoided—he often worked until three or four in the morning, only to rise again soon after dawn. Newton's housekeeper complained, for it appeared that Newton did not like her cooking. Many times all Newton ate was bread and water—just enough nourishment to keep his mental machinery functioning. He was so preoccupied with his work that he seldom left his room. And when he did, he appeared in the college halls with wig awry and stockings hanging down. He seemed completely detached from worldly problems, as though his mind were billions of miles from earth—which indeed it was.

By Easter time, 1685, Newton had finished one volume of what was to be a three-volume work. By late summer, the second volume was nearly completed. He spent a great deal of time revising and polishing his material, so that it was not until springtime, 1686, that he allowed the first volume to be sent to the Royal Society in London.

The manuscript was titled *Philosophiae Naturalis Principia Mathematica* ("Mathematical Principles of Natural Philosophy"), but has been referred to consistently as simply the *Principia*. No sooner had it been delivered to the society and been read by several members than another controversy was in the making. Newton's old antagonist Robert Hooke claimed to be the discoverer of the inverse-square law and accused Newton of including it in the *Principia* without acknowledging Hooke's priority.

The Royal Society was embarrassed, and several key members openly expressed their dismay. They knew that this kind of controversy could halt Newton's work. But Hooke remained adamant. It was true, of course, that Hooke had guessed that gravity could be measured by an inverse-square law. Other scientists, including Halley and Wren, had guessed it too, but no one until Newton could supply mathematical proof that the

law was valid. Hooke was demanding credit for a discovery he had never been able to substantiate.

Halley was determined to be the first to inform Newton of the difficulty, and he did so immediately. He knew that if the news came from another source, or if it were to become distorted, Newton would probably cease communicating with the society and that the remainder of the *Principia* might never be written. Appended to the lengthy letter that he sent to Newton was this important passage:

There is one thing more that I ought to inform you of, viz., that Mr. Hooke has some pretensions upon the invention of the rule of decreases of gravity being reciprocally as the squares of the distances from the center ... How much of this is so, you know best; as likewise what you have to do in this matter. Only Mr. Hooke seems to expect you should make some mention of him in the preface, which it is possible you may see reason to prefix.

Newton, of course, was furious. He thanked Halley for the explanatory letter and made clear his intention of ignoring Hooke's claim. "In the papers in your hands," he wrote, "there is not one proposition to which he can pretend, and so I had no proper occasion of mentioning him there...."

In the midst of this delicate exchange of letters, the society decided that Newton's book should be published. Discovering that its treasury had insufficient funds for such an undertaking, the society sought independent financing. According to an entry in the minutes of the June 2, 1686, meeting, "It was ordered that Mr. Newton's book be printed, and that Mr. Halley undertake the business of looking after it and printing it at his own charge...."

Edmund Halley was not a man of means. Though once wealthy, his family had lost its fortune, and he had to live on a relatively small income. Still, he recognized the importance of Newton's work, and he knew that it had to be published. So he accepted the responsibility—and the obligation. It was his act of generosity, freely made, that determined the outcome of the dispute between Newton and Hooke.

Halley had been the first to bring Hooke's claim to Newton's attention, but he was not the last. After Newton had heard from a number of sources that Hooke felt slighted, he wrote to Halley that he was disgusted with the entire project. He had finished two of the three volumes, and though he had begun volume three, it was still unfinished at the time volume one was going to press. Now he said he planned to withhold the third volume.

Halley was stunned. He replied to Newton as diplomatically as possible, in the hope of calming the angry genius. His tact proved equal to the task, for with his letter, Newton's resistance softened. Newton appreciated Halley's encouragement. And aware that the young astronomer was paying the costs of publication, he could see that the sale of the completed work would be less if the third volume were omitted. Once his anger had abated, Newton wrote to Halley, "I am very sensible of the great kindness of the gentlemen of your society to me, far beyond what I could ever expect or deserve, and know how to distinguish between their favor and another's humor. . . ."

Newton was feeling so conciliatory, in fact, that he even agreed to acknowledge that several other scientists, including Hooke, had deduced the inverse-square relationship between distance and gravitational force.

The *Principia* was published in the summer of 1687, and though it established Newton's fame for all times, most people who were interested in it found it impossible to fathom. The book was written in Latin, the language of scholars at that time. Also, for the benefit of scholars, the mathematical proofs were laboriously worked out in classic geometry rather than calculus. By doing this, Newton hoped that scientists would be able to understand his new ideas through the proofs. Still, few were able to shake themselves free of traditional beliefs and even glimpse the universe that Newton had accurately analyzed. For he had seen it as no man before him had been able to do.

The *Principia* did not win acceptance immediately; it took some time before men could adjust their thinking to Newton's radically new view of science. Building on Galileo's belief in thorough experimentation and on his shrewd deductions, Newton established experiment as the way of science and the only correct method of examining nature.

In the field of physics, he exposed errors that had existed—and had been compounded—for two thousand years. He forced scientists to probe deeper into nature and to discard hasty, superficial observation that so often leads to false conclusions. For example, prior to publication of the *Principia*, men believed that an object's natural state was motionlessness, or rest. In everyday observation all things seemed eventually to stop moving. Water flowed to the sea and ceased flowing; a stone rolled down a hillside and finally stopped; any object set in motion by man eventually stopped moving. Rest was seen, therefore, as an ideal state that every object sought.

By his discoveries, Newton showed that man lives in a universe in which motion is the natural state of things. His three laws of motion, set forth in the *Principia*, provide the basis of the science of mechanics. They state:

I. Every material body persists in its state of rest or uniform motion in a straight line if, and only if, it is not acted upon by an external force.

II. The net external force acting on a material object is directly and linearly proportional to, and in the same direction as, the acceleration of the object.

III. To every action there is always opposed an equal reaction.

In his first law, Newton used the word "rest" to describe the condition of an object, but for him the word had special connotations. An object lying on a table in a quiet room may be considered at rest—but only if consideration is made for the fact that the room and the object are turning as the earth turns on its axis, and both are moving with the earth in its rotation about the sun.

Newton's *Principia* contains the main body of knowledge that composes the science of mechanics, one of the major divisions of physics. Essentially, mechanics involves forces—and motions produced by forces. It is the most precise science yet conceived by man, and it is fundamental to modern life. Newtonian principles enable men to design machines and to calculate accurately the amount of energy needed to do specific jobs. They make it possible for men to build and launch a rocket so that it will blast free from the earth's gravitational influence and remain in an orbit around the earth.

The mighty seed that had grown into Isaac Newton's *Principia* was the law of universal gravitation, and that law remained a generative force in Newton's mind and in the minds of others. By means of it he presented the first satisfactory explanation of the earth's tides. But, more significantly, he established for all times a relationship between the forces on earth and those throughout the universe.

2.9   *Axioms, or Laws of Motion*

Isaac Newton

LAW I

Every body continues in its state of rest, or of uniform motion in a right line, unless it is compelled to change that state by forces impressed upon it.

Projectiles continue in their motions, so far as they are not retarded by the resistance of the air, or impelled downwards by the force of gravity. A top, whose parts by their cohesion are continually drawn aside from rectilinear motions, does not cease its rotation, otherwise than as it is retarded by the air. The greater bodies of the planets and comets, meeting with less resistance in freer spaces, preserve their motions both progressive and circular for a much longer time.

LAW II

The change of motion is proportional to the motive force impressed; and is made in the direction of the right line in which that force is impressed.

If any force generates a motion, a double force will generate double the motion, a triple force triple the motion, whether that force be impressed altogether and at once, or gradually and successively. And this motion (being always directed the same way with the generating force), if the body moved before, is added to or subtracted from the former motion, according as they directly conspire with or are directly contrary to each other; or obliquely joined, when they are oblique, so as to produce a new motion compounded from the determination of both.

LAW III

To every action there is always opposed an equal reaction: or, the mutual actions of two bodies upon each other are always equal, and directed to contrary parts.

Whatever draws or presses another is as much drawn or pressed by that other. If you press a stone with your finger, the finger is also pressed by the stone. If a horse draws a stone tied to a rope, the horse (if I may so say) will be equally drawn back towards the stone; for the distended rope, by the same endeavor to relax or unbend itself, will draw the horse as much towards the stone as it does the stone towards the horse, and will obstruct the progress of the one as much as it advances that of the other. If a body impinge upon another, and by its force change the motion of the other, that body also (because of the equality of the mutual pressure) will undergo an equal change, in its own motion, towards the contrary part. The changes made by these actions are equal, not in the veloci-

SOURCE. From *Mathematical Principles of Natural Philosophy* (or, the *Principia*), Motte-Cajori translation (Berkely, Cal.: University of California Press, 1962) Vol. 1; reprinted by permission of the Regents of the University of California.

ties but in the motions of bodies; that is to say, if the bodies are not hindered by any other impediments. For, because the motions are equally changed, the changes of the velocities made towards contrary parts are inversely proportional to the bodies. This law takes place also in attractions.

.   .   .   .

2.10   *Inertial Reference Frames*

Edwin F. Taylor and John A. Wheeler

Less than a month after the surrender at Appomattox ended the American Civil War (1861–1865), the French author Jules Verne began writing *A Trip from the Earth to the Moon* and *A Trip around the Moon.* Eminent American cannon designers, so the story goes, cast a great cannon in a pit dug in the earth of Florida with the cannon muzzle pointing skyward. From this cannon is fired a 10-ton projectile containing three men and several animals. As the projectile coasts outward in unpowered flight toward the moon after leaving the cannon, its passengers walk normally inside the projectile on the side nearer the earth. As the trip continues, the passengers find themselves pressed less and less against the floor of the space ship until finally, at the point where the earth and moon exert equal but opposite gravitational attraction for all objects, the passengers float free of the floor. Later, as the ship nears the moon, they walk around once again, but now against the side of the space ship nearer the moon. Early in the trip one of the dogs in the ship had died from injuries sustained at takeoff. The passengers had disposed of the remains of the dog through a scuttle in the side of the space ship, only to find that the corpse continues to float outside the window during the entire trip.

This story leads to a paradox of crucial importance to relativity. Verne thought it reasonable that the gravitational attraction of the earth would keep a passenger pressed against the earth side of the space ship during the early part of the trip. He also thought it reasonable that the dog should remain next to the ship, since both ship and dog independently follow the same path through space. But if the dog floats *outside* the space ship during the entire trip, why doesn't the passenger float around *inside* the space ship? If the ship were sawed in half would the passenger, now "outside," float free of the floor?

Our experience with actual space flights enables us to resolve this paradox. Jules Verne was in error about the motion of the passenger inside the space ship. Like the dog outside the ship, the passenger inside independently follows the same part through space as the space ship itself. Therefore he floats freely relative to the ship during the entire trip. It is true that the gravitational field of the earth acts on the passenger. But it also acts on the space ship. In fact, with respect to the earth, the acceleration of the *space ship* in the gravitational field of the earth is just *equal* to the acceleration of the *passenger* in the gravitational field of the earth. Because of the equality of these accelerations there will be no *relative* acceleration between passenger and space ship. Thus the space ship serves as a reference frame ("inertial reference frame") relative to which the passenger does not experience an acceleration.

To say that the acceleration of the passenger relative to the space ship is zero is not to say that his velocity relative to it is necessarily also zero. He may have jumped from the floor or sprung from the side—in which case he will hurtle across the space and strike the opposite wall. However, when he has zero initial velocity relative to the ship the situation is particularly interesting,

SOURCE. From *Spacetime Physics* by Edwin F. Taylor and John Archibald Wheeler. W. H. Freeman and Company, Copyright © 1966.

for he will also have zero velocity relative to it at all later times. He and the ship will follow identical paths through space. How remarkable that the passenger who cannot see the outside nevertheless moves on this deterministic orbit. Without a way to control his motion and even with his eyes closed he will not touch the wall. How could one do better at eliminating gravitational influences!

A modern space ship carrying a passenger is shot vertically from the earth, rises, and falls back towards the earth. [Figure 2.10–1]. (The passenger of an elevator can experiences a close approximation to this fall when the elevator cable is cut!) Choose this freely falling space ship as the best possible reference frame in which to do physics. This reference frame is best because, among other things, the laws of motion of a particle are simple in a falling space vehicle. A free particle at rest in the vehicle remains at rest in the vehicle. When the particle is given a gentle push, it moves across the vehicle in a straight line with constant speed. Further experiments show that *all* the laws of mechanics can be expressed simply with respect to a falling space ship. We call such a space ship that rises or falls freely—or more generally moves freely in space—an *inertial reference frame*.

Look at the freely falling space ship from the surface of the earth. There is a simple reason why the free particle at rest relative to the space ship remains at rest in the space ship. This reason is that, with respect to the surface of the earth, the particle and the space ship both fall with the same acceleration [Figure 2.10–1]. It is because of this equal acceleration that the *relative* positions of the particle and the space ship do not change if the particle is originally at rest in the space ship.

The definition of an inertial frame requires that *no gravitational forces will be felt in it*. If such a reference frame is to be a space ship near the earth, it cannot be a very large one because widely separated particles within it will be differently affected by the *nonuniform* gravitational field of the earth. For example, particles released side by side will each be attracted toward the center of the earth, so they will move closer together as observed from the falling space ship [Figure 2.10–2]. As another example, think of the two particles being released far apart vertically but directly above one another [Figure 2.10–3]. Their gravitational accelerations toward the earth will be in the same direction. However the particle nearer the earth will slowly leave the other one behind: the two particles will move farther apart as the space ship falls. In either of these instances the laws of

Figure 2.10–1. Space ship in free fall near the earth.

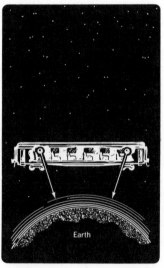

Figure 2.10–2. Railway coach in free fall in *horizontal* position near the earth.

Figure 2.10–3.   Railway coach in free fall in *vertical* position near the earth.

tion of free particles at opposite ends of the vehicle too slight to be detected? Analyzing the conditions inside one vehicle will serve to illustrate these considerations. A railway coach 25 meters long is dropped in a *horizontal* position from a height of 250 meters onto the surface of the earth [Figure 2.10–2]. The time from release to impact is about 7 seconds, or about 21×10^8 meters of light-travel time.[1] Let tiny ball bearings be released initially from rest—and in mid-air—at opposite ends of the coach. Then, during the time of fall, they will move *toward* each other a distance of 10^{-3} meters—the thickness of 9 pages of this book—because of the difference in *direction* of the earth's gravitational pull upon them. As another example, assume that the same railway coach is dropped in a *vertical* position, and that the lower end of the coach is initially 250 meters from the surface of the earth [Figure 2.10–3]. Again two tiny ball bearings are released from rest at opposite ends of the coach. In this case, during the time of fall, the ball bearings will move *apart* by a distance of 2×10^{-3} meters because of the greater gravitational acceleration of the one nearer the earth. In either of these examples let the measuring equipment in use in the coach be just short of the sensitivity required to detect the relative motion of the ball bearings. Then, with equipment of this degree of sensitivity, and with the limited time of observation, the railway coach—or, to use an earlier example, the freely falling space ship—serves as an inertial reference frame. When the sensitivity of the measuring equipment is increased, then the space ship will not serve as an inertial reference frame unless changes are made. Either the 25-meter domain in which observations are made must be shortened, or the time given to the observations must be decreased. Or, better, some appropriate combination of the space and time dimensions of the region under observation must be cut down. Or, as a final alternative, the whole apparatus must be shot by a rocket up to a region of space where one

mechanics will not be simple in a very large space ship: the large space ship will not be an inertial frame.

Now, we want the laws of mechanics to look simple in the space ship. Therefore we want to eliminate all relative accelerations produced by external causes—"eliminate" meaning to reduce these accelerations below the limit of detection so that they will not interfere with the more important accelerations we wish to study, such as those produced when two particles collide. This can be done by choosing a space ship that is sufficiently small. The smaller the space ship, the smaller will be the relative accelerations of objects at different points in the space ship. Let someone have instruments for the detection of relative accelerations with any given degree of sensitivity. No matter how fine that sensitivity, the space ship can always be made so small that these perturbing relative accelerations are too small to be detectable. Within these limits of sensitivity the space ship is then an *inertial reference frame*.

When is a space ship or any other vehicle small enough to be called an inertial reference frame? Or when is the relative accelera-

[1]The "length" of the time dimension is equal to the time (7 sec) multiplied by the speed of light (3×10^8 m/sec). [Ed.]

cannot detect the "differential in the gravitational acceleration," between one side of the coach and another—to use one way of speaking. In another way of speaking, the accelerations of the particles *relative to the coach* must be too small to be perceived. These relative accelerations can be measured from inside the coach without observing anything external. Only when these relative accelerations are too small to be detected is there a reference frame with respect to which the laws of motion are simple—an *inertial reference frame.*

A reference frame is said to be inertial in a certain region of space and time when, throughout that region of spacetime, and within some specified accuracy, every test particle that is initially at rest remains at rest, and every test particle that is initially in motion continues that motion without change in speed or in direction. An inertial reference frame is also called a *Lorentz reference frame.* In terms of this definition, inertial frames are necessarily always *local* ones, that is, inertial in a limited region of spacetime.

Region of spacetime. What is the precise meaning of this term? The long-narrow railway coach in the example served as a means to probe spacetime for a limited stretch of time and in one or another single direction in space. It can be oriented north-south, or east-west, or up-down. Whatever the orientation, the relative acceleration of the tiny ball bearings released at the two ends can be measured. For all three directions—and for all intermediate directions—let it be found by calculation that the relative drift of the two test particles is half the minimum detectable amount or less. Then throughout a cube of space 25 meters on an edge and for a lapse of time of 7 seconds, test particles moving every which way depart from straight-line motion by undetectable amounts. In other words, the reference frame is inertial in a region of spacetime with dimensions (25 meters × 25 meters × 25 meters of space) × (21×10^8 meters of time).

Test Particle. How small must a particle be to qualify as a test particle? It must have so little mass that, within some specified accuracy, its presence will not affect the motion of other nearby particles. In terms

of Newtonian mechanics the gravitational attraction of the test particle for other particles must be negligible within the accuracy specified. As an example, consider a particle of mass 10 kilograms. A second and less massive particle placed one-tenth meter from it and initially at rest will, in less than three minutes, undergo a displacement of 10^{-3} meters. Thus the 10-kilogram object is not—in this sense—a test particle. A test particle *responds* to gravitational forces but it does not itself *produce* any significant gravitational force.

It would be impossible to define an inertial reference frame if it were not for a remarkable feature of nature. Particles of different size, shape, and material in the same location all fall with the *same acceleration* toward the earth. If this were not so, an observer inside a falling space ship would notice a relative acceleration among different particles even when they are close together; at least some of the particles initially at rest would not remain at rest; that is, the space ship would not be an inertial reference frame according to the definition. How sure are we that particles in the *same* location but of *different* substances all fall toward the earth with the same acceleration? According to legend Galileo dropped balls made of different materials from the Leaning Tower of Pisa in order to verify this assumption. In 1922 Baron Ronald von Eötvös checked to an accuracy of five parts in 10^9 that the earth imparts the same acceleration to wood and to platinum. More recently Robert H. Dicke has pointed out that the sun is more suitable than the earth as source for the gravitational acceleration that one will measure. The alternation in direction of the sun's pull every 12 hours lends itself to fantastic amplification by resonance. Cylinders of aluminium and gold experience accelerations due to the sun (0.59×10^{-2} meters per second per second) that are the same to three parts in a hundred thousand million (3 in 10^{11}), according to R. H. Dicke and Peter G. Roll. This is one of the most sensitive checks of a fundamental physical principle in all of physics: the identity of the acceleration produced by gravity in every kind of test particle.

It follows from this principle that a particle made of *any* material can be used as a test particle to determine whether a given reference frame is inertial. A reference frame that is inertial for one kind of test particle will be inertial for all kinds of test particles.

2.11   *Gravitation*

Richard P. Feynman

UNIVERSAL GRAVITATION

What else can we understand when we understand gravity? Everyone knows the earth is round. Why is the earth round? That is easy; it is due to gravitation. The earth can be understood to be round merely because everything attracts everything else and so it has attracted itself together as far as it can! If we go even further, the earth is not *exactly* a sphere because it is rotating, and this brings in centrifugal effects which tend to oppose gravity near the equator. It turns out that the earth should be elliptical, and we even get the right shape for the ellipse. We can thus deduce that the sun, the moon, and the earth should be (nearly) spheres, just from the law of gravitation.

What else can you do with the law of gravitation? If we look at the moons of Jupiter we can understand everything about the way they move around that planet. Incidentally, there was once a certain difficulty with the moons of Jupiter that is worth remarking on. These satellites were studied very carefully by Roemer, who noticed that the moons sometimes seemed to be ahead of schedule, and sometimes behind. (One can find their schedules by waiting a very long time and finding out how long it takes on the average for the moons to go around.) Now they were *ahead* when Jupiter was particularly *close* to the earth and they were *behind* when Jupiter was *farther* from the earth. This would have been a very difficult thing to explain according to the law of gravitation—it would have been, in fact,

the death of this wonderful theory if there were no other explanation. If a law does not work even in *one place* where it ought to, it is just wrong. But the reason for this discrepancy was very simple and beautiful: it takes a little while to *see* the moons of Jupiter because of the time it takes light to travel from Jupiter to the earth. When Jupiter is closer to the earth the time is a little less, and when it is farther from the earth, the time is more. This is why moons appear to be, on the average, a little ahead or a little behind, depending on whether they are closer to or farther from the earth. This phenomenon showed that light does not travel instantaneously, and furnished the first estimate of the speed of light. This was done in 1656.

If all of the planets push and pull on each other, the force which controls, let us say, Jupiter in going around the sun is not just the force from the sun; there is also a pull from, say, Saturn. This force is not really strong, since the sun is much more massive than Saturn, but there is *some* pull, so the orbit of Jupiter should not be a perfect ellipse, and it is not; it is slightly off, and "wobbles" around the correct elliptical orbit. Such a motion is a little more complicated. Attempts were made to analyze the motions of Jupiter, Saturn, and Uranus on the basis of the law of gravitation. The effects of each of these planets on each other were calculated to see whether or not the tiny deviations and irregularities in these motions could be completely understood from this one law. Lo and behold, for Jupiter and Saturn, all was well,

SOURCE. From Feynman, Leighton, and Sands, *The Feynman Lectures on Physics* (Reading, Mass.: Addison-Wesley, 1963) Chapter 7.

but Uranus was "weird." It behaved in a very peculiar manner. It was not traveling in an exact ellipse, but that was understandable, because of the attractions of Jupiter and Saturn. But even if allowance were made for these attractions, Uranus *still* was not going right, so the laws of gravitation were in danger of being overturned, a possibility that could not be ruled out. Two men, Adams and Leverrier, in England and France, independently, arrived at another possibility: perhaps there is another planet, dark and invisible, which men had not seen. This

That the law of gravitation is true at even bigger distances is indicated in Figure 2.11–1. If one cannot see gravitation acting here, he has no soul. This figure shows one of the most beautiful things in the sky—a globular star cluster. All of the dots are stars. Although they look as if they are packed solid toward the center, that is due to the fallibility of our instruments. Actually, the distances between even the centermost stars are very great and they very rarely collide. There are more stars in the interior than farther out, and as we move outward there are fewer and

Figure 2.11–1.   A globular star cluster. (Mount Wilson and Palomar Observatories.)

planet, N, could pull on Uranus. They calculated where such a planet would have to be in order to cause the observed perturbations. They sent messages to the respective observatories, saying, "Gentlemen, point your telescope to such and such a place, and you will see a new planet." It often depends on with whom you are working as to whether they pay any attention to you or not. They did pay attention to Leverrier; they looked, and there planet N was! The other observatory then also looked very quickly in the next few days and saw it too.

.   .   .   .   .

fewer. It is obvious that there is an attraction among these stars. It is clear that gravitation exists at these enormous dimensions, perhaps 100,000 times the size of the solar system. Let us now go further, and look at an *entire* galaxy, shown in Figure 2.11–2. The shape of this galaxy indicates an obvious tendency for its matter to agglomerate. Of course we cannot prove that the law here is precisely inverse square, only that there is still an attraction, at this enormous dimension, that holds the whole thing together. One may say, "Well, that is all very clever but why is it not just a ball?" Because it is *spinning* and has *angular*

Figure 2.11–2.   A galaxy. (Mount Wilson and Palomar
Observatories.)

Figure 2.11–3.   A cluster of galaxies. (Mount Wilson
and Palomar Observatories.)

momentum which it cannot give up as it contracts; it must contract mostly in a plane.
(Incidentally, if you are looking for a good
problem, the exact details of how the arms
are formed and what determines the shapes of
these galaxies has not been worked out.) It is,
however, clear that the shape of the galaxy
is due to gravitation even though the complexities of its structure have not yet allowed
us to analyze it completely. In a galaxy we
have a scale of perhaps 50,000 to 100,000
light years. The earth's distance from the
sun is $8\frac{1}{3}$ light *minutes*, so you can see how
large these dimensions are.

Gravity appears to exist at even bigger
dimensions, as indicated by Figure 2.11–3,
which shows many "little" things clustered
together. This is a *cluster of galaxies*, just like a
star cluster. Thus galaxies attract each other
at such distances that they too are agglom-
erated into clusters. Perhaps gravitation exists
even over distances of *tens of millions* of light
years; so far as we now know, gravity seems to
go out forever inversely as the square of the
distance.

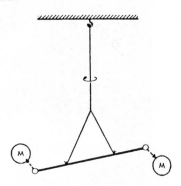

Figure 2.11–4. A simplified diagram
of the apparatus used by Cavendish to
verify the law of universal gravitation
for small objects and to measure the
gravitational constant G.

CAVENDISH'S EXPERIMENT

Gravitation, therefore, extends over enormous distances. But if there is a force between
any pair of objects, we ought to be able to
measure the force between our own objects.
Instead of having to watch the stars go around
each other, why can we not take a ball of lead
and a marble and watch the marble go toward
the ball of lead? The difficulty of this experiment when done in such a simple manner is
the very weakness or delicacy of the force. It
must be done with extreme care, which means
covering the apparatus to keep the air out,
making sure it is not electrically charged,
and so on; then the force can be measured. It
was first measured by Cavendish with an
apparatus which is schematically indicated in
Figure 2.11–4. This first demonstrated the
direct force between two large, fixed balls of
lead and two smaller balls of lead on the ends
of an arm supported by a very fine fiber, call-

ed a torsion fiber. By measuring how much
the fiber gets twisted, one can measure the
strength of the force, verify that it is inversely
proportional to the square of the distance,
and determine how strong it is. Thus, one
may accurately determine the coefficient G
in the formula

$$F = G\,\frac{mm'}{r^2}.$$

All the masses and distances are known. You
say, "We knew it already for the earth." Yes,
but we did not know the *mass* of the earth. By
knowing G from this experiment and by
knowing how strongly the earth attracts, we
can indirectly learn how great is the mass
of the earth! This experiment has been called
"weighing the earth." Cavendish claimed
he was weighing the earth, but what he was
measuring was the coefficient G of the gravity
law. This is the only way in which the mass
of the earth can be determined. G turns out
to be

$$6.670 \times 10^{-11} \text{ newton} \cdot \text{m}^2/\text{kgm}^2.$$

It is hard to exaggerate the importance of
the effect on the history of science produced
by this great success of the theory of gravitation. Compare the confusion, the lack of confidence, the incomplete knowledge that

prevailed in the earlier ages, when there were endless debates and paradoxes, with the clarity and simplicity of this law—this fact that all the moons and planets and stars have such a *simple rule* to govern them, and further that man could *understand* it and deduce how the planets should move! This is the reason for the success of the sciences in following years, for it gave hope that the other phenomena of the world might also have such beautifully simple laws.

WHAT IS GRAVITY?

But is this such a simple law? What about the machinery of it? All we have done is to describe *how* the earth moves around the sun, but we have not said *what makes it go*. Newton made no hypotheses about this; he was satisfied to find *what* it did without getting into the machinery of it. *No one has since given any machinery*. It is characteristic of the physical laws that they have this abstract character. The law of conservation of energy is a theorem concerning quantities that have to be calculated and added together, with no mention of the machinery, and likewise the great laws of mechanics are quantitative mathematical laws for which no machinery is available. Why can we use mathematics to describe nature without a mechanism behind it? No one knows. We have to keep going because we find out more that way.

Many mechanisms for gravitation have have suggested. It is interesting to consider one of these, which many people have thought of from time to time. At first, one is quite excited and happy when he "discovers" it, but he soon finds that it is not correct. It was first discovered about 1750. Suppose there were many particles moving in space at a very high speed in all directions and being only slightly absorbed in going through matter. When they *are* absorbed, they give an impulse to the earth. However, since there are as many going one way as another, the impulses all balance. But when the sun is nearby, the particles coming toward the earth through the sun are partially absorbed, so fewer of them are coming from the sun than are coming from the other side. Therefore, the earth feels a net impulse toward the sun and it does not take one long to see that it is inversely as the square of the distance—because of the variation of the solid angle that the sun subtends as we vary the distance. What is wrong with that machinery? It involves some new consequences which are *not true*. This particular idea has the following trouble: the earth, in moving around the sun, would impinge on more particles which are coming from its forward side than from its hind side (when you run in the rain, the rain in your face is stronger than that on the back of your head!). Therefore there would be more impulse given the earth from the front, and the earth would feel a *resistance to motion* and would be slowing up in its orbit. One can calculate how long it would take for the earth to stop as a result of this resistance, and it would not take long enough for the earth to still be in its orbit, so this mechanism does not work. No machinery has ever been invented that "explains" gravity without also predicting some other phenomenon that does *not* exist.

Next we shall discuss the possible relation of gravitation to other forces. There is no explanation of gravitation in terms of other forces at the present time. It is not an aspect of electricity or anything like that, so we have no explanation. However, gravitation and other forces are very similar, and it is interesting to note analogies. For example, the force of electricity between two charged objects looks just like the law of gravitation: the force of electricty is a constant, with a minus sign, times the product of the charges, and varies inversely as the square of the distance. It is in the opposite direction—likes repel. But is it still not very remarkable that the two laws involve the same function of distance? Perhaps gravitation and electricity are much more closely related than we think. Many attempts have been made to unify them; the so-called unified field theory is only a very elegant attempt to combine electricity and gravitation; but, in comparing gravitation and electricity, the most interesting thing is the *relative strenghts* of the forces. Any theory that contains them both must also deduce how strong the gravity is.

If we take, in some natural units, the repulsion of two electrons (nature's universal charge) due to electricity, and the attraction of two electrons due to their masses, we can measure the ratio of electrical repulsion to the gravitational attraction. The ratio is independent of the distance and is a fundamental constant of nature. The gravitational attraction relative to the electrical repulsion between two electrons is 1 divided by 4.17×10^{42}! The question is, where does such a large number come from? It is not accidental, like the ratio of the volume of the earth to the volume of a flea. We have considered two natural aspects of the same thing, an electron. This fantastic number is a natural constant, so it involves something deep in nature. Where could such a tremendous number come from? Some say that we shall one day find the "universal equation," and in it, one of the roots will be this number. It is very difficult to find an equation for which such a fantastic number is a natural root. Other possibilities have been thought of; one is to relate it to the age of the universe. Clearly, we have to find *another* large number somewhere. But do we mean the age of the universe in *years*? No, because years are not "natural"; they were devised by men. As an example of something natural, let us consider the time it takes light to go across a proton, 10^{-24} second. If we compare this time with the *age of the universe*, 2×10^{10} years, the answer is 10^{-42}. It has about the same number of zeros going off it, so it has been proposed that the gravitational constant is related to the age of the universe. If that were the case, the gravitational constant would change with time, because as the universe got older the ratio of the age of the universe to the time which it takes for light to go across a proton would be gradually increasing. Is it possible that the gravitational constant *is* changing with time? Of course the changes would be so small that it is quite difficult to be sure.

One test which we can think of is to deter-mine what would have been the effect of the change during the past 10^9 years, which is approximately the age from the earliest life on the earth to now, and one-tenth of the age of the universe. In this time, the gravity constant would have increased by about 10 percent. It turns out that if we consider the structure of the sun—the balance between the weight of its material and the rate at which radiant energy is generated inside it— we can deduce that if the gravity were 10 percent stronger, the sun would be much more than 10 percent brighter— by the *sixth power* of the gravity constant! If we calculate what happens to the orbit of the earth when the gravity is changing, we find that the earth was then *closer in*. Altogether, the earth would be about 100 degrees centigrade hotter, and all of the water would not have been in the sea, but vapor in the air, so life would not have started in the sea. So we do *not* now believe that the gravity constant is changing with the age of the universe. But such arguments as the one we have just given are not very convincing, and the subject is not completely closed.

It is a fact that the force of gravitation is proportional to the *mass*, the quantity which is fundamentally a measure of *inertia*—of how hard it is to hold something which is going around in a circle. Therefore two objects, one heavy and one light, going around a larger object in the same circle at the same speed because of gravity, will stay together because to go in a circle *requires* a force which is stronger for a bigger mass. That is, the gravity is stronger for a given mass in *just the right proportion* so that the two objects will go around together. If one object were inside the other it would *stay* inside; it is a perfect balance. Therefore, Gagarin or Titov would find things "weightless" inside a space ship; if they happened to let go of a piece of chalk, for example, it would go around the earth in exactly the same way as the whole space ship, and so it would appear to remain suspended before them in space. It is very interesting that this force is *exactly* proportional to the mass with great precision, because if it were not exactly proportional there would by some effect by which inertia and weight would differ. The absence of such an effect has been checked with great accuracy by an experiment done first by Eötvös in 1909 and more recently by Dicke. For all substances tried, the masses and weights are exactly proportional within 1 part in 1,000,000,000, or less. This is a remarkable experiment.

2.12

Conservation of Energy
Richard P. Feynman

There is a fact, or if you wish, a law, governing all natural phenomena that are known to date. There is no known exception to this law—it is exact so far as we know. The law is called the *conservation of energy*. It states that there is a certain quantity, which we call energy, that does not change in the manifold changes which nature undergoes. That is a most abstract idea, because it is a mathematical principle; it says that there is a numerical quantity which does not change when something happens. It is not a description of a mechanism, or anything concrete; it is just a strange fact that we can calculate some number and when we finish watching nature go through her tricks and calculate the number again, it is the same. (Something like the bishop on a red square, and after a number of moves—details unknown—it is still on some red square. It is a law of this nature.) Since it is an abstract idea, we shall illustrate the meaning of it by an analogy.

Imagine a child, perhaps "Dennis the Menace," who has blocks which are absolutely indestructible, and cannot be divided into pieces. Each is the same as the other. Let us suppose that he has 28 blocks. His mother puts him with his 28 blocks into a room at the beginning of the day. At the end of the day, being curious, she counts the blocks very carefully, and discovers a phenomenal law—no matter what he does with the blocks, there are always 28 remaining! This continues for a number of days, until one day there are only 27 blocks, but a little investigating shows that there is one under the rug—she must look everywhere to be sure that the number of blocks has not changed. One day, however, the number appears to change— there are only 26 blocks. Careful investigation indicates that the window was open, and upon looking outside, the other two blocks are found. Another day, careful count indicates that there are 30 blocks! This causes considerable consternation, until it is realized that Bruce came to visit, bringing his blocks with him, and he left a few at Dennis' house. After she has disposed of the extra blocks, she closes the window, does not let Bruce in, and then everything is going along all right, until one time she counts and finds only 25 blocks. However, there is a box in the room, a toy box, and the mother goes to open the toy box, but the boy says "No, do not open my toy box," and screams. Mother is not allowed to open the toy box. Being extremely curious, and somewhat ingenious, she invents a scheme! She knows that a block weighs three ounces, so she weighs the box at a time when she sees 28 blocks, and it weighs 16 ounces. The next time she wishes to check, she weighs the box again, subtracts sixteen ounces and divides by three. She discovers the following:

$$\left(\begin{array}{c}\text{number of} \\ \text{blocks seen}\end{array}\right) + \frac{(\text{weight of box}) - 16 \text{ ounces}}{3 \text{ ounces}} = \text{constant}. \quad [1]$$

SOURCE. From Feynman, Leighton, and Sands, *The Feynman Lectures on Physics* (Reading, Mass.: Addison-Wesley, 1963) Chapter 4.

There then appear to be some new deviations, but careful study indicates that the dirty water in the bathtub is changing its level. The child is throwing blocks into the water, and she cannot see them because it is so dirty, but she can find out how many blocks are in the water by adding another term to her formula. Since the original height of the water was 6 inches and each block raises the water a quarter of an inch, this new formula would be:

First, when we are calculating the energy, sometimes some of it leaves the system and goes away, or sometimes some comes in. In order to verify the conservation of energy, we must be careful that we have not put any in or taken any out. Second, the energy has a large number of *different forms*, and there is a formula for each one. These are: gravitational energy, kinetic energy, heat energy, elastic energy, electrical energy, chemical energy, radiant energy, nuclear

$$\left(\begin{array}{c} \text{number ot} \\ \text{blocks seen} \end{array}\right) + \frac{(\text{weight of box}) - 16 \text{ ounces}}{3 \text{ ounces}}$$

$$+ \frac{(\text{height of water}) - 6 \text{ inches}}{1/4 \text{ inch}} = \text{constant.} \qquad [2]$$

In the gradual increase in the complexity of her world, she finds a whole series of terms representing ways of calculating how many blocks are in places where she is not allowed to look. As a result, she finds a complex formula, a quantity which *has to be computed*, which always stays the same in her situation.

What is the analogy of this to the conservation of energy? The most remarkable aspect that must be abstracted from this picture is that *there are no blocks*. Take away the first terms in [1] and [2] and we find ourselves calculating more or less abstract things. The analogy has the following points.

energy, mass energy. If we total up the formulas for each of these contributions, it will not change except for energy going in and out.

It is important to realize that in physics today, we have no knowledge of what energy is. We do not have a picture that energy comes in little blobs of a definite amount. It is not that way. However, there are formulas for calculating some numerical quantity, and when we add it all together it gives "28"— always the same number. It is an abstract thing in that it does not tell us the mechanism or the *reasons* for the various formulas.

2.13   *The Nature of Heat*

Benjamin Thompson (Count Rumford)

[In 1798 Benjamin Thompson reported the results of an extensive series of experiments conducted in connection with the boring of cannons for the Bavarian Army. At the end of his paper, Thompson comments as follows:]

By meditating on the results of all these experiments, we are naturally brought to that great question which has so often been the subject of speculation among philosophers; namely,—

What is Heat? Is there any such thing

SOURCE. From *The Collected Works of Count Rumford*, edited by Sanborn C. Brown (Cambridge, Mass.: Belknap Press, Harvard University Press, 1968); Read before the Royal Society, January 25, 1798. Copyright for Rumford by the President and Fellows of Harvard College.

as an igneous *fluid?* Is there anything that can with propriety be called *calorie?*

We have seen that a very considerable quantity of Heat may be excited in the friction of two metallic surfaces and given off in a constant stream or flux in all *directions* without interruption or intermission and without any signs of diminution or exhaustion.

From whence comes the Heat which was continually given off in this manner in the foregoing experiment? Was it furnished by the small particles of metal detached from the larger solid masses on their being rubbed together? This, as we have already seen, could not possibly have been the case.

Was it furnished by the air? This could not have been the case, for in three of the experiments the machinery, being kept immersed in water, the access of the air of the atmosphere was completely prevented.

Was it furnished by the water which surrounded the machinery? That this could not have been the case is evident: first because this water was continually receiving heat from the machinery and could not at the same time be giving *to* and receiving heat *from* the same body, and secondly because there was no chemical decomposition of any part of this water. Had any such decomposition taken place (which, indeed, could not reasonably have been expected), one of its component elastic fluids (most probably inflammable air) must at the same time have been set at liberty, and in making its escape into the atmosphere would have been detected. But though I frequently examined the water to see if any air bubbles rose up through it and had ever made preparations for catching them in order to examine them if any should appear, I could perceive none, nor was there any sign of decomposition of any kind whatever or other chemical processes going on in the water.

Is it possible that the heat could have been supplied by means of the iron bar to the end of which the blunt steel borer was fixed or by the small neck of gun metal by which the hollow cylinder was united to the cannon? These suppositions appear more improbable even than either of those before mentioned, for Heat was continually going off or out of the *machinery* by both these passages during the whole time the experiment lasted.

And, in reasoning on this subject we must not forget to consider the most remarkable circumstance that the source of heat generated by friction in these experiments appeared evidently to be *inexhaustible*.

It is hardly necessary to add that anything which any *insulated* body or system of bodies can continually be furnished without *limitation* cannot possibly be a material *substance*, and it appears to me to be extremely difficult if not quite impossible to form any distinct ideas of anything capable of being excited and communicated in the manner the Heat was excited and communicated in these experiments except it be MOTION.

I am very far from pretending to know how or by what means or mechanical contrivance that particular kind of motion in bodies which has been supposed to constitute Heat is excited, continued and propagated, and I shall not presume to trouble the [reader] with mere conjectures particularly on a subject which, during so many thousand years, the most enlightened philosophers have endeavored, but in vain, to comprehend.

But, although the mechanism of heat should, in fact, be one of those mysteries of nature which are beyond the reach of human intelligence, this ought by no means to discourage us or even lessen our ardour in our attempts to investigate the laws of its operation. How far can we advance in any of the paths which science has opened to us before we find ourselves enveloped in those thick mists which on every side bound the horizon of the human intellect? But how ample and how interesting is the field that is given us to explore.

2.14   *Heat and Energy*
                                  Norman Feather

EXPERIMENTS OF JOULE AND OTHERS

In 1798 Benjamin Thompson, Count Rumford (1753–1814), presented to the Royal Society an *Enquiry concerning the Source of Heat which is excited by Friction*. A man of many parts, born in America, Fellow of the Royal Society at the age of twenty-six, at one time under-secretary of state for the British colonies, in 1798 he had just recently returned to London after eleven years service with the elector of Bavaria—as minister of war and commander-in-chief of the Bavarian army. His "enquiry" had been an experimental one, and it had been carried out in the arsenal at Munich. Being at one time responsible for the boring of cannon, Rumford had been well placed to make a large-scale experiment on the heating produced by friction. He arranged for a blunt boring tool to be used on a mass of some 60 kg. of gun-metal, covered with flannel to minimize loss of heat. When only 54 g. of metal had been ground off, the temperature of the mass had risen from 15° C to 54° C. He determined the specific heat of the abraded metal, and found it to be the same as that of gun-metal in bulk.

Rumford regarded these results as showing that the pulverization of the metal was in no direct way responsible for the appearance of heat, in the sense that the heat which appeared in the large block had not previously been associated with the fragments (it was the fashionable belief at the time that heat was a subtle material fluid permeating the pores of bodies), but that it had been produced in the boring—and that it would have been produced, in roughly the same amount, even if no metal had been abraded.

Rumford's main reasons for this con-

clusion were, first, that the thermal properties of the abraded fragments were the same as those of the metal in bulk, as attested by specific heat determinations, and, secondly, that such a large quantity of heat had been produced and such a small mass of fragments. He added the luminous comment, "and it appears to me to be extremely difficult, if not quite impossible, to form any distinct idea of anything capable of being excited and communicated in these experiments except it be motion." There is an echo here of a wise judgment of Francis Bacon, Baron Verulam (1561–1626), Lord Chancellor of England, delivered two centuries earlier, "the very essence of heat, or the substantial self of heat, is motion and nothing else." But that was only a philosopher's dictum.

The year after he presented his *Enquiry* to the Royal Society, Rumford was instrumental in promoting the foundation of the Royal Institution. For four years from its incorporation in 1800 he resided in the Institution's premises in Albemarle Street, and, in his capacity as patron and manager, on 16 February, 1801, he appointed Humphry Davy (1778–1829) as assistant lecturer in chemistry and director of the laboratory. Davy was self-taught; less than four years previously he had begun his study of experimental science, but during that time he had carried out an experiment (1799) which is our only reason for referring to him here—and which possibly was the determining reason for his appointment by Rumford. Davy had shown that ice may be melted simply by friction between one piece of ice and another. Here certainly heat had been produced which was not previously associated with the bodies concerned: the

SOURCE. From *An Introduction to the Physics of Mass, Length and Time* (Edinburgh: Edinburgh University Press, 1959) Chapter 14, copyright by Edinburgh University Press.

specific heat of water is roughly twice as great as the specific heat of ice.

The experiments of Rumford and Davy, at least as we have described them, were largely qualitative experiments; those of Joule, which we are now to describe, were rigorously quantitative. James Prescott Joule (1818–1889) began his scientific career as a self-taught electrical engineer in Manchester. Before he was twenty-two he had established the quantitative relationship, now known as Joule's law, between the rate of production of heat in a wire carrying an electric current and the measures of the current and the difference of potential between the ends of the wire. He began to see that this relationship connected an amount of heat measured in calories with the measures of two electrical quantities ultimately defined in dynamical terms—in terms of force and work. After further electrical experiments, Joule became convinced that what he had been investigating was a special example of a much more general phenomenon, and in 1843 he introduced the term "mechanical value of heat" (or, as he later wrote, "mechanical equivalent of heat") to express the view that it is possible to produce heat merely by doing mechanical work, and that when this occurs the amount of work which is not otherwise accounted for can be accounted for by supposing that it has been "converted" into heat. Within the restricted field of dynamics we have hitherto recognized only the conversion of work into energy—kinetic energy or potential energy, according to the circumstances of the case. According to Joule's view, mechanical work can be converted either into energy or into heat. For long periods during the next thirty-five years, he devoted himself to many series of careful experiments which in their overall result left no doubt of the essential validity of this assertion.

.   .   .   .   .

Certain of Joule's experiments were repeated in the following year by H. A. Rowland (1848–1901) in Baltimore, U.S.A., with even greater refinement. The general conclusion was essentially the same as before, but Joule's accuracy was improved upon, and Rowland was able to demonstrate the variation of the specific heat of water with temperature, in terms of the variation with initial temperature of the amount of work necessary to raise the temperature of a mass of water by one degree by the process of stirring. From that date until the present time, other experiments, by many different methods, have merely added to the volume of evidence for the belief, now held to be incontrovertible, that whenever heat appears which cannot be accounted for in terms of the simple conservation law, then mechanical work has been done, either at the macroscopic or at the molecular level, and that for every calorie of heat so appearing 4.1855×10^7 ergs of work have been performed. This conversion factor, 4.1855×10^7 ergs per calorie (or 4.1855 joules per calorie), is the mechanical equivalent of heat, according to the best modern determinations.

THE CONSERVATION OF ENERGY

Joule's experiments, as we have seen, spanned a period of some forty years, from 1840 to 1880. By the end of that period their full significance had come to be appreciated by scientists in general, but it is fitting to record that for a young Prussian army surgeon, Hermann Ludwig Ferdinand von Helmholtz (1821–1894), the evidence which was available as early as 1847 was sufficient to engender in his mind notions which, far ahead of their time, were found as the years passed to bear the hall-mark of genius. On July 23, 1847 Helmholtz presented his ideas to the Physical Society of Berlin in a paper entitled *On the Conservation of Force* (Thomas Young's "energy" had at that time no counterpart in the German language). Helmholtz had previously made an important discovery in physiology, but this was his first real contribution to physics. Indeed, he was to hold, in succession, three university chairs of physiology before he became officially recognized as a professional physicist, on his appointment to the chair of physics at Berlin in 1871. However, Helmholtz's genius was many sided, and in his later life he returned to reinforce the grand sweep of his early conception—of a law of conservation of energy which was universal in its scope—with modern instances. Ever since his death that process of proof by justification has continued.

Work done by a force when its point of application moves in its own direction, may appear as kinetic energy, as potential energy or as heat, depending upon whether there is resistance to motion and, if so, upon the nature of the resistance—that is the conclusion which we have drawn from the experiments described in the last section. Clearly,

it makes sense only if the various forms under which work may appear are of the same kind, only, that is, if the concepts energy and heat refer essentially to the same attribute of matter. Historically, the origins of the concepts are separate and distinct; logically, one of them, however convenient, is superfluous. Logically, there is no reason for the retention of the calorie as a separate unit in the c.g.s. system, once a distinctive name of convenience has been given to a unit of comparable size (1 joule = 10^7 ergs, 1 calorie = 4.1855 joules), but logic alone has a poor power of persuasion even with scientists.

.   .   .   .   .

2.15

Time's Arrow

Stanley W. Angrist and Loren G. Hepler

Rivers could flow uphill if the river beds in which they flow were to cool slightly, giving up thermal energy to the river water and thus conserving the energy of the universe. But they don't.

The water in the river might dissociate spontaneously into hydrogen and oxygen at the expense of the thermal energy of the surroundings. But it doesn't.

The air above the river might spontaneously liquefy, liberating thermal energy to its surroundings; or it might separate, spontaneously, into pure oxygen and nitrogen. But it doesn't.

Furthermore, an old man sitting on the river bank watching the river flow by might grow young— his wrinkles might disappear, his hair might grow thick and black, and his muscles might become taut and resilient. But, alas, none of these things happen.

In each case the first law of thermodynamics[1] could be satisfied without difficulty. Each event hypothesized above is the reverse of an actual happening. Rivers flow downhill, hydrogen and oxygen form water (sometimes

explosively) and liberate thermal energy to the surroundings, liquid air absorbs thermal energy from its surroundings (unless they are very cold) and evaporates. Man inevitably grows old, and in the process he frequently acquires wrinkles, gray hair, and a flabby paunch. The reverse of these events doesn't take place. In each case the answer comes down to the statement that the second law of thermo-dynamics says that while each of the events described is conceivable, each is so highly unlikely that it is practically impossible.

Nowadays people have no difficulty in accepting the first law of thermodynamics, which often seems to be a pretty obvious bit of common sense. But the second law is different. It sometimes seems to be a "True, but so what?" law, and at other times seems to run contrary to "common sense." We shall see, however, that this law leads us into many strange and wondrous areas where the first law alone is of little interest or importance.

Because the second law of thermodynamics may be stated in so many different forms, it may be construed by many to be an ill-

SOURCE. From *Order and Chaos* (New York: Basic Books, 1967), Chapter 7.
[1]Conservation of energy. [Ed.]

defined law and, therefore, not a *really* important law of nature. This is not so. It is just
because it is statable in so many forms that
it has become a thread that weaves its way
through all aspects of science and life. A. S.
Eddington, a noted British astronomer,
stated it succinctly in the Gifford Lectures
of 1927 when he said:

The law that entropy always increases—the
second law of thermodynamics—holds, I think,
the supreme position among the laws of Nature.
If someone points out to you that your pet theory
of the universe is in disagreement with Maxwell's
equations—then so much the worse for Maxwell's
equations. If it is found to be contradicted by
observation—will, these experimentalists do
bungle things sometimes. But if your theory is
found to be against the second law of thermodynamics, I can give you no hope; there is nothing
for it but to collapse in deepest humiliation.

Scientists have found it both necessary
and convenient when discussing heat to express themselves quantitatively. To do so
adequately, at least two numbers must be
used: one to measure the quantity of energy,
the other to measure the quantity of disorder.
The calorie has proved suitable for measuring the quantity of energy. The quantity of
disorder is measured in terms of the *entropy*.[2]

The idea of entropy is bound inextricably
with all our theoretical ideas about heat and
certain other phenomena, including information. One way of defining entropy is in
terms of the number of states that are possible in a system in a given situation. The
disorder arises because we do not know which
state the system is in. Disorder[3] is then essentially the same thing as ignorance, which is

how entropy is related to information theory.
In unsophisticated terms, entropy is a measure of the number of independent degrees
of freedom that a system has. A high entropy
system is free to be in many different states.
We might make this point a little clearer by
examining the value of entropy per mole for
some different substances at room temperature and atmospheric pressure.

Substance	Entropy
	[cal/(°K-mole)]
Diamond	0.6
Platinum	10.0
Lead	15.5
Water (liquid)	16.7
Laughing gas	62.6

Examination of this table reveals that
entropy is closely related to hardness. In fact,
hard, abrasive materials such as diamond,
garnet, topaz, and silicon carbide in which
individual atoms are bound together in
three-dimensional lattices by chemical bonds
that severely limit thermal motion of the
atoms have small entropies. It is also apparent that soft substances such as gases
contain large amounts of thermal disorder
(their molecules are shooting about in every
direction with a wide variety of speeds) at
room temperature and thus have large entropies.

Since entropy is a measure of the molecular
randomness in a system, we expect that there
will be an increase in entropy associated
with melting a solid (little randomness) to a
liquid (more randomness) or evaporating a
liquid to a gas (most randomness). The

---

[2]Clausius gave us this name after observing that the ratio of heat to temperature had special properties. He first
called it *Verwandlungsinhalt* (transformation-content) because it measured the transformability of heat. Later he called
it entropy, which comes from two Greek stems meaning "turning into." We believe that considering both words, on
balance, it was a fortunate change. Stephen Leacock commented on this quantity: "All physicists sooner or later
say, 'Let us call it entropy,' just as a man says, when you get to know him, 'Call me Charlie.'"

[3]Disorder is a tricky fellow to pin down. Everyone knows what it means to shuffle a pack of cards and would
be willing to claim that a deck had been well shuffled after observing the shuffling and resulting hands dealt by a
good (and honest) card player. Is it possible to describe the shuffle as good when only the shuffling operation is
specified or is it necessary to know the results of a shuffle before one can say it is good? G. N. Lewis, an American
chemist, has pointed out correctly that it would be possible to formulate the rules of some card game so that any
arrangement of the cards whatever would be a "regular" arrangement from the point of view of that game. (Is
there a poker player alive who has not played at least one game of 5-card draw with twos, threes, fives, and one-
eyed jacks wild?) Disorder is not an absolute concept; rather it is relative and has meaning only in context.

increase in entropy for melting ice is 5.3 cal/ (°K-mole) and for boiling water is 25.7 cal/ (°K-mole).

Entropy has another useful function in that it can provide a rigorous definition of temperature. Temperature (T) may be defined as the ratio of the change in energy of a system (ΔE) to its change in entropy (ΔS) for a constant volume (zero work) process:

$$T = \frac{\text{increase in energy}}{\text{increase in entropy}} = \frac{\Delta E}{\Delta S}$$

This way of looking at the absolute temperature says it is the quantity of energy that must be added to a system to alter its entropy one unit. For example, if we have to add 375 calories of thermal energy to a system to alter its entropy by one entropy unit (1 cal/°K), then the temperature of the system is 375 calories per entropy unit; thus the temperature of the system is 375°K as shown by:

$$T = \frac{\Delta E}{\Delta S} = \frac{375 \text{ cal}}{1 \text{ cal/°K}} = 375°K$$

The increase in entropy of a substance when thermal energy is added to it is proportional to the amount of thermal energy added. The entropy of a cold substance is altered more by the addition of one calorie of thermal energy than is the entropy of a hot substance. This definition of temperature has the virtue of being precise even though it may not convey much information in a qualitative sense about temperature. It is also a definition of temperature that does not depend on the particular properties of the thermometers used to measure it. A mercury thermometer, of course, depends on the properties of mercury and the glass that make up the thermometer. Unfortunately, we do not have convenient "entropy meters" which would allow us to make everyday use of this definition of temperature.

Now that we have a little background in the subjects of order, disorder, entropy, and the like we can move on to the second law

of thermodynamics.

.   .   .   .   .

Consider a piece of ice floating in a glass of water resting on a table, the whole arrangement (except for the ice) being at room temperature. The ice melts and in so doing it cools the water in the glass and the portion of the table under the glass. As this process is carried out, the ice changes from the ordered solid state to the disordered liquid state. The entropy increase of the ice is greater than the entropy decrease of the water in the glass and the table, which are both cooled by the melting of the ice.

We have not observed (and do not ever expect to observe) the reverse of this process whereby a portion of the water and the table spontaneously absorb energy from another portion of the water and produce a piece of ice in the glass. Once again we observe that microscopic disorder (entropy) does not spontaneously decrease.

We can go through a similar argument for a closed can (called a bomb by chemists) containing one mole of hydrogen gas and one-half mole of oxygen gas at room temperature. If a slight activation energy (a match or an electric arc) is supplied to this mixture, it will release a great deal of energy (68 kilocalories) to its surroundings and produce one mole of liquid water. We have never observed the reverse reaction in which the surroundings spontaneously give up 68 kilocalories of energy and dissociate water into hydrogn and oxygen. *Microscopic disorder (entropy) of a system and its surroundings (all of the relevant universe) does not spontaneously decrease.*

We call this sentence a statement of the second law of thermodynamics. And we know that it is correct.[4]

.   .   .   .   .

We now have a test to use to determine which way a process will proceed. For example, the test says that thermal energy will flow[5] to produce a net increase in entropy.

.   .   .   .   .

[4]Richard Bellman said, "What I tell you three times is true."

[5]The second law, like the other laws of thermodynamics, is silent about the rate of natural processes. It says thermal energy will flow from the hot to the cold reservoir; a bouncing ball will come to rest, eventually. But how long is eventually? Seconds or eons? The second law doesn't say. Its NO is emphatic; its YES is only permissive.

Let us carry out a simple series of thought experiments that will illuminate another aspect of the second law of thermodynamics. Consider a box divided into two sections by a partition with a small hole in it. Assume that we have available to us equipment that will allow us to introduce identical molecules one at a time into the box; these molecules are free to go to either side of the box. First we place just one molecule in the box. The number of possible states for the molecule in the box is two—that is, the molecule is either in the left or in the right portion of the box:

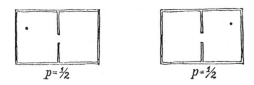

The probability that the molecule is in one side or the other is one-half; both states are equally likely to occur. The probabilities for both states must add up to one, since it is a certainty that the molecule is in the box somewhere.

We now consider the case of two molecules in the box. There are four possible states: two molecules on the left, two molecules on the right, and two states with a molecule on each side. We note that in this last case we can get a molecule on each side two different

ways because we can always interchange the left molecule with the right and the right with the left (since they are identical). Thus,

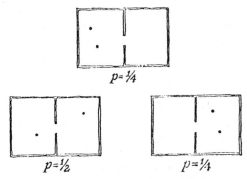

We observe that the center arrangement is the most likely to occur because it has a probability twice that of the other arrangements. Once again we note that the probabilities add to unity since it is a certainty that the molecules will take up one of these arrangements.

The probability that n events will occur simultaneously is simply the product of the probability of the individual events occurring.[6] Since the probability that one molecule will be on a give side $1/2$, the probability that all n molecules in a system will simultaneously be on one side is $(\frac{1}{2})^n$. Thus for a system of four molecules the probability of all four being on one side is $\frac{1}{2} \times \frac{1}{2} \times \frac{1}{2} \times \frac{1}{2} = \frac{1}{16}$. Probabilities of other arrangements are as follows:

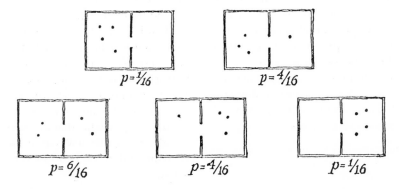

[6]Simple probability calculations can also be quite revealing in the examination of certain occupations. Consider the football parlay card where the point spreads are chosen by an expert. On such cards the probability of picking a single winner is about one-half. Then the probability of choosing three winners out of three picks is $(\frac{1}{2})^3 = \frac{1}{8}$, and the probability of choosing ten winners out of ten picks is $(\frac{1}{2})^{10} = \frac{1}{1024}$. Who gets rich when parlay cards offer odds of four to one for picking three winners out of three attempts and one hundred to one for picking ten winners out of ten attempts?

The probability of forming the second state is $\frac{4}{16}$ since there are four different ways to form each of these states. Similarly, the probability of forming the middle state is $\frac{6}{16}$.

We summarize some conclusions that can be reached from considerations of these and similar illustrations for larger numbers of molecules as follows: (1) As the number of molecules involved increases, the probability that all will be simultaneously in one side of the box becomes very small (with n equal only to ten the probability becomes less than one in a thousand); (2) the molecules are most likely to be distributed evenly between the two sides of the box. Here it is important to recognize that the relatively great probability of the evenly divided distribution as compared to the distribution with all molecules on one side arises from the large number of ways in which an even distribution can be obtained as compared to the single way in which one can obtain a distribution with all molecules on one side. We associate a large number of ways in which a given distribution can be obtained with randomness or molecular disorder and thence high entropy. Order and low entropy are associated with states that can be achieved in only a small number of ways.

We conclude that if we introduce a number of molecules into one side of the box and uncover the small hole between the two parts, it is highly unlikely that the molecules will stay on that side. They will arrange themselves from a state of low probability (low entropy) to the state of maximum probability (high entropy)—that is, they will soon distribute themselves as uniformly as possible throughout the box. Two American chemists, Gilbert N. Lewis and M. Randall, generalized this observation for all natural physical phenomena by writing:

Every system which is left to itself will, on the average, change toward a condition of maximum probability.[7]

This last statement is a general statement of the second law of thermodynamics and is seen to be equivalent to our earlier statement about the impossibility of microscopic disorder decreasing spontaneously. Man creates local and temporary islands of decreasing entropy in a world in which the entropy as a whole certainly increases, and it is the existence of these islands that enables some of us to assert the existence of progress.

The reader must not be discouraged if his understanding of the second law and entropy do not appear crystal-clear at this point. Many scientists and engineers regard entropy as a convenient mathematical function—simply a ratio of heat to temperature, nothing more and nothing less. They regard all attempts to give it a physical meaning as futile. As Morton Mott-Smith has pointed out, velocity could also be regarded as merely a useful mathematical function, the ratio of distance to time. However, because of our direct experience with motion, velocity means more to us than a mere ratio. But we can have no direct experience with entropy; we cannot feel it like temperature or see its effects like heat. Our knowledge of it is necessarily roundabout and our conception will always be a little vague. Boltzmann, who discovered the mathematical relation between entropy and probability, once said, "How awkward is the human mind in divining the nature of things, when forsaken by the analogy of what we see and touch directly."

All other variables with which science is concerned can be increased or decreased—but entropy and time always increase. Entropy can only be decreased temporarily and then only in a localized region at the expense of a greater increase elsewhere. It is a one-way variable that marks the universe as older today than it was yesterday. Entropy, as Arthur Eddinton expressed it, is "Time's Arrow."

[7]Sometimes paraphrased as: If you think things are mixed up now, just wait.

CHAPTER 3
ELECTRICITY AND LIGHT

3.1   *The One Fluid Theory of Electricity*

Benjamin Franklin

1. A person standing on wax, and rubbing a tube, and another person on wax drawing the fire; they will both of them, provided they do not stand so as to touch one another, appear to be electrified to a person standing on the floor; that is, he will perceive a spark on approaching each of them with his knuckle.

2. But if the persons on wax touch one another during the exciting of the tube, neither of them will appear to be electrified.

3. If they touch one another after the exciting the tube and drawing the fire as aforesaid, there will be a stronger spark between them than was between either of them and the person on the floor.

4. After such a strong spark neither of them discover any electricity.

These appearances we attempt to account for thus:

We suppose, as aforesaid, that electrical fire is a common element, of which every one of these three persons has his equal share before any operation is begun with the tube. A, who stands upon wax, and rubs the tube, collects the electrical fire from himself into the glass; and his communication with the common stock being cut off by the wax, his body is not again immediately supplied. B, who stands upon wax likewise, passing his knuckle along near the tube, receives the fire which was collected by the glass from A; and his communication with the common stock being cut off, he retains the additional quantity received. To C standing on the floor, both appear to be electrified; for he, having only the middle quantity of electrical fire, receives a spark upon approaching B, who has an over quantity, but gives one to A, who has an under quantity. If A and B approach to touch each other, the spark is stronger; because the difference between them is greater. After such touch, there is no spark between either of them and C, because the electrical fire in all is reduced to the original equality. If they touch while electrizing, the equality is never destroyed, the fire only circulating. Hence have arisen some new terms among us. We say, B (and bodies alike circumstanced) is electrized positively; A, negatively; or rather, B is electrized *plus*, A, *minus*. And we daily in our experiments electrize *plus* or *minus*, as we think proper. To electrize *plus* or *minus*, no more needs be known than this; that the parts of the tube or sphere that are rubbed, do in the instant of the friction

SOURCE. From a letter dated June 1, 1747. Published in *Philosophical Transactions* **45**:98 (1750).

attract the electrical fire, and therefore take it from the thing rubbing. The same parts immediately, as the friction upon them ceases, are disposed to give the fire, they have received, to any body that has less. Thus you may circulate it, as *Mr. Watson* has shown; you may also accumulate or subtract it upon or from any body, as you connect that body with the rubber, or with the receiver, the communication with the common stock being cut off.

3.2   *Law of Electrical Force*
Charles-Augustin Coulomb

Construction and use of an electric balance-based on the properties of metallic wires of having a force of reaction of torsion proportional to the angle of torsion.

Experimental determination of the law according to which the elements of bodies electrified with the same kind of electricity repel each other.

In a memoir presented to the Academy in 1784, I determined by experiment the laws of the force of torsion of a metallic wire, and I found that this force was in a ratio compounded of the angle of torsion, of the fourth power of the diameter of the suspended wire, and of the reciprocal of its length, all being multiplied by a constant coefficient which depends on the nature of the metal and which is easy to determine by experiment.

I showed in the same memoir that by using this force of torsion it was possible to measure with precision very small forces, as for example, a ten thousandth of a grain. I gave in the same memoir an application of this theory, by attempting to measure the constant force attributed to adhesion in the formula which expresses the friction of the surface of a solid body in motion in a fluid.

I submit today to the Academy an electric balance constructed on the same principle; it measures very exactly the state and the electric force of a body however slightly it is charged.

Construction of balance. Although I have learned by experience that to carry out several electric experiments in a convenient way I should correct some defects in the first balance of this sort which I have made; nevertheless as it is so far the only one that I have used I shall give its description, simply remarking that its form and size may be and should be changed according to the nature of the experiments that one is planning to make. The first figure represents this balance in perspective and the details of it are as follows:

On a glass cylinder *ABCD* (Figure 3.2–1) 12 inches in diameter and 12 inches high is placed a glass plate 13 inches in diameter, which entirely covers the glass vessel; this plate is pierced with two holes of about twenty lines[1] in diameter, one of them in the middle, at *f*, above which is placed a glass tube 24 inches high; this tube is cemented over the hole *f* with the cement ordinarily used in electrical apparatus: at the upper end of the tube at *h* is placed a torsion micrometer which is seen in detail in figure 2. The upper part, No. 1, carries the milled head *b*, the index *io*, and the clamp *q*; this piece fits into the hole *G* of the piece No. 2; this piece No. 2 is made up of a circle *ab* divided on its edge into 360 degrees and of a copper tube *Φ* which fits into the tube *H*, No. 3, sealed to the interior of the upper end of the glass tube or column *fh* of figure 1. The clamp *q* (Figure 3.2–1, 2, No. 1), is shaped much like the end of a solid crayon holder, which is closed by means of the ring *q*. In this holder

SOURCE. From *Mémoires de l'Académie Royale des Sciences*, 1785 (published, 1788). See W. L. Magie, *A Source Book in Physics* (Cambridge Mass.: Harvard University Press, 1963) p. 408.

[1] 1 line = 1/40 inch [Ed.]

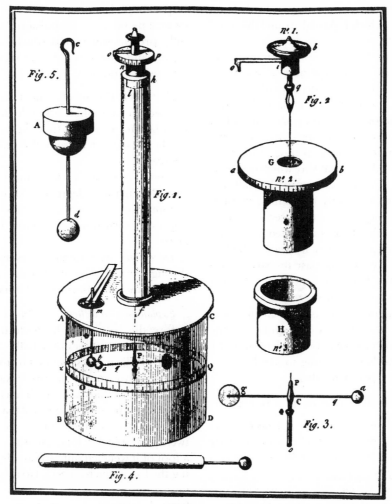

Figure 3.2–1.

is clamped the end of a very fine silver wire; the other end of the silver wire (Figure 3.2–1, 3) is held at P in a clamp made of a cylinder Po of copper or iron with a diameter of not more than a line, whose upper end P is split so as to form a clamp which is closed by means of the sliding piece Φ. This small cylinder is enlarged at C and a hole bored through it, in which can be inserted (Figure 3.2–1, 1) the needle ag: the weight of this little cylinder should be sufficiently great to keep the silver wire stretched without breaking it. The needle that is shown (Figure 3.2–1, 1) at ag suspended horizontally about half way up in the large vessel which encloses it, is formed either of a silk thread soaked in

Spanish wax or of a straw likewise soaked in Spanish wax and finished off from q to a for eighteen lines of its length by a cylindrical rod of shellac; at the end a of this needle is carried a little pith ball two or three lines in diameter; at g there is a little vertical piece of paper soaked in terebinth, which serves as a counterweight for the ball a and which slows down the oscillations.

We have said that the cover AC was pierced by a second hole at m. In this second hole there is introduced a small cylinder $m\Phi t$, the lower part of which Φt is made of shellac; at t is another pith ball; about the vessel, at the height of the needle, is described a circle zQ divided into 360 degrees: for greater simplicity I use a strip of paper divided into 360 degrees which is pasted around the vessel at the height of the needle.

To arrange this instrument for use I set on the cover so that the hole m practically corresponds to the first division of the circle zoQ traced on the vessel. I place the index oi of the micrometer on the point o or the first division of this micrometer; I then turn the micrometer in the vertical tube fh until, by looking past the vertical wire which suspends the needle and the center of the ball, the needle ag corresponds to the first division of the circle zoQ. I then introduce through the hole m the other ball t suspended by the rod $m\Phi t$, in such a way that it touches the ball a and that by looking past the suspension wire and the ball t we encounter the first division o of the circle zoQ. The balance is then in condition to be used for all our operations; as an example we go on to give the method which we have used to determine the fundamental law according to which electrified bodies repel each other.

FUNDAMENTAL LAW OF ELECTRICITY

The repulsive force between two small spheres charged with the same sort of electricity is in the inverse ratio of the squares of the distances between the centers of the two spheres.

Experiment. We electrify a small conductor, (Figure 3.2–1, 4) which is simply a pin with a large head insulated by sinking its point into the end of a rod of Spanish wax; we introduce this pin through the hole m and with it touch the ball t, which is in contact with the ball a; on withdrawing the pin the two balls are electrified with electricity of the same sort and they repel each other to a distance which is measured by looking past the suspension wire and the center of the ball a to the corresponding division of the circle zoQ; then by turning the index of the micrometer in the sense pno we twist the suspension wire lp and exert a force proportional to the angle of torsion, which tends to bring the ball a nearer to the ball t. We observe in this way the distance through which different angles of torsion bring the ball a toward the ball t, and by comparing the forces of torsion with the cor-

responding distances of the two balls we determine the law of repulsion. I shall here only present some trials which are easy to repeat and which will at once make evident the law of repulsion.

First Trial. Having electrified the two balls by means of the pin head while the index of the micrometer points to o, the ball a of the needle is separated from the ball t by 36 degrees.

Second Trial. By twisting the suspension wire through 126 degrees as shown by the pointer o of the micrometer, the two balls approach each other and stand 18 degrees apart.

Third Trial. By twisting the suspension wire through 567 degrees the two balls approach to a distance of 8 degrees and a half.

Explanation and Result of This Experiment. Before the balls have been electrified they touch, and the center of the ball a suspended by the needle is not separated from the point where the torsion of the suspension wire is zero by more than half the diameters of the two balls. It must be mentioned that the silver wire lp which formed this suspension was twenty-eight inches long and was so fine that a foot of it weighed only $\frac{1}{16}$ grain. By calculating the force which is needed to twist this wire by acting on the point a four inches away from the wire lp or from the center of suspension, I have found by using the formulas explained in a memoir on the laws of the force of torsion of metallic wires, printed in the Volume of the Academy for 1784, that to twist this wire through 360 degrees the force that was needed when applied at the point a so as to act on the lever an four inches long was only $\frac{1}{340}$ grains: so that since the forces of torsion, as is proved in that memoir, are as the angles of torsion, the least repulsive force between the two balls would separate them sensibly from each other.

We found in our first experiment, in which the index of the micrometer is set on the point o, that the balls are separated by 36 degrees, which produces a force of torsion of $36° = \frac{1}{3400}$ of a grain; in the second trial the distance between the balls is 18 degrees, but as the micrometer has been turned through 126 degrees it results that at a distance of 18 degrees the repulsive force was equivalent to 144 degrees; so at half the first distance the repulsion of the balls is quadruple.

In the third trial the suspension wire was twisted through 567 degrees and the two balls are separated by only 8 degrees and a half. The total torsion was consequently 576 degrees, four times that of the second trial, and the distance of the two balls in this third trial lacked only one-half degree of being reduced to half of that at which it stood in the second trial. It results then from these three trials that the repulsive action which the two balls exert on each other when they are electrified similarly is in the inverse ratio of the square of the distances.

In the remarks which Coulomb adds to this account he points out first that when the very fine wire which he describes is used there is some uncertainty about the natural position of the zero. This uncertainty can be corrected by a suitable modification of the method of observation but he suggests that it is generally better to use a thicker wire. He also calls attention to the possible loss of electricity during the experiment, and suggests a way by which this can be observed and

allowed for. He also points out that the repulsive action actually takes place along the chord of the arc by which the distances are measured, but shows that the errors introduced by measuring the distance by the degrees at least partly compensate each other, and that the errors are unimportant when the deflections do not exceed 25 to 30 degrees. He also shows how the instrument can be used to detect exceedingly small quantities of electricity.

Second Memoir on Electricity and Magnetism in which there are determined the laws according to which the magnetic fluid as also the electric fluid act either by repulsion or by attraction.

The first section of this memoir calls attention to the difficulties encountered in the use of the torsion balance when the electrical force is attractive, and describes the precautions which it was found necessary to take in order to obtain satisfactory results. The law of inverse squares was found to hold also in the case of attraction.

Second experimental method to determine the law with which a sphere one or two feet in diameter attracts a small body charged with electricity of a different sort from its own.

The method which we shall follow is analogous to that which we have used in the seventh volume of the *Savans Étrangers* to

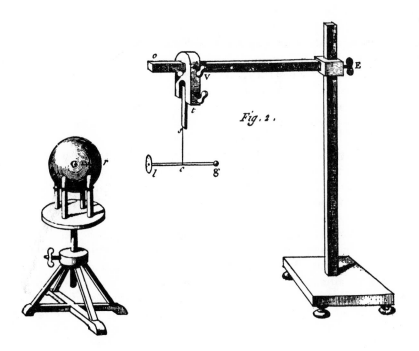

Fig. 2.

Figure 3.2–2.

determine the magnetic force of a steel plate in relation to its length, its thickness, and its width. It consists in suspending a needle horizontally, of which the end only is electrified and which, when brought to a certain distance from a sphere, electrified with the other sort of electricity, is attracted and oscillates because of the action of the sphere: we determine then by calculation from the number of oscillations in a given time the attractive force at different distances, just as we determine the force of gravity by the oscillations of an ordinary pendulum.

We shall first consider some observations which have guided us in the experiments which are to follow. A silk fiber taken from a cocoon which can sustain 80 grains without breaking yields so readily to torsion that if we suspend horizontally to such a fiber three inches long in vacuum a small circular plate of which the weight and diameter are known we shall find from the period of oscillation of this little plate, using the formulas explained in a memoir on the force of torsion printed in the Volume of the Academy for 1784, that when we use a lever of 7 or 8 lines long to twist the fiber about its axis of suspension we shall need for a complete rotation to use usually not more than the force of a sixty thousandth of a grain; and if the suspended fiber is twice as long there will be needed only a hundred and twenty-thousandth of a grain. Therefore if we suspend a needle horizontally on this fiber, when the needle has come to rest and the fiber is entirely untwisted, and if by means of any force we set the needle in oscillations whose amplitude does not depart from the line in which the torsion is zero by more than 20 or 30 degrees, the force of torsion will have no sensible effect on the period of the oscillation, even when the force that produces the oscillations is not more than a hundredth of a grain. Premising this much, let us see how we proceed to determine the law of electrical attraction.

We suspend, (Figure 3.2–2, 2) a needle *lg* made of shellac by a silk thread *sc* 7 to 8

inches long of a single fiber such as is drawn from the cocoon; at the end *l* we fix perpendicularly to the needle a little disc 8 or 10 lines in diameter, made very light and cut from a sheet of gilt paper; the silk thread is attached at *s* to the lower end of a little rod *st* dried in a furnace and coated with shellac or with Spanish wax; this rod is held at *t* by a clamp which slides along a ruled rod *oE* and can be placed anywhere we desire by means of the screw *V*.

G is a globe of copper or of cardboard covered with tin. It is carried on four uprights of glass coated with Spanish wax, and terminated, in order to make the insulation more perfect, by four rods of Spanish wax three or four inches long. The lower ends of these four uprights are set in a base which is placed on a little movable table that, as the figure shows, can be set at the height which is most convenient for the experiment; the rod *Eo* may also, by means of the screw *E*, be set at a convenient height.

When everything is ready we adjust the globe *G* in such a way that its horizontal diameter *Gr* is opposite the center of the plate *l*, which is some inches away from it. We give an electric spark to the sphere from a Leyden jar[2], we then touch the place *l* with a conductor and the action of the electrified sphere on the electric fluid of the unelectrified plate gives to the plate a charge of the other sort from that of the sphere; so that when the conducting body is removed the sphere and the plate act on each other by attraction.

Experiment. The sphere *G* was a foot in diameter; the plate *l* was 7 lines in diameter; the shellac needle *lg* was 15 lines long; the suspension fiber *sc* was a silk fiber taken from the cocoon and 8 lines long: when the slider was at the point *o* the plate *l* touched the sphere at *r*, and as the slider was moved toward *E* the plate was removed from the center of the sphere by the quantity given by the divisions 0, 3, 6, 9, 12 inches, and when the sphere was electrified with what is called positive electricity and the plate with negative

---

[1]A type of capacitor in which a considerable electrical charge can be stored [Ed.]

electricity by the method which has been described, we had:

Trial 1—The plate *l* being at ...... 3 inches from the surface of the sphere or 9 inches from its center gave .... 15 oscillations in 20″.

Trial 2—The plate *l* distant by .... 18 inches from the center of the sphere gave .... 15 oscillations in 40″.

Trial 3—The plate *l* distant by ....24 inches from the center of the sphere gave .... 15 oscillations in 60″.

Explanation of This Experiment and its Result. When all the points of a spherical surface act by an attractive or repulsive force which varies inversely as the square of the distance on a point placed at any distance from this surface it is known that the action is the same as if all the spherical surface were concentrated at the center of the sphere.

As in our experiment the plate *l* was only 7 lines in diameter and as in the trials its least distance from the center of the sphere was 9 inches, we may, without sensible error, suppose that all the lines which are drawn from the center of the sphere to a point of the plate are parallel and equal; and in consequence the total action of the plate can be supposed to be united at its center just as in the case of the sphere; so that for the small oscillations of the needle, the action which makes the needle oscillate will be a constant quantity for a given distance and will act along the line which joins the two centers. Therefore if we call the force φ and the time of a certain number of oscillations T we shall have T proportional to $1/\sqrt{\varphi}$ but if d is the distance *Gl* from the center of the sphere to the center of the plate and if the attractive forces are proportional to the reciprocal of the square of the distances or to $1/d^2$, it follows that T will be proportional to d or to the distance; so that when we make our trials and change the distance, the time of the same number of oscillations ought to be proportional to the distance from the center of the plate to the center of the sphere: let us compare this theory with experiment.

Trial 1—Distance between centers ....9 inches ....15 oscillations in 20″.

Trial 2— .... 18 inches .... in 41″.

Trial 3— .... 24 inches .... in 60″.

The distances are as the numbers 3, 6, 8,
The times of the same number of oscillations 20, 41, 60,
By theory they ought to have been 20, 40, 54,

Thus in these three trials the difference between theory and experiment is $\frac{1}{10}$ for the last trial compared with the first, and almost nothing for the second trial compared with the first; but it should be remarked that it took almost four minutes to make the three trials; that although the electricity held pretty well on the day this experiment was tried, it nevertheless lost $\frac{1}{40}$ of its amount each minute. We shall see, in a memoir which will follow the one which I am presenting today, that when the electric density is not very great, the electric action of two electrified bodies diminishes in a given time exactly as the electric density or as the intensity of the action; therefore, since our trials lasted four minutes and since the electric action lost $\frac{1}{40}$ each minute from the first to the last trial, the action arising from the intensity of the electric density independently of the distance should be diminished by almost a tenth; consequently, to have the corrected time of the 15 oscillations in the last trial, we must set $\sqrt{(10)}:\sqrt{(9)}::60″$: the quantity required, which will be found to be 57 seconds, which differs only by $\frac{1}{20}$ from the 60 seconds found by experiment.

We have thus come, by a method absolutely different from the first, to a similar result; we may therefore conclude that the mutual attraction of the electric fluid which is called positive on the electric fluid which is ordinarily called negative is in the inverse ratio of the square of the distances; just as we have found in our first memoir, that the mutual action of the electric fluid of the same sort is in the inverse ratio of the square of the distance.

3.3   *Treatise on Light (1678)*

Christian Huyghens

It is inconceivable to doubt that light consists in the motion of some sort of matter. For whether one considers its production, one sees that here upon the Earth it is chiefly engendered by fire and flame which contain without doubt bodies that are in rapid motion, since they dissolve and melt many other bodies, even the most solid; or whether one considers its effects, one sees that when light is collected, as by concave mirrors, it has the property of burning as a fire does, that is to say it disunites the particles of bodies. This is assuredly the mark of motion, at least in the true Philosophy, in which one conceives the causes of all natural effects in terms of mechanical motions. This, in my opinion, we must necessarily do, or else renounce all hopes of ever comprehending anything in Physics.

And as, according to this Philosophy, one holds as certain that the sensation of sight is excited only by the impression of some movement of a kind of matter which acts on the nerves at the back of our eyes, there is here yet one reason more for believing that light consists in a movement of the matter which exists between us and the luminous body.

Further, when one considers the extreme speed with which light spreads on every side, and how, when it comes from different regions, even from those directly opposite, the rays traverse one another without hindrance, one may well understand that when we see a luminous object, it cannot be by any transport of matter coming to us from this object, in the way in which a shot or an arrow traverses the air; for assuredly that would too greatly impugn these two properties of light, especially the second of them.

It is then in some other way that light spreads; and that which can lead us to comprehend it is the knowledge which we have of the spreading of Sound in the air.

We know that by means of the air, which is an invisible and impalpable body, Sound spreads around the spot where it has been produced, by a movement which is passed on successively from one part of the air to another; and that the spreading of this movement, taking place equally rapidly on all sides, ought to form spherical surfaces ever enlarging and which strike our ears. Now there is no doubt at all that light also comes from the luminous body to our eyes by some movement impressed on the matter which is between the two; since, as we have already seen, it cannot be by the transport of a body which passes from one to the other. If, in addition, light takes time for its passage—which we are now going to examine—it will follow that this movement, impressed on the intervening matter, is successive; and consequently it spreads, as Sound does, by spherical surfaces and waves: for I call them waves from their resemblance to those which are seen to be formed in water when a stone is thrown into it, and which present a successive spreading as circles, though these arise from another cause, and are only in a flat surface.

To see then whether the spreading of light takes time, let us consider first whether there are any facts of experience which can convince us to the contrary. As to those which can be made here on the Earth, by striking lights at great distances, although they prove that light takes no sensible time to pass over these distances, one may say with good reason that they are too small, and that the only conclusion to be drawn from them is that the

SOURCE. From *Moments of Discovery: The Origins of Science*, G. Swartz and P. W. Bishope, (Eds.) (New York: Basic Books, 1958); translated by S. P. Thompson (1912).

passage of light is extremely rapid.

It is true that we are here supposing a strange velocity that would be a hundred thousand times greater than that of Sound. For Sound, according to what I have observed, travels about 180 Toises[1] in the time of one Second, or in about one beat of the pulse. But this supposition ought not to seem to be an impossibility; since it is not a question of the transport of a body with so great a speed, but of a successive movement which is passed on from some bodies to others. I have then made no difficulty, in meditating on these things, in supposing that the emanation of light is accomplished with time, seeing that in this way all its phenomena can be explained, and that in following the contrary opinion everything is incomprehensible. For it has always seemed to me that even Mr. Des Cartes, whose aim has been to treat all the subjects of Physics intelligibly, and who assuredly has succeeded in this better than any one before him, has said nothing that is not full of difficulties, or even inconceivable in dealing with Light and its properties.

But that which I employed only as a hypothesis, has recently received great seemingness as an established truth by the ingenious proof of Mr. Römer which I am going here to relate, expecting him himself to give all that is needed for its confirmation. It is founded as is the preceding argument upon celestial observations, and proves not only that Light takes time for its passage, but also demonstrates how much time it takes, and that its velocity is even at least six times greater than that which I have just stated.

For this he makes use of the Eclipses suffered by the little planets which revolve around Jupiter, and which often enter his shadow: and see what is his reasoning [Figure 3.3–1]. Let A be the Sun, $BCDE$ the annual orbit of the Earth, F Jupiter, GN the orbit of the nearest of his Satellites, for it is this one which is more apt for this investigation than any of the other three, because of

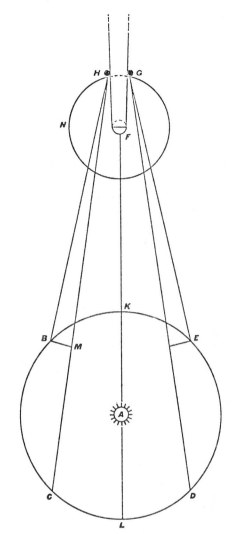

Figure 3.3–1.

the quickness of its revolution. Let G be this Satellite entering into the shadow of Jupiter, H the same Satellite emerging from the shadow.

Let it be then supposed, the Earth being at B some time before the last quadrature, that one has seen the said Satellite emerge from the shadow; it must needs be, if the Earth remains at the same place, that, after $42\frac{1}{2}$ hours, one would again see a similar emergence, because that is the time in which

[1]About 1150 feet. (1 Toise = 76.74 inches.) [Ed.]

it makes the round of its orbit, and when it would come again into opposition to the Sun. And if the Earth, for instance, were to remain always at B during 30 revolutions of this Satellite, one would see it again emerge from the shadow after 30 times $42\frac{1}{2}$ hours. But the Earth having been carried along during this time to C, increasing thus its distance from Jupiter, it follows that if Light requires time for its passage the illumination of the little planet will be perceived later at C than it would have been at B, and that there must be added to this time of 30 times $42\frac{1}{2}$ hours that which the Light has required to traverse the space MC, the difference of the spaces CH, BH. Similarly at the other quadrature when the earth has come to E from D while approaching toward Jupiter, the immersions of the Satellite ought to be observed at E earlier than they would have been seen if the Earth had remained at D.

Now in quantities of observations of these Eclipses, made during ten consecutive years, these differences have been found to be very considerable, such as ten minutes and more; and from them it has been concluded that in order to traverse the whole diameter of the annual orbit KL, which is double the distance from here to the sun, Light requires about 22 minutes of time.

The movement of Jupiter in his orbit while the Earth passed from B to C, or from D to E, is included in this calculation; and this makes it evident that one cannot attribute the retardation of these illuminations or the anticipation of the eclipses, either to any irregularity occurring in the movement of the little planet or to its eccentricity.

If one considers the vast size of the diameter KL, which according to me is some 24 thousand diameters of the Earth, one will acknowledge the extreme velocity of Light. For, supposing that KL is no more than 22 thousand of these diameters, it appears that being traversed in 22 minutes this makes the speed a thousand diameters in one minute, that is $16\frac{2}{3}$ diameters in one second or in one beat of the pulse, which makes more than 11 hundred times a hundred thousand Toises; since the diameter of the Earth contains 2,865 leagues, reckoned at 25 to the degree, and each league is 2,282 Toises, according to the exact measurement which Mr. Picard made by order of the King in 1669. But Sound, as I have said above, only travels 180 toises in the same time of one second: hence the velocity of Light is more than six hundred thousand times greater than that of Sound.[2] This, however, is quite another thing from being instantaneous, since there is all the difference between a finite thing and an infinite. Now the successive movement of Light being confirmed in this way, it follows, as I have said, that it spreads by spherical waves, like the movement of Sound.

But if the one resembles the other in this respect, they differ in many other things; to wit, in the first production of the movement which causes them; in the matter in which the movement spreads; and in the manner in which it is propagated. As to that which occurs in the production of Sound, one knows that it is occasioned by the agitation undergone by an entire body, or by a considerable part of one, which shakes all the contiguous air. But the movement of the Light must originate as from each point of the luminous object, else we should not be able to perceive all the different parts of that object, as will be more evident in that which follows. And I do not believe that this movement can be better explained than by supposing that all those of the luminous bodies which are liquid, such as flames, and apparently the sun and the stars, are composed of particles which float in a much more subtle medium which agitates them with great rapidity, and makes them strike against the particles of the ether which surrounds them, and which are much smaller than they. But I hold also that in luminous solids such

---

[2]The value obtained in this way for the speed of light is about $\frac{2}{3}$ of the true value, 3×10^{10} cm/sec. [Ed.]

as charcoal or metal made red hot in the fire, this same movement is caused by the violent agitation of the particles of the metal or of the wood; those of them which are on the surface striking similarly against the ethereal matter. The agitation, moreover, of the particles which engender the light ought to be much more prompt and more rapid than is that of the bodies which cause sound, since we do not see that the tremors of a body which is giving out a sound are capable of giving rise to Light, even as the movement of the hand in the air is not capable of producing sound.

Now if one examines what this matter may be in which the movement coming from the luminous body is propagated, which I call Ethereal matter, one will see that it is not the same that serves for the propagation of Sound. For one finds that the latter is really that which we feel and which we breathe, and which being removed from any place still leaves there the other kind of matter that serves to convey Light. This may be proved by shutting up a sounding body in a glass vessel from which the air is withdrawn by the machine which Mr. Boyle has given us, and with which he has performed so many beautiful experiments. But in doing this of which I speak, care must be taken to place the sounding body on cotton or on feathers, in such a way that it cannot communicate its tremors either to the glass vessel which encloses it, or to the machine; a precaution which has hitherto been neglected. For then after having exhausted all the air one hears no Sound from the metal, though it is struck.

One sees here not only that our air, which does not penetrate through glass, is the matter by which Sound spreads; but also that it is not the same air but another kind of matter in which Light spreads; since if the air is removed from the vessel the Light does not cease to traverse it as before.

And this last point is demonstrated even more clearly by the celebrated experiment of Torricelli, in which the tube of glass from which the quicksilver has withdrawn itself, remaining void of air, transmits Light just the same as when air is in it. For this proves that a matter different from air exists in this tube, and that this matter must have penetrated the glass or the quicksilver, either one or the other, though they are both impenetrable to the air. And when, in the same experiment, one makes the vacuum after putting a little water above the quicksilver, one concludes that the said matter passes through glass or water, or through both.

As regards the different modes in which I have said the movements of Sound and of Light are communicated, one may sufficiently comprehend how this occurs in the case of Sound if one considers that the air is of such a nature that it can be compressed and reduced to a much smaller space than that which it ordinarily occupies. And in proportion as it is compressed the more does it exert an effort to regain its volume; for this property along with its penetrability, which remains notwithstanding its compression, seems to prove that it is made up of small bodies which float about and which are agitated very rapidly in the ethereal matter composed of much smaller parts. So that the cause of the spreading of Sound is the effort which these little bodies make in collisions with one another, to regain freedom when they are a little more squeezed together in the circuit of these waves than elsewhere.

But the extreme velocity of Light, and other properties which it has, cannot admit of such a propagation of motion, and I am about to show here the way in which I conceive it must occur. For this, it is needful to explain the property which hard bodies must possess to transmit movement from one to another.

When one takes a number of spheres of equal size, made of some very hard substance, and arranges them in a straight line, so that they touch one another, one finds, on striking with a similar sphere against the first of these spheres, that the motion passes as in an instant to the last of them, which separates itself from the row, without one's being able to perceive that the others have been stirred [Figure 3.3–2]. And even that one which was used to strike remains motionless with them.

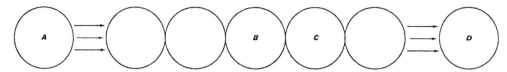

Figure 3.3–2.

Whence one sees that the movement passes with an extreme velocity which is the greater, the greater the hardness of the substance of the spheres.

But it is still certain that this progression of motion is not instantaneous, but successive, and therefore must take time. For if the movement, or the disposition to movement, if you will have it so, did not pass successively through all these spheres, they would all acquire the movement at the same time, and hence would all advance together; which does not happen. For the last one leaves the whole row and acquires the speed of the one which was pushed. Moreover there are experiments which demonstrate that all the bodies which we reckon of the hardest kind, such as quenched steel, glass, and agate, act as springs and bend somehow, not only when extended as rods but also when they are in the form of spheres or of other shapes. That is to say they yield a little in themselves at the place where they are struck, and immediately regain their former figure. For I have found that on striking with a ball of glass or of agate against a large and quite thick piece of the same substance which had a flat surface, slightly soiled with breath or in some other way, there remained round marks, of smaller or larger size according as the blow had been weak or strong. This makes it evident that these substances yield where they meet, and spring back: and for this time must be required.

I have then shown in what manner one may conceive Light to spread successively, by spherical waves, and how it is possible that this spreading is accomplished with as great a velocity as that which experiments and celestial observations demand. Whence it may be further remarked that although the particles are supposed to be in continual movement (for there are many reasons for

this) the successive propagation of the waves cannot be hindered by this; because the propagation consists nowise in the transport of those particles but merely in a small agitation which cannot help communicating to those surrounding, notwithstanding any movement which may act on them causing them to be changing positions amongst themselves.

But we must consider still more particularly the origin of these waves, and the manner in which they spread. And first, it follows from what has been said on the production of Light, that each little region of a luminous body, such as the Sun, a candle, or a burning coal, generates its own waves of which that region is the center. Thus in the flame of a candle, having distinguished the points A, B, C, concentric circles described about each of these points represent the waves which come from them. And one must imagine the same about every point of the surface and of the part within the flame.

But as the percussions at the centers of these waves possess no regular succession, it must not be supposed that the waves themselves follow one another at equal distance: and if the distances marked in the figure [Figure 3.3–3] appear to be such, it is rather to mark the progression of one and the same wave at equal intervals of time than to represent several of them issuing from one and the same center.

After all, this prodigious quantity of waves which traverse one another without confusion and without effacing one another must not be deemed inconceivable; it being certain that one and the same particle of matter can serve for many waves coming from different sides or even from contrary directions, not only if it is struck by blows which follow one another closely but even for those which act on it at the same instant. It can do so because

Figure 3.3–3.

And if these contrary movements happen to meet one another at the middle sphere, *B*, or at some other such as *C*, that sphere will yield and act as a spring at both sides, and so will serve at the same instant to transmit these two movements.

But what may at first appear full strange and even incredible is that the undulations produced by such small movements and corpuscles, should spread to such immense distances; as for example from the Sun or from the Stars to us. For the force of these waves must grow feeble in proportion as they move away from their origin, so that the action of each one in particular will without doubt become incapable of making itself felt to our sight. But one will cease to be astonished by considering how at a great distance from the luminous body an infinitude of waves, though they have issued from different points of this body, unite together in such a way that they sensibly compose one single wave only, which, consequently, ought to have enough force to make itself felt. Thus this infinite number of waves which originate at the same instant from all points of a fixed star, big it may be as the Sun, make practically only one single wave which may well have force enough to produce an impression on our eyes. Moreover from each luminous point there may come many thousands of waves in the smallest imaginable time, by the frequent percussion of the corpuscles which strike the Ether at these points: which further contributes to rendering their action more sensible.

the spreading of the movement is successive. This may be proved by the row of equal spheres of hard matter, spoken of above. If against this row there are pushed from two opposite sides at the same time two similar spheres *A* and *D*, one will see each of them rebound with the same velocity which it had in striking, yet the whole row will remain in its place, although the movement has passed along its whole length twice over.

3.4   *Michael Faraday and the Physics of 100 Years Ago*

L. Pearce Williams

Michael Faraday was born on September 22, 1791, in Newington, Surrey, near London. His father, a journey-man blacksmith, had left the North-country to try to better his lot in the metropolis as a depression gradually settled over England. His mother had worked as a maidservant before her marriage to James Faraday and had

SOURCE. From *Science* **156**: 1335–1342 (9 June 1967).

already borne her husband a daughter, Elizabeth, and a son, Robert. A second daughter, Margaret, soon followed Michael. The family was desperately poor. James Faraday was in almost constant ill health and could work only sporadically. As prices rose as a result of England's involvement in the French Revolutionary Wars, simple subsistence became a major problem. In later years Faraday told of having been given a single loaf of bread which was to serve him as his main course for a week.

What sustained the Faradays throughout their hardships was a simple but extraordinarily powerful religious faith. James Faraday was a Sandemanian. The Sandemanian Church rejected what it considered to be all the false trappings of the Church of England and sought to recapture both the letter and the spirit of the early Church. The congregation was a true brotherhood in which all helped one another both materially and spiritually. Life within the Sandemanian community was often hard, but it was never desperate. With the exception of the influence of his family, about which we know very little, the Sandemanian Church was the most important factor in Michael Faraday's education. It was at a church "school" that he learned the three R's—the total extent of his formal education. More importantly, the Sandemanian religion provided him with two convictions that were essential elements of his later scientific career. According to Judeo-Christian tradition, the universe was, literally, made for man. The philosophical result of this view is the belief in final causes which, as innumerable science texts assure us, was banished from science by the Scientific Revolution. They were not banished from Faraday's mind. He had a deep conviction of the ultimate harmony of the world which led him onward in his physical pursuits. This harmony, he also believed, was designed for man's wellbeing. Thus he could state at the end of a lecture on ozone in 1859:

These are the glimmerings we have of what we are pleased to call the *second causes* by which the *one Great Cause* works his wonders and governs this earth. We flattered ourselves we knew what air was composed of, and now we discover a *new* property which is imponderable, and invisible, except through its *effects* which I shewed you in the last experiment; but while it fades the ribbon, it gives the glow of health to the cheek, and is just as necessary for the good of mankind, as the other parts of which air is composed.

From his religion Faraday also drew a deep and profound sense of his own (and everyone else's) fallibility. He *knew* that he must err and accepted this as a simple fact. He would do his utmost to minimize his errors—he would repeat his experiments hundreds of times; he would scrutinize them with the most critical eye; he would check and recheck his arguments. But he would not insist upon them, once published. If others challenged his results or his ideas, he refused to be drawn into controversy except insofar as he was willing to clarify his sometimes obscure language. Thus, in Faraday's scientific career there is none of the acerbity, belligerence, and intransigence that marked the careers of such contemporaries as Liebig and Tyndall. The truth (always with a small *t*), he felt, would ultimately emerge from the critical interplay of ideas without noisy advocacy.

The young Faraday experienced these religious influences in a world far removed from the world of science. At the age of 14 he was apprenticed to a bookbinder and seemed destined to lead the life of an ordinary tradesman in London. Then, his passion for science was aroused by reading the article on electricity in a volume of the *Encyclopaedia Britannica* brought into his master's shop to be bound. Still, he might have remained a slightly unusual bookbinder had he not been brought to Sir Humphry Davy's attention and, through a series of fortuitous circumstances, hired as Davy's assistant at the Royal Institution in 1813. Thus began his apprenticeship as a chemist.

THE FOUNDATIONS

Under Davy's tutelage Faraday rapidly assimilated the chemical knowledge of the day. More important than the mere digestion of facts, however, was the theoretical point of view he received from his famous mentor. Davy himself had lived through a personal "scientific revolution" in his youth and was able to pass on its meaning to his disciple.

At the end of the eighteenth and the beginning of the nineteenth centuries the nature of physical science underwent an important and fundamental change. By and large, the Newtonian breakthrough of the seventeenth century had focused the attention of physicists primarily upon the physics of the observable world. The publication of the *Mécanique analytique*, by Lagrange, in 1788 marked the culmination of the development of terrestrial mechanics in the eighteenth century. The appearance of the thick quartos of the *Mécanique céleste* of Pierre Simon de Laplace in the early years of the nineteenth century left no doubt of the ability of the Newtonian principles of natural philosophy to deal with celestial phenomena. By the time Faraday began his active career as a scientist, the realm of macrophysics seemed pretty well under the control of the physicist.

The triumphs of Newtonian physics also served to lay out, more or less implicitly, the rules of the game. In macrophysics, certain fundamental aspects were given. If one attacked the problem of the equilibrium of a number of heavy bodies suspended by ropes from pulleys, there was no need to worry about the reality of these bodies. They were simply there, and could be represented by their weight or by the force of gravity acting upon them. It was permissible, indeed necessary, to make certain simplifying assumptions—such as frictionless pulleys, inextensible ropes, and the concentration of mass at the center of gravity—in order to write the mathematical equations by means of which the situation could be analyzed. Similarly, in celestial mechanics, the real existence of the celestial bodies themselves was never in doubt. They were the centers of the gravitational forces which were the objects of Laplace's analysis. Thus, subtly and almost imperceptibly, there were introduced into physics certain ideas and attitudes which, in the nineteenth century, had almost

the strength of dogma. Force, for example, was always associated with body in macrophysics. Hence, wherever force appeared, it seemed only reasonable to assume the existence of a body, even if no body were perceptible. In macrophysics, all forces, were central forces, acting at a distance, between the reacting bodies. All forces, then, must be central forces, acting at a distance between the centers of gravity of the bodies from which these forces arose. Finally, the success of the application of mathematics to macrophysics led to the widespread feeling that mathematical representation was the essence of all physics. If a physical hypothesis could not be put in mathematical terms, then the chances were very high that it was a false one.

With the solution of the major problems of macrophysics, the nineteenth-century physicist turned to an area which had been relatively neglected by the physicists of the eighteenth century. Although Newton had thrown out many helpful hints on the way to approach microphysics, this field had been cultivated largely by chemists. Some of them had learned some very important lessons from their failure to achieve dramatic breakthroughs like those made by their physicist friends. By the time the physicist became seriously interested in such topics as light, heat, electricity, magnetism, and molecular forces, some chemists were on the verge of rejecting the foundations which had served so well to uphold the edifice of macrophysics. Fortunately for Faraday, Davy was one of these chemists.

THE CORPUSCULAR SYSTEM

The basic point at issue was that of the origin of force. In macrophysics, force was considered an essential quality of body. The weight of an object was clearly to be associated with the object, just as the gravitational force of the sun obviously had its origin in the sun. Thus, in microphysics, it seemed only natural to assume that the presence of a force necessarily implied the existence of some material body from which the force emanated. Although it was possible for the same body to be the seat of different kinds of force, it seemed simpler to assume different bodies for each specifically different force. By the beginning of the nineteenth century, certain "forces" were known that could be studied in the laboratory— light, heat, electricity, and magnetism. Each force, it was thought, originated in a specific kind of body. Light, as Sir Isaac Newton had strongly implied, was composed of corpuscles of different sizes, hence the different colors. The particles of heat, or "caloric," were mutually repulsive, and this explained their ability to cause ordinary objects to expand when the caloric content was increased. Most physicists on the Continent believed that the positive and negative aspects of electricity and the boreal and austral forces of magnetism required separate corpuscles. Hence, corpuscles of positive electricity, negative electricity, boreal magnetism, and austral magnetism were conceived as the underlying material basis for electrical and magnetic phenomena. If one adds to these "imponderable" substances the atoms of ordinary matter, whose relative weights John Dalton determined in the first decade of the nineteenth century, the picture is complete.

To the casual observer, there was a wonderful simplicity in this system. Seven "elementary" particles made up the totality of reality. Of these seven, five—positive electricity, negative electricity, boreal magnetism, austral magnetism, and ponderable matter—obeyed the Newtonian inverse-square law. The only mysteries left for the generation that followed Laplace were the force laws of action of light and of caloric. Otherwise, the future task of physics was that of devising more refined methods of mathematical computation so that the results of the interaction of the "elementary" particles could be calculated to the desired degree of accuracy.

OBJECTIONS

This microphysical world appealed to many—especially to the mathematical physi-

cist. But it also repelled many, among whom were some of the leading chemists of the day. To mathematical illiterates like Sir Humphry Davy, elegant equations were so much gibberish. Even had he understood them, it is doubtful that Davy would have been seduced by them; for they gave no real aid in understanding the specific interactions with which the chemist was concerned. Given the atomic weights of nitrogen, sulfur, and chlorine, for example, could the physicist provide any insight into their possible chemical combinations? Could he even predict whether they would react or simply coexist in a flask forever as nitrogen, sulfur, and chlorine?

The corpuscular system was less than seductive to the chemist for another reason. The mathematical physicist, in his mind's eye, could see the trajectories of the ultimate particles and the differential equations associated with them. Each particle had its own, personal, equation; particulate interactions involved the association of these equations. To the chemist, all seven kinds of particles were necessarily associated. For example, for a chemist, an atom of iron contained a central ponderable atom, an atmosphere of caloric (for caloric could be squeezed out of an iron bar by pounding it), an atmosphere of light (the iron will glow when hit long enough), positive and negative electricity (the voltaic current will decompose iron salts in solution), and austral and boreal magnetism. The simplicity of the corpuscular model disappears when the ultimate particles are forced into the union demanded by chemistry.

Beyond these purely chemical objections there were, as well, certain methodological aspects which could upset the thoughtful. The aim of macrophysics was to reduce observable phenomena to mathematical description. In macrophysics, the number of purely physical assumptions necessary to accomplish this could be kept to a minimum through accurate and direct observation. The Newtonian assumption that the r in the inverse-square law of gravitation should be the distance between the centers, rather than between the superficies, of two gravitating bodies could be not only proved mathematically but easily checked by observation. In the microphysical realm, hypothetical actions were not so easily evaluated. One case may serve to illustrate this. In 1820, Augustin Fresnel pointed out to Ampère that the circular electrical currents which Ampère felt must exist in permanent magnets could not be ordinary currents concentric with the axis of the magnet. If such currents existed, they should be detectable by their heating effect, but a magnet was not warmer than its surroundings. But, argued Fresnel, the same results could be obtained if electrical currents were assumed to circulate around the molecules of the magnet. The fact that molecular electrical currents did not generate heat did not perturb Fresnel for, as he put it, "our ideas on the constitution of bodies are too incomplete for us to know whether electricity ought to produce heat in this case." In short, on the molecular level, anything goes. Ad hoc hypotheses could be framed almost without limit so long as they offered a means of explaining physical phenomena or provided a basis for mathematical analysis.

One final aspect of the corpuscular universe should be noted, for it contributed to the opposition to it that arose in the early nineteenth century. In the hands of an ardent proponent like Laplace, this system of the world was overtly, even defiantly, materialistic, deterministic, and atheistic. Following hard on the heels of the French Revolution, this system was strongly tinged with political and social subversion. There were many, particularly in England, who felt that the great cataclysm of 1789 had been caused by precisely those ideas which Laplace and his followers now asserted to be the necessary foundations of physics. Any alternative system which avoided materialism and atheism would be eagerly accepted, provided, of course, it also proved useful in the pursuit of scientific truth. In the early years of the nineteenth century such a system was promulgated in England, and both Davy and Faraday were strongly influenced by it.

THE SYSTEM OF FORCES

In 1799, Samuel Taylor Coleridge, the poet and literary critic, returned from a year's journey to Germany filled with enthusiasm for the new Kantian metaphysics. He was especially impressed with the dismissal of atomism by some of Kant's disciples and the substitution of forces as the basic phenomenal reality of the world. After all, we do not directly experience atoms or the other theoretical entities of microphysics. What we observe and measure and probe are the forces with which these hypothetical corpuscles are endowed. Why not, then, do away with the material substratum entirely and consider forces to be the ultimate reality? There were a number of advantages to this approach. For Coleridge, the important thing was that it removed the duality of matter and spirit and permitted God, once again, to become a living presence in the world. For the natural philosopher, it simplified matters considerably. If force were the ultimate reality, then only the two forces of attraction and repulsion existed. The appearance of these forces as electrical and magnetic attractions and repulsions, thermal expansion, chemical affinity, and so on, depended upon the conditions under which the basic forces manifested themselves. The seven elementary particles of Laplacian physics could be replaced by two forces. Furthermore, predictions could be made of new effects which found no place in the Laplacian system. One would not, for example, expect the conversion of one *particle* into another; electricity was electricity and magnetism was magnetism and, while the two kinds of particles might interact, there was no reason to anticipate the *production* of magnetism by the electrical particles, or vice versa. If all observable forces, however, were but manifestations of attraction and repulsion under different conditions, then it was logical to assume the conversion of one force into another when the proper conditions were present. It was this logic that guided Hans Christian Oersted in his 20-year quest for the magnetic effect of an electric current.

One of Coleridge's best friends was Humphry Davy, himself a poet and ardent student of philosophy. Even before Coleridge returned to England, Davy's own questing mind had led him to a point where he could be only favorably impressed by the philosophical news from Germany. In April 1799, Davy had written to his friend and patron Davies Gilbert (*I*, p. 67): "The supposition of active powers common to all matter, from the different modifications of which all the phenomena of its changes result, appear to me more reasonable than the assumption of certain imaginary fluids alone endowed with active powers, and bearing the same relation to common matter, as the vulgar philosophy supposes spirit to bear to matter." It should be noted that, in the face of public criticism, Davy steadfastly refused to use the Laplacian language of imponderable fluids and spoke, instead, of the powers and energies of matter.

The system of forces was certainly seductive, but it, too, involved certain difficulties. For the physicist, it offered a considerable simplification of the basic hypotheses of his science, but, for the chemist, it was also a source of specifically chemical difficulties. The conversions of the two basic forces offered little to the man interested in the singular qualities of the chemical elements. Given the view that chemical affinity and electricity, for example, were somehow connected, to what did one appeal for an understanding of the specific chemical differences between sodium, potassium, and chlorine? There was no place for such singularities within the physics of forces. Once again, it would appear to be Davy who saw the way to reconcile the system of forces with another system in which chemical singularities could exist. In the eighteenth century an atomic theory had been formulated by Father Roger Joseph Boscovich, in which only forces figured. The atom was a mathematical point, surrounded by alternating zones of repulsive and attractive forces [Figure 3.4–1]. These "atoms," in combination with one another, made up the moleculae of the chemical elements. Chemical qualities were the result of the different patterns of force produced by the different combinations of the point atoms.

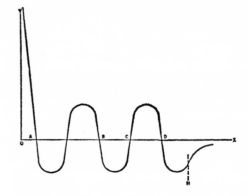

Figure 3.4–1. The Boscovichean point atom has its center at O. The x-axis represents the distance from this mathematical point. The part of the curve above the x-axis represents repulsive force; the part below the x-axis, attractive force. If a test particle is brought in from infinity, it will follow the hyperbola required by the inverse-square law until it reaches some microscopic distance such as that indicated at H. The attractive force then varies in accordance with the curve, turning into a repulsive force at D, back to an attractive force at C, and so on. Between A and OY, the relationship between the y-axis and the curve representing the repulsive force becomes asymptotic, thus the property of impenetrability for the point atom is preserved.

Thus, even the most basic questions of the chemist could be answered in terms of forces, and, it should again be noted, forces were not hypothetical but real. They were experimentally determinable facts, not metaphysical hypotheses. To someone who was suspicious of hypotheses, except insofar as they suggested experiments, the system of force and of point atoms had a clear advantage over that of material particles. If to this scientific superiority is added a religious dimension, the advantage becomes compelling.

Faraday was aware of all these factors. In his early years as an apprentice chemist he refused to commit himself to any system, simply suspending judgment on the nature of ultimate reality. But, as he delved deeper into the nature of matter and force, the system of forces and point atoms gradually received his allegiance until, by the 1830's, his theoretical ideas were consistently expressed in its terms. Not until the 1850's did he abandon some of its tenets to rise to a height of abstraction unmatched by any of his contemporaries.

THE TRANSMISSION OF FORCE

Faraday's first scientific paper, "Analysis of Native Caustic Lime of Tuscany," was published in 1816 and was followed in the next 5 years by papers of a similar nature. It was as an analytical chemist that he discovered and described benzene, in 1825. His intense interest in electricity and magnetism was aroused in the spring of 1821, when his friend Richard Phillips, editor of the *Annals of Philosophy*, asked him to write a history of the new field of electromagnetism. Hans Christian Oersted's announcement of the magnetic effect of an electric current, in the summer of 1820, had touched off such a flurry of experiment and theorizing that many people were confused as to what actually did occur in the neighborhood of a current-carrying wire. Phillips knew that his friend would repeat the experiments, examine the theoretical systems built upon them, and give his readers a sober and critical account of both. What he did not know was that his simple request would lead to Faraday's first important discovery in electromagnetism and to a new concept of the transmission of force.

In September of 1821, Faraday suddenly realized that his experimental investigation of the magnetic effects to be found near a current-carrying wire led to a startling prediction. If a single magnetic pole were free to move, it would travel around the wire in a circle. He immediately devised a simple apparatus to illustrate this effect and thus invented the first electric motor in which electrical force was converted into mechanical motion.

The theoretical implications of his discovery were equally dramatic. The "pattern" of force around a current-carrying wire was obviously circular. To a man trained in classical, mathematical physics (as André-Marie Ampère was, for example), this fact had to be explained in terms of central forces acting in straight lines between some kind of current elements in both the wire and the magnetic detector. To Faraday, the experimental fact sufficed; the magnetic force *was* circular. He

even went on to show how the attractions and repulsions of ordinary magnetic poles could be deduced from his circular line of magnetic force [Figure 3.4–2]. Thus magnetic central forces were shown to be the resultant of the circular force. Faraday's total lack of mathematical training here stood him in good stead. Had he viewed physics as Laplace or Ampère did, he would have been forced (as Ampère was) to decompose his circular magnetic force into central forces and reduce electromagnetism to neat, Laplacian terms. Instead, he defended his circular force and thereby gave birth to the idea of the line of force, which was to be central to his thinking throughout his life. Central forces require no emphasis on the line of force: the line is always straight, connecting the centers of the two interacting bodies. The believer in central forces also need not trouble himself with the

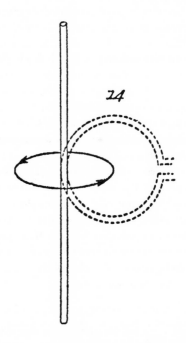

Figure 3.4–2. Faraday's illustration of the "polar" nature of the circular line of force. When the current-carrying wire is bent into a circle, the line of force enters on one side and exits on the other side of the plane of the circle. The lines of force will be crowded together within the circle and dispersed outside it, giving the appearance of polarity. If the circle is repeated many times (that is, if there is a helix) the "polarity" will become obtrusive.

mechanism of the interaction. By the nineteenth century, "action at a distance" was accepted by all but the most finicky physicists, and it seemed only logical to assume that this action had to be in straight lines. Faraday's line of force, however, by being curved, almost required a consideration of the mechanism of transmission of the force. If it were simply "action at a distance," the action certainly took the long way round in its manifestation. From 1821 on, Faraday was to agonize over the way in which force was transmitted. Out of his agony were to come his great discoveries.

Almost from the moment of his first discovery, Faraday thought of the electromagnetic force as a strain imposed upon the molecules of the surrounding medium. The theory of point atoms lent itself paticularly well to this view, although Faraday did not, at this time, explicitly call upon it. The hypothesis of such an intermolecular strain did permit him to envision electromagnetic effects in a rather unorthodox way. Instead of assuming currents of positive and negative fluids passing by one another in some complex fashion, might it not be possible to explain the phenomena of electrodynamics in terms of the vibration of strained molecules? A current might be a wave passing down a wire; the magnetic effect of the wave was a state of tension induced in the surrounding medium by the wave. The ability of waves to transmit "force" without the transmission of an independent body in which the force was inherent had only recently been triumphantly demonstrated in Fresnel's theory of the undulatory nature of light. Why should not electricity act in the same way?

ELECTROMAGNETIC INDUCTION

This appears to have been the thinking behind Faraday's famous induction-ring experiment of August 1831. He expected a momentary wave to pass through the primary coil. The magnetic effects would be intensified by the iron ring and, thereby, react upon the secondary coil, producing a momentary current in it. What he did not expect was a second current in the secondary when the primary circuit was broken. Yet, this was not too surprising, after all, for if the first pulse of the electric "wave" set up a magnetic strain in the surrounding medium, then the momentary current in the secondary marked the creation of the strain. The collapse of the strain, it could be argued, should be indicated by another momentary current, in the opposite direction. In between, the medium must exist in a state of constant strain, and Faraday christened this condition the electrotonic state. The embarrassing thing about this state to Faraday, the experimentalist, was the fact that it was totally indetectable! In 1822 Faraday had attempted to detect a strain in a decomposing electrolytic solution by shining a ray of plane-polarized light through it to see if its plane was altered, but to no avail. Now, he tried other methods, but still could get no effect. He was convinced, however, that the electronic state must exist, and, over the years, he kept devising experimental tests for detecting it.

In 1831 his attention could not be allowed to wander far from his new discovery of electromagnetic induction. Immediately after this discovery he substituted a moving permanent magnet for the iron ring. Then, this effect was broadened by introducing a rotating circular copper disk between the poles of a permanent horseshoe magnet. In the *First Series of the Experimental Researches in Electricity* he reported the discovery of electromagnetic induction, the invention of the dynamo, and the law connecting the lines of force with the current generated. Not bad for a start!

A start was all it was, for Faraday realized the vast new territory he had opened up. But, as he pushed ahead, reporting his path in the *Second Series*, he became aware of the necessity of preparing a base camp from which further trips into the wilderness could begin.

INTERMOLECULAR ACTION

Faraday's discoveries had already stirred the speculative fancies of many who saw in the electromagnetic current a separate "fluid" to be added to the other imponderables. Faraday was intent upon reducing, not multiplying, the theoretical entities of his science, so he paused in 1832 to prove to himself and to his contemporaries that all electrical manifestations—electrostatic, electrodynamic, galvanic, animal, and thermoelectric—involved precisely the same force of nature. Faraday was particularly concerned to prove that electrostatic discharge could cause electrochemical decomposition, for there were those who insisted upon a separate galvanic fluid generated by the voltaic cell. It was in these experiments that Faraday was led to his famous two laws of electrolysis and to his discoveries in electrochemistry. More important, however, was his realization that electrochemical decomposition could take place without the presence of electrical poles. The older theories had viewed electrochemical decomposition as the action, at a distance, of the poles of the electrochemical cell upon the decomposing molecules. The poles were the centers from which this decomposing force emanated to tear the molecules apart. To Faraday's surprise, he found that decomposition took place when an electrostatic generator was discharged into the air *through* a piece of blotting paper soaked in potassium iodide. Here there were no poles; the mere passage of the electricity was sufficient. It was at this moment that Faraday conceived a daring thought. Perhaps the electrical forces did not act at a distance, as everyone since Franklin, or at least Coulomb, had assumed. Perhaps they were transmitted from particle to particle. Decomposition occurred when the interparticulate strain shifted the forces of chemical affinity from the constituents of one molecule to those of its neighbors on each side. The strain, then, would be accompanied by a physical migration of the molecular constituents without these constituents ever existing in the free state in the solution. This immediately solved a basic problem that had plagued Faraday's predecessors, for, if molecules were torn apart by the poles, why could the "fragments" not be detected by ordinary chemical means?

In his mind's eye Faraday could now picture the electrochemical lines of intermolecular strain. The volume of the molecules would necessarily make these lines curves.

These curves accounted for the fact that electrochemical deposition upon the electrodes was uniform and not concentrated on the sides facing each other. The curves also looked suspiciously like magnetic lines of force, but Faraday was too excited over their implications for the theory of electricity to be drawn away into the theory of magnetism. What if *all* electrical action were intermolecular instead of action at a distance? To Faraday, and to physicists of his generation nourished upon the concept of action at a distance, such a heresy was almost unthinkable. Yet, there were precedents for it. As Faraday was to point out later, Newton himself had rejected action at a distance and, perhaps, all Faraday was doing was restoring physics to its proper Newtonian foundations. In any case, the idea of intermolecular action offered some intriguing possibilities for experiment, and Faraday was quick to exploit them.

UNIFIED THEORY OF ELECTRICITY

We can deal with only one of them here, but it was a fundamental one for Faraday. For the orthodox physicist, the forces between two charged bodies depended solely upon the quantities of the charges and the distance between the two bodies. If the action, however, depended upon the transmission of electrostatic force by the molecules of the intervening medium, then the "amount" of force transmitted might bear some relation to the nature of these molecules. Thus Faraday was led to the discovery of specific inductive capacity which, like chemical affinity, bore a definite relation to the particles involved. But, it might be objected, where are these molecules *in vacuo* where electrostatic action still takes place? Point atoms, it should be remembered, are infinite, so the problem does not arise. Few of Faraday's contemporaries could see this, and most felt him to be completely muddled. He *was* muddled, for the forces of the point atoms *do* act at a distance, or, as Faraday saw them, simply *were* in space. Disturbance of these forces led to electrical phenomena. Upon this confused idea, Faraday constructed a unified theory of electricity.

Insulators were bodies which could withstand a great deal of electrostatic strain; electrolytes were bodies whose "breaking point" was exactly determined by the chemical affinities of their constituents. The stronger the affinities were, the easier it was to distort the molecule and permit the transfer of partners. When the "slippage" took place, the strain was momentarily relaxed, only to be built up immediately again. This buildup and breakdown of strain constituted the electric "wave" or current. Good conductors would take up little strain, and so the buildup and breakdown were extremely rapid and the "wave" was easily generated and renewed.

In 1838, after 7 years of concentrated effort, Faraday presented this theory to the world. The relief of the mental strain under which he had labored for 7 years was too sudden and his intellectual faculties collapsed. From 1839 until 1845 he was able to work only fitfully between bouts of giddiness, headache, and loss of memory. His condition was not improved by the coolness with which his theory was received. It would take more than a host of experiments and fundamental discoveries to convince his orthodox brethren. What was needed was mathematics to make the theorists sit up and take notice. This Faraday was unable to provide.

Fortunately, there were some younger men in England to whom Faraday's ideas had the appeal of novelty and heterodoxy. The young William Thomson, later Lord Kelvin, was just testing his powers as a theorist in 1845 when he encountered Faraday's papers on electricity. Unlike his older contemporaries, he was attracted rather than repelled by Faraday's daring and sometimes incredible hypotheses of electrical action. Nothing would do but that he reduce Faraday's often obscure language to the purity and elegance of mathematics and *then* see what could be made of his theories. The results were reported to Faraday in a letter from Thomson dated August 6, 1845. Not only did Thomson take Faraday seriously, he was even able to suggest some consequences of Faraday's theory which might be capable of experimental verification. In particular, he suggested that a state of

electrostatic tension ought to be detectable by plane-polarized light. The electrotonic state, then, did not exist solely in Faraday's imagination but was deducible from Thomson's mathematics.

ILLUMINATION OF LINES OF FORCE

With Thomson's analysis to spur him on, Faraday plunged back into his experimental search for the electrotonic state. It still resisted all his efforts to detect it. Finally he abandoned the purely electrical road. Perhaps electrostatic forces were simply too feeble to produce a detectable effect. Magnetic forces, on the other hand, were something else again; where an electrostatically charged body could lift milligrams of chaff, a powerful electromagnet could hold hundredweights in its embrace. Furthermore, magnetic lines of force, like their electrostatic cousins, were curved, indicating to Faraday that the transmission of the magnetic force was likewise intermolecular. Might not a "magnetotonic" state be more easily detectable, then, than the electrotonic? Should not a transparent body placed in the powerful magnetic field between the poles of an electromagnet be strained? And should not this strain be detectable by polarized light? The experiment was tried, but with no effect. Faraday now was not to be put off by Nature's reluctance to reveal herself. The effect *must* exist! Air, flint glass, iceland spar—all were examined to no avail. Finally Faraday hit upon a piece of borate-of-lead glass of very high refractive index that he had made back in the 1820's for optical researches. The plane of the polarized light was now rotated sufficiently to be easily detected. The state of strain for which he had looked for so many years was now an experimental fact.

Yet, as was so often the case in Faraday's experimental investigations, more evidence was forthcoming than was necessary simply to prove the point he sought. The rotation of the plane of polarized light revealed a strain, but it was a peculiar kind of strain. The direction of the rotation depended *solely* upon the polarity of the magnetic field; the glass seemed merely to make the effect perceptible. The effect, as Faraday somewhat poetically put it, amounted to the "illumination of lines of force." The emphasis had shifted subtly from the condition of the particles of the heavy glass to the peculiar nature of the magnetic line of force.

DIAMAGNETICS AND PARAMAGNETICS

Before he could investigate the nature of the magnetic line of force, however, Faraday realized that he had opened a way into a new territory. If the heavy glass served to illuminate the line of force, it could not be indifferent, as a body, to the magnetic force itself. When the glass was freely suspended in the intense magnetic field between the poles of the Royal Institution's powerful electromagnet, it moved as if it were trying to escape from the field. Its long axis also turned perpendicular to the lines of magnetic force. The action of the glass was precisely opposite to the action of a bar of iron. All bodies, Faraday now found, reacted to the magnet either as the glass did or as an iron bar would. The two classes of bodies were christened diamagnetics and paramagnetics, and the science of magnetism was now extended to include all matter.

It was one thing to classify, it was another to understand. To most of Faraday's contemporaries, the problem did not appear a difficult one to solve. Since the action of diamagnetics was opposite to the action of paramagnetics, it seemed to follow that the polarity of diamagnetics must simply be opposite to that of paramagnetics. If one accepted Ampère's theory of magnetism, this meant that somehow the currents circulating around the ultimate molecules of diamagnetics and those circulating around the molecules of paramagnetics must be moving in opposite directions. Both Edmond Becquerel and Wilhelm Weber adopted modifications of this theory and insisted upon its necessary consequence—that diamagnetic polarity was simply the opposite of paramagnetic polarity.

Faraday, who had never accepted Ampère's physical model, was not so sure. His doubts led him to seek experimental evidence of diamagnetic polarity unrelentingly for 5 years, but to no avail. The polarity was *not* in the particles but was in the line of force. From 1850 on, Faraday shifted the focus of his attention from the manifestations of force in matter to the line of force in space itself. In many ways these last researches were to be the subtlest and most abstract and fundamental of them all. Out of them were to come the foundations of classical field theory.

The magnetic line of force differed from the electrostatic in two important respects. The electrostatic line of force depended upon molecular strains for its propagation, so it always had "ends." These were, so to speak, the "poles" which could be labeled positive and negative. The electrostatic line of force, then, could never originate and terminate upon the same conducting body. But this was precisely what the magnetic line of force did. The magnetic "poles," however, appeared to be perfectly arbitrary points in a homogeneous bar magnet. Since there was no detectable diamagnetic polarity, might it not be possible to eliminate the poles themselves? What, then, would become of the lines of force in such a view? Must they not be continuous, closed curves which passed *through* the magnet?

In a series of brilliantly simple experiments Faraday showed this to be the case. Using the law of electromagnetic induction from his *First Series of Experimental Researches,* he showed that the number of lines of force external to the magnet was the same as the number of lines that passed through the magnet. The magnet seemingly served only to concentrate the lines of force, or, as Faraday put it, a magnet was "the habitation of lines of force."

The concentration of the lines of force in the magnet implied that the lines of force were more easily conducted through the soft iron than through the surrounding medium. It was this implication that Faraday made explicit in defining the difference between paramagnetics and diamagnetics. Paramagnetics conducted the lines of force easily, so the lines converged upon paramagnetic bodies; diamagnetic bodies were poor conductors, so the lines avoided them. Each type of body, therefore, produced characteristic patterns of lines of force. It was by using these patterns that Faraday was able to refute those who claimed reverse paramagnetic polarity for diamagnetics. In Figure 3.4–3 the case for the diamagnetic is *not* the reverse of the case for the paramagnetic. If poles be defined as places of maximum concentration of the lines of force, then it may be easily seen that there are no poles at all associated with the diamagnetic body.

SPATIAL STRAIN

The magnetic "conductibility" of bodies determined their para- or diamagnetic condition, but what happened when no bodies were around? Faraday had long known that

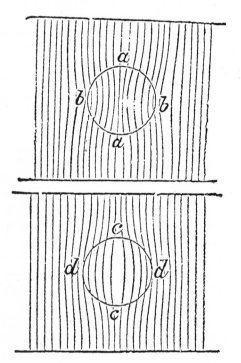

Figure 3.4–3. Diagrammatic representations of a paramagnetic substance (top) and a diamagnetic substance (bottom) in a uniform magnetic field. The "polarity" of the paramagnetic substance is represented by the compression of the lines of force at *aa*. There is no such compression in the diamagnetic substance; *cc* does not represent polarity opposite to that at *aa*.

the magnetic lines of force existed in the best vacuum he could obtain. Ordinary bodies merely served to concentrate or diffuse them; empty space, it appeared, could conduct them. Conduction of the line of force involved the presence of a strain, and this presence forced the question of what carried the strain in empty space. Faraday was almost alone, in the nineteenth century, in refusing to accept the ether as the basis for magnetic strains. The ether, it must be remembered, was considered to be atomic, and, if that were the case, it would have to exhibit the kind of polarities which Faraday had proved did not exist. He seems quietly to have adopted a phenomenalistic point of view. If there was a strain but no substance to be strained, then so be it. He had, after all, argued before that we can only know force, not substance, and if the force was manifest, then that was as far as we can go. The lines of force simply were strains. As far back as 1846 he had even suggested, in a wild speculation, that vibrations of lines of force might be light waves. Then he had quite explicitly denied the existence of the ether.

By 1855 the system was complete. Matter itself was but a peculiar kind of spatial strain with which the magnetic and electrostatic lines of force were associated. The energy of the universe was to be found in these strains. The fundamental postulate of field theory had been laid down.

Unfortunately, very few of Faraday's contemporaries were aware of the fact that a revolution was going on under their very noses. A typical reaction was that of Sir George Biddell Airy, the Astronomer Royal, who, when asked what he thought of Faraday's work on magnetism, replied, "The effect of a magnet upon another magnet may be represented *perfectly* by supposing that certain parts act just as if they pulled by a string, and that certain other parts act just as if they pushed with a stick. And the representation is not vague, but is a matter of strict numerical calculation. . . . I can hardly imagine anyone who practically and numerically knows this agreement, to hesitate an instant in the choice between this simple and precise action, on the one hand, and anything so vague and varying as lines of force, on the other hand."

One of the few who did hesitate was James Clerk Maxwell, who saw what Airy and others missed. The lines of force could be represented mathematically, they could be given all the precision that Airy demanded, and the concept of the field might lead to new and exciting discoveries. In the 1860's, as Faraday slowly sank into senility, Maxwell's eager mind began to explore the electromagnetic field. When Faraday died on August 25, 1867, Maxwell had already begun to lay the foundations for his great treatise. It would have pleased Faraday to know that his idea of the line of force was, in Maxwell's hands, to become the unifying thread in a physical system encompassing the cosmos. It was the line of force that tied all together into a *Uni*verse worthy of the God he had worshipped all his life.

3.5   *The Spectrum of Electromagnetic Radiation*
                        Norman Feather

The interrelation of the two subjects light and electricity, through the identification of light as an electromagnetic wave process, remained speculative for more than twenty years after Maxwell's theory was published. On the other hand, the assimilation of the subjects light and heat had already been effected—some twenty years previously. At the intuitive level, the recognition of a common behavior must have occurred at a very early stage in human evolution: the light and heat of the sun appear in permanent association. At the scientific level, however, the definitive experiments were not made until the first half of the nineteenth century. There had been speculation, it is true: indeed in 1717 Newton had written (*Optiks*, book 3, query 18): "Is not the heat . . . conveyed through the vacuum by the vibrations of a much subtiler medium than air. . . . And is not this medium the same by which light is refracted or reflected. . . . And is it not . . . expanded through the heavens?" But it was not until 1800 that F. W. Herschel, the astronomer, having produced a spectrum of sunlight with a prism, found that a thermometer placed in the spectrum registered a higher temperature in the red than in the violet, and a higher temperature still when it was situated outside the visible spectrum, beyond the red. A new field of observation was clearly awaiting a more sensitive instrument of detection.

The necessary instrument was realized in the thermo-multiplier (thermopile) of Leopoldo Nobili (1784–1835) and Macedonio Melloni (1798–1854) in 1831. In that year, however, Melloni, having been involved in the abortive revolution of 1830, had to leave his professorship at Parma and flee to France. But he was able to continue his work, and in his country of exile he carried out the first detailed demonstration that heat radiation is reflected and refracted according to the same laws as apply to visible light. He also made an extensive study of the process of absorption. As a result, he became a convinced advocate of the view that radiant heat is essentially "dark" or "obscure" light, different only in respect of wavelength from such radiation. But the conclusive verification of this point of view came from the work of others in the years which followed.

In a brilliant series of investigations, accounts of which were published over the period 1835 to 1838, J. D. Forbes demonstrated the polarization of heat, by double refraction as well as by reflection: then in 1847 A. H. L. Fizeau and Jean Bernard Léon

Foucault (1819–1868) observed and studied its diffraction by a grating. By 1850 the statement might well have been made that light is merely visible heat. Such a statement is at least consonant with the more modern view—that energy in whatever form must ultimately degenerate into heat through interaction with matter. Moreover, long before 1850, it was known that there was also "obscure" light of shorter wavelength than the visible. J. W. Ritter (1776–1810) had discovered the photochemical action of this "ultra-violet" radiation in 1801, and in 1804 Young succeeded in obtaining a "photograph" of Newton's rings making use of it. By 1862, therefore, Maxwell's inference, that light is propagated as an electromagnetic wave process, would naturally be understood to refer not only to visible light, but also to "obscure" light of longer and shorter wavelength—to radiations of the infra-red and ultra-violet regions of the spectrum, of altogether indefinite range. Against this background of speculation, the obvious supposition would be that if demonstrably electromagnetic radiation were to be produced in the laboratory it would be radiation in the extreme infra-red, rather than in the ultra-violet, radiation of wavelength measured in centimetres, not in a unit ten thousand times smaller than this.

Success in this direction was not achieved until 1887. The successful experimenter was Heinrich Rudolf Hertz (1857–1894). Hertz had gone to Munich to study engineering, but almost immediately, at the age of twenty, he abandoned his intention of entering the profession of architecture, for which that study was prescribed, and set himself to master the work of the French mathematical physicists. Then, in October 1878, he presented himself in the laboratories of Kirchhoff and Helmholtz in Berlin. He was at once accepted for research, and became assistant to Helmholtz in 1880. His experimental investigations in the field of Maxwell's theory were started in 1885, when he was appointed professor of physics at Karlsruhe Polytechnic, and they virtually ceased when he moved to the university chair at Bonn in 1889. During these four years he

had provided full justification for Maxwell's speculations, in almost every particular. Hertz had been thinking about these matters since the problem was first put to him by Helmholtz in 1879, but it was in fact a chance observation that put him on the right path to their experimental demonstration. He was operating an induction coil with a spark gap, and he noticed that, when a nearly closed loop of wire was connected, by one metallic conductor only, to any point in the secondary circuit of the activated coil, "sympathetic" sparking occurred across the small gap in the loop. Having found that he could obtain similar effects even though there was no metallic connection between the loop and the coil, he had a rudimentary detector of electrical effects at a distance ready-made in the nearly closed loop.

Hertz's oscillator (transmitter) eventually consisted of a pair of rectangular metal plates, one in connection with each knob of the spark gap of an induction coil. The connections were symmetrical and the plates were arranged to be co-planar. His "improved" detector was a single circular loop of copper wire, closed except for an adjustable gap of small separation. With this equipment, suitably modified to suit the occasion, he was able to show that there is a radiation (as distinct from an induction) effect from such a source, and to demonstrate that, in respect of reflection, refraction and polarization, this radiation has properties similar to those of visible light. In addition, he was able to determine the velocity of propagation, at least as to order of magnitude. "Hertzian" waves, waves indubitably electromagnetic in character, were in this way shown to be precisely of the type that Maxwell had predicted—and the inference that light, visible and "obscure," is propagated as an electromagnetic disturbance in an insulating medium was at last vindicated.

This brief history of vindication would not be unbiased if reference were not made to the work of David Edward Hughes (1831–1900). Born in London, by the age of nineteen Hughes was professor of music at a small college in Kentucky. He had, however, the talents of an inventor rather than a musician. He gave up his teaching post in 1854, and in the following year was granted a US patent for a type-printing telegraph. He invented the carbon microphone, and, having improved on his original design, he noticed that such an instrument, when used in a receiving circuit, was sensitive to "sudden electrical impulses, whether given out . . . through the extra current from a coil or from a frictional machine," up to a distance of 500 yards or more. This was in 1879. Here was a purely empirical discovery, not preceded by a detailed theoretical study of Maxwell's theory, or conditioned by preconceived notions on a grand scale. Hughes demonstrated his experiments in London, but the opinion of Stokes and others (1880) was that Hughes's own idea that he was dealing with electric waves was unjustified. As a result the matter was not pursued farther, and an account of the observations was not published until 1899.

Nowadays the term "Hertzian waves" has become old-fashioned: what was a frontier research in fundamental physics has given rise, in the field of large-scale engineering practice, to a world-wide network of "radio" stations crowding the spectrum of wavelengths from several kilometers to a centimeter or so. Over a frequency range of almost twenty octaves, electromagnetic waves are in daily use for the peaceful—and less peaceful—activities of mankind. Another fourteen octaves, or thereabouts, bridges the gap between the centimeter waves of the special-purpose radio transmitter and visible light. Physicists have studied the properties of this infra-red radiation ("near," "far" and "extreme" infra-red) in sufficient detail over this whole range of frequencies to know that there is merely a continuous gradation of effect, that there are no discontinuities of character. The traditional nomenclature reflects differences in modes of production or detection, not differences in properties. In 1917, E. F. Nichols and J. D. Tear were able to detect, and examine spectroscopically, a constituent of heat radiation (in the extreme-infra-red) with a wavelength of 4.2×10^{-2} cm—and also to produce, electrically, radiation

of shorter wavelength (2.2×10^{-2} cm), using a suspension of fine metallic particles in an insulating oil as a system of multiple spark gaps. In this way an "overlap" was effected in the region of the spectrum of wavelength around $\frac{1}{3}$ mm. The overlap could equally well have been made elsewhere: it is little more than an accident of history that it was first made precisely where it was.

Visible light, having wavelengths from about 7.7×10^{-5} cm (red to 3.9×10^{-5} cm (violet), accounts for almost exactly one octave of the spectrum of electromagnetic radiations. In the survey just completed, we have traversed some thirty-three octaves of lower frequency. In respect of the board features of propagation phenomena, we have been unable to record any fundamental discontinuity in behavior over this enormous range. The modern physicist will assert that over a range of higher frequencies almost equally great—some twenty-eight octaves according to present reckoning—there is observable radiation of the same general character. He will say that from the near ultra-violet, through the spectral region of the X-rays, through that of the γ-rays, and beyond, there is no abrupt change in propagation behavior. To make our survey complete, therefore, we must add, briefly, the history of discovery in these latter fields.

In 1895 Wilhelm Konrad Röntgen discovered a "dark" radiation, to which all bodies appeared in some degree transparent, emitted from the glass wall of a discharge tube under the particular conditions of discharge in which "cathode rays" show faintly in the tube and produce a patch of bright phosphorescence on the glass. He named this radiation the X-radiation. In the following year the radioactivity of uranium was discovered, and was exhibited as a characteristic atomic property of the element, by Antoine Henri Becquerel (1852–1908). In 1899 Ernest Rutherford (1871–1937) made the first analysis of the radiations spontaneously emitted by uranium preparations, and recognized two components which he designated the α- and β-rays. In 1900 a third more penetrating component was found by Paul Villard. Rutherford confirmed this finding, and applied the term γ-rays to the new component.

From the outset it appeared that the γ-radiation from radioactive substances and the X-radiation from discharge tubes had very closely similar properties. Among the more obvious of these was the action on a photographic plate in the manner of visible (or ultra-violet) light. As early as 1896, indeed, Stokes had given good reasons why X-rays should have the character of pulses of electromagnetic radiation simulating the behavior of light of the extreme ultra-violet. But to obtain experimental confirmation of this point of view was very difficult, technically, at the time. The significant criteria, of course, are the diffraction and polarization effects should be found under appropriate conditions of experiment. Early attempts (1899–1901) to demonstrate the diffraction of X-rays in passing through narrow slits were inconclusive. Then, in 1905, polarization by reflection (scattering) appeared to be proved by the experiments of C. G. Barkla in Liverpool. However, at about the same time, Barkla (1877–1944), later professor at King's College, London, and at Edinburgh, was discovering other effects of an entirely different character, and the simple issue tended to become confused. We cannot discuss these other effects here.

.   .   .   .   .   .

The difficulties of interpretation to which they gave rise remained largely unresolved when, in 1912, Max von Laue (1879–1960) and his pupils found incontrovertible evidence for diffraction, using the naturally ordered array of atoms in a crystal as a three-dimensional grating. After 1912, therefore, it was at least certain that the X-rays from a discharge tube exhibit the two effects of diffraction and polarization— and that they are produced in an electrical process (the stopping of negatively charged electrons in the metal "target" of the tube). The conclusion could no longer be avoided: whatever other properties the radiation may

possess, it exhibits these three features, which enable wavelengths to be assigned, and allow the assigned wavelengths to be interpreted as those of a transverse electromagnetic wave process of the Maxwellian type. As far as these features are concerned, there is no discontinuity of behavior as the wavelength decreases. Two years later Rutherford and E. N. da C. Andrade succeeded in observing the diffraction of certain of the less penetrating γ-rays, using a crystal "grating" in a modified arrangement, thereby extending the range of validity of this statement.

The experiments of von Laue and Rutherford and Andrade, to which we have just referred, established the essential identity of two types of radiation to which physicists had already given distinctive names. In this respect the result was entirely analogous to that achieved by Nichols and Tear three years later. As in this later case, therefore, we conclude that the distinction of nomenclature is significant only in respect of the mode of origin of the radiation: X-rays are radiations produced when charged particles (in the original experiments, electrons) are suddenly accelerated or decelerated—and γ-rays are similar radiations having their origin in processes occurring in the nuclei of atoms (originally, the nuclei of atoms of the radioactive substances). Strictly, we should not use these terms with other connotations. But we note, in passing, that the spectral regions to which the terms may properly be applied now overlap very considerably.

We have stated that some twenty-eight octaves of electromagnetic radiation of frequency greater than that of visible light have been the subject of experimental investigation. This statement implies a spectrum extending, in wavelength, continuously from about 4×10^{-5} cm to 1.5×10^{-13} cm. The X-rays with which Barkla and von Laue worked occupy roughly the middle octave or two of this range. Pedantically, we should refuse to accept a radiation as a wave radiation unless we can determine its wavelength by a direct comparison with a standard length, or its frequency, effectively, by direct counting against a standard of time. The procedure of wavelength determination for X-rays and the less penetrating γ-rays, in terms of the grating spacing of a crystal, fulfils these requirements completely, but a crystal grating spacing is the smallest comparison length that matter-in-bulk provides us with, and the crystal diffraction method reaches its natural limit of usefulness when the wave-length of the radiation is about 1.5×10^{-10} cm. Obviously, our statement that a further ten octaves of the sectrum beyond this limit have been the subject of study requires further justification. We do not withdraw the statement—but the pedant may look askance at our attempt to justify it.

The argument in justification is as follows. We have recorded the fact that "other effects of an entirely different character" began to appear in Barkla's experiments, so that the simple fact of polarization was overlaid with observations for which no explanation was forthcoming on the basis of a classical wave-thory interpretation. Gradually, however, these other effects were brought within the compass of a new theory in which the frequency of the wave picture was retained as a characteristic parameter descriptive of the "quality" of the radiation. It was then found that, in parallel with the continuity of behavior in relation to wave-process phenomena, as wavelength decreased and frequency increased, there was continuity of behavior in relation to these other effects, which became the more pronounced as the wave-process effects became difficult to demonstrate experimentally. Ultimately, for radiations of the highest frequencies, it was no longer possible to observe the wave-like properties directly (crystal gratings were too coarse), but the other effects provided an entirely acceptable basis for the calculation of frequencies—and of wavelengths, too, for the velocity of propagation is the velocity of light in all cases. It was a far-reaching extrapolation, it is true, but there is as yet no reason to suppose that in relation to empty space the idea of an electromagnetic wave of wavelength however small is a physically meaningless idea. So we do not withdraw the statement that electromagnetic radiation has been studied over the whole spectral range down to wavelengths of the order of 1.5×10^{-13} cm (which is, in fact, of of the order of the radius of a proton, the nucleus of the hydrogen atom)—but we have obviously reached the stage at which we should enquire more fully regarding these "other effects" which have bulked so largely in our latest argument.

3.6

Scientific Imagination
Richard P. Feynman

I have asked you to imagine these electric and magnetic fields. What do you do? Do you know how? How do I imagine the electric and magnetic field? What do I actually see? What are the demands of scientific imagination? Is it any different from trying to imagine that the room is full of invisible angels? No, it is not like imagining invisible angels. It requires a much higher degree of imagination to understand the electromagnetic field than to understand invisible angels. Why? Because to make invisible angels understandable, all I have to do is to alter their properties *a little bit*—I make

SOURCE. From Feynman, Leighton, and Sands, *The Feynman Lectures on Physics* (Reading, Mass.: Addison-Wesley, 1963) Chapter 20.

them slightly visible, and then I can see the shapes of their wings, and bodies, and halos. Once I succeed in imagining a visible angel, the abstraction required—which is to take almost invisible angels and imagine them completely invisible—is relatively easy. So you say, "Professor, please give me an approximate description of the electromagnetic waves, even though it may be slightly inaccurate, so that I too can see them as well as I can see almost invisible angels. Then I will modify the picture to the necessary abstraction."

I'm sorry I can't do that for you. I don't know how. I have no picture of this electromagnetic field that is in any sense accurate. I have known about the electromagnetic field a long time—I was in the same position 25 years ago that you are now, and I have had 25 years more of experience thinking about these wiggling waves. When I start describing the magnetic field moving through space, I speak of the E and B fields and wave my arms and you may imagine that I can see them. I'll tell you what I see. I see some kind of vague shadowy, wiggling lines—here and there is an E and B written on them somehow, and perhaps some of the lines have arrows on them—an arrow here or there which disappears when I look too closely at it. When I talk about the fields swishing through space, I have a terrible confusion between the symbols I use to describe the objects and the objects themselves. I cannot really make a picture that is even nearly like the true waves. So if you have some difficulty in making such a picture, you should not be worried that your difficulty is unusual.

Our science makes terrific demands on the imagination. The degree of imagination that is required is much more extreme than that required for some of the ancient ideas. The modern ideas are much harder to imagine. We use a lot of tools, though. We use mathematical equations and rules, and make a lot of pictures. What I realize now is that when I talk about the electromagnetic field in space, I see some kind of a superposition of all of the diagrams which I've ever seen drawn about them. I don't see

little bundles of field lines running about because it worries me that if I ran at a different speed the bundles would disappear. I don't even always see the electric and magnetic fields because sometimes I think I should have made a picture with the vector potential and the scalar potential, for those were perhaps the more physically significant things that were wiggling.

Perhaps the only hope, you say, is to take a mathematical view. Now what is a mathematical view? From a mathematical view, there is an electric field vector and a magnetic field vector at every point in space; that is, there are six numbers associated with every point. Can you imagine six numbers associated with each point in space? That's too hard. Can you imagine even *one* number associated with every point? I cannot! I can imagine such a thing as the temperature at every point in space. That seems to be understandable. There is a hotness and coldness that varies from place to place. But I honestly do not understand the idea of a *number* at every point.

So perhaps we should put the question: Can we represent the electric field by something more like a temperature, say like the displacement of a piece of jello? Suppose that we were to begin by imagining that the world was filled with thin jello and that the fields represented some distortion—say a stretching or twisting—of the jello. Then we could visualize the field. After we "see" what it is like we could abstract the jello away. For many years that's what people tried to do. Maxwell, Ampère, Faraday, and others tried to understand electromagnetism this way. (Sometimes they called the abstract jello "ether.") But it turned out that the attempt to imagine the electromagnetic field in that way was really standing in the way of progress. We are unfortunately limited to abstractions, to using instruments to detect the field, to using mathematical symbols to describe the field, etc. But nevertheless, in some sense the fields are real, because after we are all finished fiddling around with mathematical equations—with or without making pictures and drawings or trying to visualize the thing—we can still make the instruments detect the signals from Mariner II and find out about galaxies a billion miles away, and so on.

The whole question of imagination in science is often misunderstood by people in other disciplines. They try to test our imagination in the following way. They say, "Here is a picture of some people in a situation. What do you imagine will happen next?" When we say, "I can't imagine," they may think we have a weak imagination. They overlook the fact that whatever we are *allowed* to imagine in science must be *consistent with everything else we know*: that the electric fields and the waves we talk about are not just some happy thoughts which we are free to make as we wish, but ideas which must be consistent with all the laws of physics we know. We can't allow ourselves to seriously imagine things which are obviously in contradiction to the known laws of nature. And so our kind of imagination is quite a difficult game. One has to have the imagination to think of something that has never been seen before, never been heard of before. At the same time the thoughts are restricted in a strait jacket, so to speak, limited by the conditions that come from our knowledge of the way nature really is. The problem of creating something which is new, but which is consistent with everything which has been seen before, in one of extreme difficulty.

While I'm on this subject I want to talk about whether it will ever be possible to imagine *beauty* that we can't *see*. It is an interesting question. When we look at a rainbow, it looks beautiful to us. Everybody says, "Ooh, a rainbow." (You see how scientific I am. I am afraid to say something is beautiful unless I have an experimental way of defining it.) But how would we describe a rainbow if we were blind? We are blind when we measure the infra-red reflection coefficient of sodium chloride, or when we talk about the frequency of the waves that are coming from some galaxy that we can't see—we make a diagram, we make a plot. For instance, for the rainbow, such a plot would be the intensity of radiation vs.

wavelength measured with a spectrophotometer for each direction in the sky. Generally, such measurements would give a curve that was rather flat. Then some day, someone would discover that for certain conditions of the weather, and at certain angles in the sky, the spectrum of intensity as a function of wavelength would behave strangely; it would have a bump. As the angle of the instrument was varied only a little bit, the maximum of the bump would move from one wavelength to another. Then one day the physical review of the blind men might publish a technical article with the title "The Intensity of Radiation as a Function of Angle under Certain Conditions of the Weather." In this article there might appear a graph such as the one in Figure 3.6–1. The author would perhaps remark that at the larger angles there was more radiation at long wavelengths, whereas for the smaller angles the maximum in the radiation came at shorter wavelengths. (From our point of view, we would say that the light at 40° is predominantly green and the light at 42° is predominantly red.)

Now do we find the graph of Figure 3.6–1 beautiful? It contains much more detail than we apprehend when we look at a rainbow, because our eyes cannot see the exact details in the shape of a spectrum. The eye, however, finds the rainbow beautiful. Do we have enough imagination to see in the spectral curves the same beauty we see when we look directly at the rainbow? I don't know.

But suppose I have a graph of the reflection

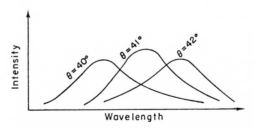

Figure 3.6–1.   The intensity of electromagnetic waves as a function of wavelength for three angles (measured from the direction opposite the sun), observed only with certain meteorological conditions.

coefficient of a sodium chloride crystal as a function of wavelength in the infra-red, and also as a function of angle. I would have a representation of how it would look to my eyes if they could see in the infra-red— perhaps some glowing, shiny "green," mixed with reflections from the surface in a "metallic red." That would be a beautiful thing, but I don't know whether I can ever look at a graph of the reflection coefficient of NaCl measured with some instrument and say that it has the same beauty.

On the other hand, even if we cannot see beauty in particular measured results, we *can* already claim to see a certain beauty in the equations which describe general physical laws. For example, in the wave equation [3.6–1], there's something nice about the regularity of the appearance of the x, the y, the z, and the t. And this nice symmetry in appearance of the x, y, z and t suggests to the mind still a greater beauty which has to do with the four dimensions, the possibility that space has four-dimensional symmetry, the possibility of analyzing that and the developments of the special theory of relativity. So there is plenty of intellectual beauty associated with the equations.

$$\frac{\partial^2 \psi}{\partial x^2} + \frac{\partial^2 \psi}{\partial y^2} + \frac{\partial^2 \psi}{\partial z^2} - \frac{1}{c^2}\frac{\partial^2 \psi}{\partial t^2} = 0 \quad [3.6 - 1]$$

PART TWO

THE MODERN ERA

Looking Backward

Paul R. Heyl

Boston, it used to be said, was not a place, but a state of mind; likewise we may fairly say that the most remarkable thing about late nineteenth century physics was the state of mind of the physicists.

The nineteenth century had seen as great an advance in the science of physics as in other lines of human activity. The story may be briefly summarized here. The physics of the eighteenth century had been a rather loose and disjointed affair, consisting mainly of uncorrelated facts about half a dozen entities as then recognized—matter, heat (caloric), light corpuscles, electricity, magnetism, and phlogiston; but by 1890 these had been correlated and consolidated, along with newly discovered material, into a closely woven fabric based upon three central concepts: matter, energy, and ether.

The physicists of that day contemplated this well-knit structure with much satisfaction. It was generally held that at last theory was on the right track, and that the future would but develop and extend the lines already laid down. There was, in fact, every reason for scientific optimism. Lodge, in an address delivered in 1889, said:

The present is an epoch of astounding activity in physical science. Progress is a thing of months and weeks, almost of days. The long line of isolated ripples of past discovery seem blending into a mighty wave, on the crest of which one begins to discern some oncoming magnificent generalization. The suspense is becoming feverish, at times almost painful. One feels like a boy who has been long strumming on the silent keyboard of a deserted organ, into the chest of which an unseen power begins to blow a vivifying breath. Astonished, he now finds that the touch of a finger elicits a responsive note, and he hesitates, half delighted, half affrighted, lest he be deafened by the chords which it would seem he can now summon forth almost at will.

But by 1895 this attitude of mind had changed, a strange and widespread pessimism having displaced the previous optimism. It was believed by many that all (or nearly all) the possible great discoveries had been made. Professors of physics warned their students that other lines held out more promise, and that the future of physics was to be one of residuals, second order effects, and the never ending quest for "one more decimal place."

This state of mind was the more remarkable because there was no assignable reason for it; yet its reality can be attested by many physicists now living, who were students during the early eighteen-nineties. Without the testimony of tradition nothing would be known of it today, as it seems to have left no record in the scientific literature of the time.

It was but the darkest hour before the dawn, for late in 1895 Röntgen discovered the X-rays, and the crest of the wave was upon us. Ten years more saw the discovery of radioactivity, the genesis of the quantum theory, and the publication of Einstein's first papers on relativity. The ice jam was broken, and the river of discovery flowed on once more in full flood.

A correct perspective of present-day physics must be based upon a clear picture of the science as it was just prior to Röntgen's discovery. Let us therefore examine in more detail the three fundamental concepts of that period—matter, energy, and ether.

Matter was generally recognized as atomic in its structure, no change having taken place in this point of view during the century previous, though there were a few belated individuals who persisted in refusing their assent to this doctrine. The last notable opponent of the atomic theory, Ernst Mach of Vienna, died as recently as 1916.

Matter was rather a sharply defined

SOURCE. From *New Frontiers of Physics* (New York: Appleton, 1930).

concept. It had two outstanding characteristics, inertia and gravitation, which served
to distinguish and identify it, and which form
the basis of the ordinary tests which any
person naturally applies in deciding whether
a closed barrel is full or empty, that is,
whether there is any "matter" contained in
it. One may kick the barrel slightly to see if
it will easily start rolling, or he may try to
lift it. If it resists being set in motion, that
is, in scientific parlance, if it exhibits inertia,
he concludes that it is full of something; and
if it is heavy to lift, that is, if it gravitates
strongly, he comes to the same conclusion. By
these two tests it was decided whether anything was material or immaterial. If the latter,
it must find a place either as a form of energy
or as an ethereal manifestation.

The early years of the nineteenth century
had seen established the doctrine that matter
was unalterable in amount by physical or
chemical changes. This was known as the
principle of the conservation of matter.

But matter in itself was but a lifeless concept. Its principal interest lay in the fact
that it was a vehicle for energy.

Energy was distinctly a nineteenth century
concept. Into it were merged the "imponderables" of the century before. Energy was
broadly defined as the power of doing work;
and as it was obviously necessary that power
should be exercised by something, energy
came into the structure of physical theory
as a dependent or ancillary concept. As the
beauty of a rose is its color, and as this color
would be nothing without the rose (or some
other material basis), so energy was the life
of matter; but to assert its independent
existence would have been regarded as meaningless.

Energy, unlike matter, was not regarded
as atomic in its structure, but as capable
of indefinite subdivision. Thus the energy
content of a given body could increase or
decrease by smooth and gradual change.
It was recognized that there were many forms
of energy: heat, light, electrification, and
magnetization, as well as the purely mechanical forms of kinetic and potential energy,
and that all of these different forms of energy

were mutually interconvertible. This principle was called the correlation of energy.
Moreover, it had been shown that in these
various changes there was no loss or gain
of energy, which principle was called the
conservation of energy. The establishment
of this latter principle was regarded as one
of the great triumphs of the nineteenth
century.

But if energy could have no separate
existence and could not be created or annihilated, how then could the heat energy
of the sun reach us? For millions of miles
and during a period of some eight minutes,
it must traverse a vacuum far more perfect
than anything that we can realize in the
laboratory. What could carry this energy to
us?

This and other similar considerations
led to the establishment of the ether of space
as one of the fundamentals of physical
science of the period of which we are speaking. The concept of an ether is as old as
Newton and Huygens, but came into its own
only with the displacing of the corpuscular
theory of light by the wave theory in the
first half of the nineteenth century. Once
light was admitted to be a wave motion, the
question naturally arose, a motion of what?
This brought but a half humorous definition
of the ether as the nominative of the verb
to undulate.

But most physicists took the ether far more
seriously than that. It was a question of either
postulating the existence of a medium filling
all space or admitting action at a distance, and
when this issue was clearly raised there were
few, indeed, who hesitated. The ether, though
invisible and intangible, was regarded as an
entity as real as matter, and for the same
reason. "We know of matter only by its properties," said the physicists of 1895, "and
we are, in fact, as well acquainted with the
properties of the ether as we are with those
of iron. We know that iron transmits waves
of sound with a certain speed, and we know
that the ether transmits waves of light with
a certain speed. We know that iron is opaque
and that ether is transparent. We know that
iron exhibits friction, but that ether must

be absolutely frictionless, else the motion of the heavenly bodies would be retarded. We know that iron has a density of 8, and Lord Kelvin has recently calculated that the ether has a density of 10^{18}. What more can one want?"

Whether or not the ether was atomic in its structure was a matter that gave rise to some difference of opinion. It was generally agreed that if there was any such granular structure as exists in ordinary matter, it must be on an almost infinitely finer scale.

With the advent of Maxwell's electromagnetic theory of light the necessity for the recognition of the ether grew, if possible, more imperative, for this theory provided an answer to the long discussed question as to the nature of electricity—was it matter or energy? Maxwell's theory provided a classification for electricity and magnetism in the scheme of things, definitely removing them from the category of matter, where the eighteenth century had placed them, and transferring them to the newly erected domain of energy, recognizing them as forms of energy of the ether.

Thus the concept of the ether flourished famously and for a time even attempted to annex matter as a tributary state. For this Lord Kelvin was responsible. He suggested in 1867 that the atom of matter might be a vortex ring in the ether, something like a smoke ring in air. This suggestion found much favor among physicists, as is shown by Maxwell's article "Atom" in the ninth edition of the *Encyclopaedia Britannica* (1875). This hypothesis was not abandoned until the eighties, after it was found an extensive mathematical examination that such vortex rings in a perfect fluid would not show the slightest tendency to gravitate.

An excellent summary of the position which the ether concept held in the minds of physicists is found in an address delivered by Lodge in 1882:

One continuous substance filling all space: which can vibrate as light; which can be sheared into positive and negative electricity; which in whirls constitutes matter; and which transmits by continuity, and not by impact, every action and reaction of which matter is capable. This is the modern view of the ether and its functions.

Such, in brief, was the structure of physical theory as it existed in the closing years of the nineteenth century; a theory closely knit and correlated, with a pigeonhole for every known phenomenon; a theory remarkably consistent considering its range, and regarded with much satisfaction, though Kelvin remarked, in an address at the Royal Institution on April 27, 1900, that there were two clouds in the sky.

One cannot avoid the thought that in using this phrase Kelvin may have prophetically sensed the possibility of these clouds increasing as did those seen by the servant of Elijah. As a matter of fact, this is just about what happened; for one of these clouds was the negative result of the famous Michelson-Morley experiment on ether drift, which gave rise eventually to the theory of relativity; and the other was a certain inperfection in the theory of the statistical distribution of energy, similar to that which later led to the quantum theory.

And so, with complacent satisfaction as to the past and pessimism as to the future, the century approached its close.

CHAPTER 4
RELATIVITY

4.1

Albert Einstein

Antonina Vallentin

"When I was still a rather precocious young man, I already realized most vividly the futility of the hopes and aspirations that most men pursue throughout their lives," wrote Einstein in his biographical notes.

This precocious young man launched himself into what was, in fact, a mad adventure. Nothing that usually makes for success in an adventure spurred him on: neither ambition; nor taste for money; nor a desire to assert himself. The secret of his success lay elsewhere. It lay in something negative: not in what he did possess, but in what he did not possess and never would. It lay in his freedom from encumbrances: he traveled as light in the beginning as he was to travel all his life. He remained without requirements, so that one might have suspected him of a longing for austerity. On the contrary, in middle life he was not at all insensible to beauty and the good things of life. There was vitality in every inch of his great body, but he despised with all the violence that contempt aroused in him the enjoyment of mere material satisfaction. "Well-being and happiness," he wrote one day, "never appeared to me as an absolute aim. I am even inclined to compare such moral aims to the ambitions of a pig."

He has firmly refused to adapt himself in any way to the demands of his fame. He has maintained a personal, simple way of questioning the necessity of an action, or of expected behavior, or of an attitude.[1] In vain one would explain to him the customary formalities, and those who had not known him long would explain patiently, as to a backward child. They would repeat: "This is done ..." "Why is it done?" he would ask. Until you noticed his smile he seemed like a malicious child. "Tails? Why tails? I never had any and never missed them."

He defended himself against submission to conventions with obstinacy and wit. He was only really at his ease in an open-necked shirt and sandals. He had a preference for old clothes, a mended jumper, an ancient waistcoat, any material strange to the touch: a shabby dressing-gown was always more cozy than any grand new one given him as a present. Luxurious gifts had a way of disappearing from one day to the next: they were given away to some poor wretch, whom Einstein would persuade to hurry off discreetly with his parcel before the family noticed it.

Albert Einstein aimed at the Polytechnic in Zurich and hoped his mathematics would help him to pass his entrance exams. Rich relations helped Einstein with his studies by supplying a hundred Swiss francs a month, but they had no suspicion of the forces they

SOURCE. From *The Drama of Albert Einstein* (New York: Doubleday, 1954); copyright by Antonina Vallentin, reprinted by permission of Doubleday and Company, Inc.

[1]This biography was published one year before Einstein's death in 1955. [Ed.]

were helping to unleash, and one of the most remarkable scientific careers in history was launched out of family charity, exercised more from a sense of duty than any confidence in the boy's gifts or any interest in science.

His career began with a failure. His knowledge of mathematics was not enough to make up for gaps in other fields. Although he was morally mature beyond his age, the young man had to return to the school desk like any dunce or time-waster. It was the ordinary canton school in the small Swiss town of Aarau.

After passing the examinations for the Zurich Polytechnic, he decided to devote himself to teaching instead of specializing as engineer, as had been his original intention.

"There is such a thing as a passionate desire to understand, just as there is a passionate love for music. This passion is common with children, but it usually vanishes as they grow up. Without it, there would be no natural science, and no mathematics." Thus Einstein wrote in 1950, in connection with his scientific memoir on the theory of generalized gravitation. This passion remained with him forever. It was never blunted. It still predominates in the man of over seventy. It guided the young man on his chosen path with the sureness of a sleepwalker. The birth of a theory that was to convulse the world originated with this passion, the violence of this curiosity. The question that worried the youth of sixteen was: What would happen if a man tried to imprison a ray of light? The question was naturally more complex, but as scientific formulae were beyond me, it was with these simple words that Einstein explained what he himself considered to be the starting point of his life work.

Once he had asked himself the question, the problem haunted him. It was always present as he continued his studies at the Polytechnic and struggled with the material difficulties in his path. In spite of an intense interest in mathematics, he decided to study to be a teacher of physics. Mathematics, he said to himself, were divided into so many specialized fields that any one of them could absorb the short span of a man's life. "I was like Buridan's donkey, who could not decide which stack of hay to choose," he said later. His choice was dictated by his contempt for knowledge acquired automatically, which he would have had to absorb before reaching essential principles. It sums up his lasting sense of economy in all unnecessary intellectual effort. One could call it a reluctance to use a very fine instrument for coarse bits of work.

During his first visit to America in 1921, he was given a questionnaire covering all the intellectual equipment a student was supposed to carry with him through life, once his university studies ended. To one question as to the speed of sound, Einstein replied: "I don't know, I don't crowd my memory with facts that I can easily find in an encyclopaedia."

In contrast to this resistance to what he considered this dead weight of knowledge, the work in the laboratory fascinated him by its direct contact with experiment. But Einstein was forced to proclaim later the absolute priority of pure speculation. Summing up this conviction in recently published notes, he writes: "A theory can be checked by experiment, but there is no path that leads from experiment to a theory that has not yet been established."

This fascination for laboratory work arose from a characteristic one would not have expected in him. One might have thought that, having worked with principles which the man in the street could not grasp, he had arrived at a divorce from reality. His aloofness from daily chores also might have created the impression that there was a screen between him and the material side of life. In fact there is nothing of the absent-minded scientist about him. The vagueness often observed in him is put on, a kind of protection against an intrusive presence. Though concentration on his work removes him from the outside world, when not so absorbed his attention is curiously alert, arrested often in the most unexpected way

by an apparently uninteresting phenomenon. He is intensely observant: his glance
fixes on some detail and lingers over something which other people do not even notice.
His interest will be suddenly aroused by
statements made by specialists on questions
which one would have thought alien to his
world, secrets of craftsmanship and details
of engineering. Everything ingenious in the
material sphere engages his attention. If you
watch him handle some object, he seems to
take possession of it. He does not touch
things as though he wanted to push them
away as people do when their thoughts are
elsewhere; he follows the outline, and taps
the surface; he tests their faculties. He has
a taste for solid things such as steel structures;
he likes to watch a thing in the making. He
feels at home in a world of stable laws, in a
reality ruled by unalterable material facts.

The experiments in the laboratory which
so fascinated him were the most attractive
part of the Polytechnic programme. All his
life he had retained a distaste for education
when it stuffs young minds with facts, names
or formulae. He was apt to say that one need
not go to a university to learn these—they
can be got out of books. Education should
be devoted wholly to helping young people
to *think*, to give them the training which no
text-book can provide. "It is truly a miracle,
that modern education hasn't completely
stifled the sacred curiosity of research," he
said.

Constraint has always been his personal
enemy. His whole youth was a battle against
it. When he uttered the German word for it,
an abrupt word, with a particular sinister
sound, *Zwang*, everything tolerant, humorous or resigned in his expression vanished.
He spat out the word, as one does a fish bone.
While editing his biographical notes in the
peace of his Princeton study where no power
on earth threatens his absolute liberty,
Einstein remembers the havoc which it
played with his life. "The constraint was so
terrifying that after I had passed the final
examination, I found myself unable to think
of any scientific problem for almost a year."

No matter how hungry he was, Einstein

meticulously put aside twenty francs from
the hundred francs his family gave him each
month to pay for his naturalization. By the
time he passed his examination he was a Swiss
subject. His papers were in order. He asked
for what seemed to him the most obvious
post—that of an assistant, which professors
usually granted to gifted pupils. But no professor had marked his exceptional faculties,
not one had enough interest in him to forward his career.

The young man had been forced, owing to
this failure, to look for a job. He tried for
a professorship in physics at a lycée, but he
was unsuccessful, both in Zurich and in
smaller Swiss towns. Was the despised candidate not qualified for the post he desired?
Was it maybe fear of his genius which put
off the administrators who might otherwise
have hired him? Einstein still wonders and
does not know whether he was just too in-
significant; or whether there was already an
odour of sulphur about him. He began to
answer newspaper advertisements, rushing
hither and thither when he read of a vacancy,
any vacancy. He landed as temporary assistant in a technical school in Winterthur
which kept him going for several months.
With gratitude he took on a modest post
as tutor at a boarding school in Schaffhausen,
looking after two backward boys. He liked
the job, but he had his own ideas about
teaching physics. He lost the job because
he wanted to teach freely, in his own way.
He could find no other place in secondary
education.

Whenever Einstein speaks to a child one
realizes the barriers which exist in his relationship with adults. He is quick to reach
an understanding with children and they
have only to look at each other to become
accomplices. It is not the re-emergence of
the backward child or boy in Einstein:
children's minds have for him the same
fascination as the authentic material of inanimate objects: he loves their naivety, still
unaffected by conventional restraints, their
impetuous questions and their lack of
embarrassment about the gaps they reveal
in their knowledge. Above all Einstein shares

their laughter and their mysterious sense of humor which makes the grown-ups exclaim: What on earth are they laughing about? (The humorous verses Einstein composed for my daughter when she was small were never beyond her comprehension.) And so Einstein goes on regretting never having been a schoolteacher. He thinks of children as lost in a universe of physical phenomena. He knows that he can make the great laws of nature accessible to them. He loves an explanation to be both exact and clear. He once interrupted a grandiloquent statement on a scientific discovery by a visitor with: "If it is something that one can understand, one can also explain it clearly." It annoys him that the automatic way in which physics are so often taught to children is responsible for the number of adults excluded for ever from awareness of that miraculous universe. I talked to him about my own ignorance of the most elementary facts, as I had a blind spot for the elementary principles of mathematics and physics. "What nonsense!" he said aggressively. "It is merely that you haven't been taught properly." The telephone interrupted us. "You see," I said, "I have been told how a telephone works, but I still don't grasp it." "Why, it's very simple," and he explained it so clearly that it became quite obvious and also wonderful. "You see what a good teacher of elementary physics I might have been," he said, laughing at my sense of achievement.

"Every scientist ought to have a shoemaker's job," Einstein once said. He found one in 1902, after a long and anxious search, thanks to a schoolfellow who was touched by his obvious poverty and who asked his father to intervene for Einstein with the director of the Federal Office of Patents in Berne. Here again he almost met with failure. The director was looking for someone capable of judging whether a request for a patent had any justification. "What do you know about patents?" he asked. "Nothing," replied Einstein. The interview might have ended there and then. But the frankness of this brief admission intrigued the man, who had been informed of the financial straits of the candidate who faced him. A long conversation proved that the applicant did know enough to recognize the value of an invention.

On this lower level Albert Einstein, when twenty-three years old, settled down in Berne, married and founded a family. Two sons were born in succession. Albert Einstein loved his sons as much as he could love anyone. But his thoughts were elsewhere. To understand what happened in the years following his new-found security, one must consider not only the creative miracle of genius—but also the curious capacity he had for absenting himself from his immediate worldly surroundings. A separation seemed to take place between his mind and his body—not unlike the ectasy of a saint. The word "transfiguration" acquired an almost literal sense. But Einstein's "absences" were in their external aspect, as alien as possible to ecstatic conditions; they were so realistic and commonplace that a superficial observer might never ever notice them.

Albert Einstein would be there, his senses radiantly alive to the simple and sensual pleasures of life. You might be sitting alone with him, or with two or three others, or the room might even be full of people. He had just finished a sentence. You would think he was following the conversation intently. Suddenly he would fall silent and stop listening to you. He would rise to his feet without a word, or remain sitting motionless. The effect would be the same. He would be unreachable. You could start talking noisily, or lapse into an even more embarrassing silence, with everybody staring at him, but he would neither see nor hear them. This had nothing in common with the absent-mindedness of a man wrapped up in his own thoughts, who continues to circulate in a world of reality, causing absurd misunderstandings. With Einstein, the eclipse was and is total.

In his office at the Federal Patent Office during the three years that he conscientiously performed his duties there, the same thing must often have happened. As soon as he got back his interest in scientific research, and recovered from the saturation of knowledge he needed for his examination, he attacked the problem of light rays which had haunted him ever since he was sixteen. He spoke once of the "desperate efforts" pursued by him over many years to find a solution.

During this time of extraordinary creative richness Einstein attended to many other problems that came his way, edited papers on various subjects, and treated them all as of equal importance. The theory which scientists the world over agreed had caused a break with the experience of many centuries, and the creation of a new universe, first saw the light in a specialized German publication, the *Year Book of Physics*, a bulky volume filled with the most diverse contributions. Einstein published one article in it yearly after 1901, when he completed his studies. The 1905 volume contained five contributions from him. They were all on different subjects. Perhaps the author did not wish to emphasize one subject more than another. But the one entitled "A New Definition of Molecular Dimensions" was to give him his chance of a university career. Another deals with the quantum law of the emission and absorption of light and explains the phenomenon known in physics as "photo-electrical effect." The principle formulated by Einstein has influenced research in the physics of the quantum and spectroscopy everywhere. His formulation of this law gave birth later to television and other applications of the photo-electric cell. This is the contribution that later won for him the Nobel Prize. A third article took as its starting point the experiment made by the botanist Robert Brown on the movement of minute particles suspended in liquid. The fourth paper was the longest of all. Its title in no way augured the revolution that it was to create: "On the Electrodynamics of Bodies in Movement." But men with a knowledge of scientific publications note a purely superficial difference: this particular article contains no references, quotes no authority, and has few footnotes and those that there are only serve to explain the text. Later on, the experts were unanimous in declaring that what might have been but an outline of the theory of relativity was

in fact a complete paper on the subject. Einstein himself must have been conscious of the importance of the solution he had arrived at. It was learned much later that he had a total collapse after it, exhausted by his superhuman effort. He was ill for a fortnight. The shock was very violent.

Einstein never forgot the moment when the ultimate light was revealed to him. He was conscious of living through a supreme experience. He knew no greater human emotion: no violent passion or deep distress could compare in intensity with the transformation of the universe which had taken place before his eyes. In an essay written in January 1940 on the foundation of theoretical physics, after speaking of the new concepts of fields discovered by Faraday, Einstein wrote: "The precise enunciation of the laws of these fields in space-time was the work of Maxwell. Can we visualize what he must have gone through when the differential equations that he had formulated proved that the electromagnetic fields are propagated in the shape of polarized waves and with the speed of light? . . . To few people in this world has it been given to witness such an experiment," he added with one of his rare glimpses of emotion, reminded no doubt of his own happiness.

Perhaps it was the hard struggle he had to reach his final convictions, perhaps it was the very intensity of his creative powers, but he seemed (was it from birth or from that moment?) unconscious of everything life could offer him and impervious to emotion. In his affections as well as in his thoughts, he seemed to be only lending his presence which he might withdraw at any time. Spurred on by his supreme effort Albert Einstein not only reconsidered theories of physics and questioned mechanical laws that had seemed safe against attack, but in his outline of the theory of relativity, as well as in a short article—the fifth—which appeared in the *Year Book of Physics* under the title "Does the Inertia of a Body Depend on its Energy?", he gave warning of a power which was to shake the world. For the first time the formula rendering the use of atomic energy theoretically possible appeared in print, and the possibility of annihilating humanity became a subject of speculation. The old concept of our universe had been upset by one man.

Once again, as at the beginning of his career, Einstein's work was met with curious indifference. Those who were much later to grasp the immense importance of these theories seem not to have noticed the article in which he developed them for the first time. A long way from Zurich, in the University of Cracow, a Polish professor, Witkowski, exclaimed on reading the article: "A new Copernicus has been born." He roused enthusiasm in one of his pupils who later became a remarkable physicist. This young professor, Loria by name, spoke of Einstein's article to other colleagues, repeating the words of his teacher: "A new Copernicus has been born." Einstein?—the name meant nothing to his colleagues. The professor talked with such fervor to the German physicist Max Born that they went to the library to look for the 1905 *Year Book of Physics*, and Max Born was later to write: "One of the most remarkable volumes in all scientific literature is Volume 17, Series 4, of the *Year Book of Physics*, 1905."

Though the importance of his work was not at once recognized, its scope was noticed sufficiently to produce the impression that his modest position was somehow inadequate. But one could not, after all, offer a Chair to a civil servant of the Federal Office! In all the Germanic countries the line of demarcation between the secondary school and the university is strictly defined. In Latin countries a future university professor must justify himself by a work of distinction, to reach the first stage, the grade of lecturer. One of the articles published in the *Year Book of Physics*— not the one on the theory of relativity—was considered sufficient to "qualify" Einstein. A lectureship does not bring with it, however, other remuneration than the scanty contributions of the students. Einstein could not think of abandoning his meagre "livelihood." A compromise was finally found: the University of Berne let him keep the patent job and combine it with addresses to students as university lecturer.

One of his colleagues reports that Einstein, pursuing his own work in his mind, showed little interest in university matters and did not take enough trouble in preparing his lectures. The physicist who had proposed him for the Chair in Zurich, himself a professor there, came to Berne to get a report on his protégé. Disappointed, he told him that the lectures he addressed to his students would not be appropriate for the auditorium he had planned for him. "I do not specially wish to transfer to Zurich," said Einstein quickly, as a man who wishes above all to be left in peace. Still, he knew that he could not remain for ever working in Berne. He had to think of his career. It was inconceivable to refuse a Chair. A rival sprang up on his path who was already established in the University of Zurich and who had been a former Colleague in the Polytechnic. Friedrich Adler's father, Victor Adler, was leader of the Austrian Socialist Party; he was well-known and had a large following abroad. Socialist circles in Zurich were very much on his son's side. A famous name might easily have eclipsed a stranger. Once more Einstein's career was in jeopardy. But integrity was a god to Friedrich Adler. He it was who intervened with the university authorities and stood aside.

From 1909, the year of Einstein's appointment as professor extraordinary at Zurich (he was not yet holder of a Chair), his remuneration was no higher than it had been as an official in Berne, but his obligations increased. The family budget was only balanced by taking in students as lodgers.

The appreciation of the scientific world only came to Einstein gradually. A year later, when he was just over thirty, he was granted a Chair. In 1910 he was called to the German university of Prague. This nomination was only an incident in his career, but it brought him to two important turning-points in his personal life. The regulations concerning Government employment in Austria forced him to declare his religion. He had no ties with the Jewish community after he left Munich, and no more links with Jewish tradition. At that time it was sufficient to state one's religion on paper and in this way lose all connection with Jewry. Many obstacles were overcome in that way. Einstein's friends insisted that he should make what could be called a concession to prejudice. They quoted to him many "conversions" for the sake of a career, but it was precisely these opportunist arguments and these laws enforced by the Emperor Francis Joseph that brought Einstein to the reaffirmation of one of the principles of his life. The young professor filled in his questionnaire in his clear and regular handwriting: "Religion—Israelite."

A further change took place in his private life. The move, with all that it involved in the way of adaptation to the new milieu, to changed conditions in everyday life, served to accentuate the friction that existed between him and his wife.

Einstein stayed only two years in Prague. He did not know how people felt about him. He was absorbed in his work, isolated as though on a desert island and hardly noticing other people. He was documenting and widening the theory of relativity as he had expounded it in 1905. "I only became fully conscious of the fact," he said, "that the specific theory of relativity[1] was only a first step towards a necessary evolution when I tried to integrate gravitation into the framework of that theory." In 1912 six articles in the *Year Book of Physics* are devoted for the most part to the problem of gravitation. The integration he attempted proved to be difficult. "The path was more arduous than had been expected," he wrote later, "because it was in contradiction to Euclid's Geometry."

At one time Einstein tried to describe the processes of his scientific work and in doing so he exposed the workshop of his mind. Starting with primary concepts directly linked to sense experiences and theorems which are interdependent, the scientist tries to discover the logical unity in the image of the universe. He goes beyond what Einstein calls "the secondary layer" to arrive at a system of the greatest conceivable unity. He

considered it an error to designate these superimposed layers of thought as "degrees of abstraction." "I do not consider it right to conceal the logical independence of a fundamental concept from the sense experiment." And he adds, in his usual picturesque language: "The connection is not comparable to that of soup to beef, but more to that of the cloakroom ticket to the overcoat."

In his attempt to throw light on how scientific thought functions, Einstein underlines that "the fundamental concepts and theorems and the relations between them should be as narrow as possible but freely selected." But the freedom "is not much of a freedom, it is not like the freedom of a novelist, rather like that of a man trying to solve a cleverly designed crossword puzzle. He may suggest any word as a solution, but there is probably only one word which can really solve the puzzle in all its parts." Einstein at that time was only halfway on the path to his unique solution of the mystery. Following this path with characteristic tenacity, he challenged almost casually those whom he had not yet convinced, and the challenge had consequences whose spectacular repercussions ever he could not foresee. In examining the influences of gravitation on light, he declared, making another breach in Newtonian physics, that a light ray undergoes a deviation in proportion to the gravitation, so that it acquires the shape of a parabola. He wrote this from his desk, but perhaps he looked up at an exceptionally bright night sky as he did so: even if his eyes remained fixed on his paper, he had a picture of the star-studded sky in his mind's eye. By thought alone, and by application of the logic which leads to the unity of the world, he established the laws which governed this inaccessible world. In the paper he went on to prepare for the *Year Book of Physics*, he stressed that this deviation of light rays of stationary stars must be visible during an eclipse of the sun and that it would therefore be

[1]More commonly known as the *special* theory of relativity. [Ed.]

possible to verify his theory by experiment. He concluded his articles "It would be urgently desirable for astronomers to become interested in this question, even if the considerations given here appear insufficiently founded or even adventurous." Einstein's theories are thus introduced into the domain of the possible.

From now on, Einstein was no more a solitary searcher, a mere man of promise, with brilliant gifts and bold ideas which frightened his conservative colleagues.

The Zurich Polytechnic realized belatedly its mistake in not clinging to the former student who was one day to be its chief pride. It now offered a Chair to the candidate who had been rejected so coldly only a few years before. And in 1912 Einstein returned to lecture to the selfsame benches where once he had sat and listened. At first, back in the familiar atmosphere in which he had first met his wife, their relations may have improved. But their differences were too deep-rooted for the breach to be altogether healed. And their stay in Zurich was not long enough to bring together two people who were by now virtually estranged.

The year after his nomination to Zurich, Albert Einstein received the most flattering invitation that could have been given to a man of thirty-four. Germany's two most eminent physicists, Planck and Nernst, proposed him as a member of the scientific institute founded by the Emperor, the Kaiser-Wilhelm Gesellschaft. The two scientists went to Zurich to persuade Einstein of the enormous advantages offered to him. He would have a Chair at the university, without the obligation of regular lectures, and apart from this he would have more time to devote to his scientific work, with a high salary and membership of the Prussian Academy of Science—the equivalent of a marshal's baton offered to a young lieutenant. The proposal was enough to dazzle even a man as insensitive to material advantages and honors as Einstein.

But, flattered as he must have been, nevertheless Einstein was cautious. The professor of a university was a Government servant. As a civil servant he acquired with his post German nationality. Einstein had not lost the convictions of his youth which had once made him break with the country of his birth. His Swiss nationality had given him the feeling of being international rather than national, "a citizen of the world." The obstinacy with which he defended his point of view was so instinctive that it might have sprung from some presentiment of the future.

The problem he had raised was a hard one for the German scientists to swallow, but they were eager to benefit from his growing fame. Besides, there was a precedent. The faculty of letters had among its prominent personalities a Frenchman, Professor Haguenin, who before accepting his nomination had stipulated that he wished to keep his nationality. Einstein and Haguenin were the only ones in this large university body to remain foreigners while becoming Prussian civil servants.

In the autumn of 1913 Einstein left Zurich and went to Berlin alone. His separation from his wife, at first temporary, soon became final. He knew that it was inevitable when he went. He was deeply distressed when he left his sons: it was probably the only time in his life when anyone saw him cry.

The declaration of war hit Berlin as if the sky had fallen.

On the afternoon of August 1, 1914, it still seemed possible that the disaster would be averted. The crowd marched through the streets of Berlin crying "Down with the war!" The workmen who walked in close ranks, and the public orators who climbed upon improvised platforms, angrily addressing the crowd, seemed like a tidal wave which might really sweep away the arrogant military caste with their sham Lohengrin at their head. Next day the crowd, even denser and more excited, stampeded in the square in front of the palace. The Kaiser appeared. The crowd was delirious with enthusiasm. This war which Wilhelm II had declared suddenly appeared to them as a long-awaited release, a holy war: but had they any idea what it was to be a release from?

Einstein, like all those who lived through these hours in Berlin and saw this delirium, was never to forget how a crowd could be won over completely to a destructive impulse. "This blindness, incomprehensible to me has struck like an epidemic many who have always seemed to think and feel clearly up till now," he wrote to Romain Rolland. His status as a neutral protected him from the necessity of making a stand. On the strength of it he was able to refuse to sign the manifesto of the ninety-three intellectuals, the capitulation of German spiritual independence. But he did not retreat altogether into the protective shelter of neutrality. In the same month of October when this manifesto appeared, he signed a counter-manifesto, written by that indomitable German pacifist Professor G. Nicholai, who later achieved fame by his spectacular escape by aeroplane from wartime Germany.

His reputation as an eccentric saved him from the most savage attacks of his indignant colleagues. Nevertheless in their eyes he was a moral leper. He was to remember that when he came for the first time to Paris. He was invited by Madame Curie and when he sat down in the salon he glanced at the row of chairs beside him which no one dared to occupy and smiled across at young Frederic Joliot. "Come and sit next to me, for I feel exactly as if I was at a meeting of the Prussian Academy where two chairs are always left empty beside mine," he explained laughing.

It would have been more difficult for Einstein to preserve peace for his work, and stand the increasing privations that the war effort imposed upon Germany, had he been wholly alone in the foreign Prussian capital. But on arriving in Berlin he found support, and soon after shelter, with one of his uncles and his daughter. The meeting between Einstein and his cousin Elsa was the beginning of the only permanent personal relationship in his life: thenceforward their relationship dominated his private life. Einstein, distrustful of all instrusions into his private life, must have been on the defensive against every stranger. But with Elsa it was as if he rediscovered something with which he had been familiar all his life and his defences disappeared. Their marriage, when it took place, seemed the most obvious thing in the world to both of them.

Whenever he was working particularly hard, Elsa saw that he had silence around him and made his everyday life as easy as possible, so that the transition from the intellectual effort should be smooth and unobtrusive. When he emerged from his study where he had been closeted for many hours with his assistant, followed by clouds of smoke and pulling at his pipe, with his eyes shining, Elsa would slowly bring him back to reality as though awakening a sleepwalker; she would gradually bring to his attention the people around him and the food on his plate, which he was chopping with his knife like a blind man. One day, in a moment of relaxation, she asked him: "People talk a lot about your work at the moment. Everybody keeps asking me for news. I appear so stupid when I have to say that I know nothing. Couldn't you just tell me a little about it?"

"Yes," said Einstein, "it must be irritating for you." He thought for a moment. He smiled; it was a tender smile. Elsa gazed at him with her vague, serene, short-sighted glance. "Well," he began with a visible effort. He stopped, then suddenly his face lit up happily. "Well, if people ask you, you can tell them that you know all about it, but can't tell them as it is a great secret." He was delighted to have discovered this solution.

Elsa had an acute gift of observation, which was softened by the natural tolerance of her kind nature. Above all she had a sense of humor that triumphed over all the great and small difficulties of life. An incident that might have caused a nervous breakdown with anybody else made her shriek with laughter, gay laughter that swept away all irritations like so many cobwebs.

Now that he had no domestic cares, it was much easier for Einstein to exchange the peace of his home for the increasingly engrossing adventure of his work. On November 28, 1915, at the height of the war, Einstein wrote to a physicist friend, Arnold Sommerfeld, who had written him several unanswered letters: "This last month I have lived through the most exciting and the most exacting period of my life: and it would be true to say that it has also been the most fruitful. Writing letters has been out of the question. I realized that up till now my field equations of gravitation had been entirely devoid of foundation. When all my confidence in the old theory vanished, I saw clearly that a satisfactory solution could only be reached by linking it with the theory of the Riemann variations. The wonderful thing that happened then was that not only did the theory of Newton result from it, as a first approximation, but also the perihelion motion of Mercury (43″ per century) as a second approximation. For the deviation of light by the sun, I obtained twice the former amount."

In a speech on Planck, a little later, Einstein actually described his own state of mind at the time. He borrowed a sentence from Leibnitz to explain the love of research that stimulates every scientist, the "desire for a pre-established harmony." It was, according to him, mistaken to attribute (as is generally done) Planck's indefatigable tenacity and patience to his unusual strength of will and rigorous discipline, for the "emotional condition that allows similar achievements to be accomplished is more like that of a deeply religious man or of a man in love; the daily effort is not dictated either by a purpose or a programme, but by an immediate need."

In 1916 there appeared in the *Year Book of Physics*, as well as in a separate publication, the work he had mentioned to Sommerfeld: "The Foundations of the Theory of General Relativity." This work, Einstein's main work, took up sixty-four pages. Rarely, perhaps never before in the history of human thought, did so small a publication have such a disturbing effect on the world. Einstein himself was aware that a great victory had been won and he was fully conscious of his own contribution. One day, much later, his collaborator, Infeld said to him: "I believe

that the special theory of relativity would have been formulated with but little delay whether or not you had done it." "Yes, that is true," said Einstein. He thought, indeed, that a scientist like Langevin, for instance, might have developed it; for according to Einstein, Langevin had clearly realized its essential features. However, he added immediately: "But this is not true of the general theory of relativity. I doubt whether it would have been known yet."

This awareness of his achievement and his feeling of a spiritual triumph had nothing to do with self-satisfaction. It was more like the joy of a believer in seeing a miracle accomplished in front of him, an almost humble joy. Einstein described later this faith that possessed him: "In a certain sense I hold it to be true that pure thought can grasp reality as the ancients dreamed it."

But the moment when this work of Einstein's first saw the light could not have been more ill-chosen, and more unpropitious for the diffusion of his ideas. The war in Germany was at its height and the atmosphere was tense. Zeppelin bombs broke the silence of London nights. Only one human voice managed to penetrate the noise of war: and it was the voice of a scientist who spoke the language of the enemy.

The English scientists listened and what they heard deeply disturbed them. They pored with passionate interest over the article Einstein published in the reports of the Prussian Academy of Science: "Cosmological Notes on the Theory of General Relativity."

"Modern cosmology was born in that year of 1917," Einstein's collaborator, Professor Infeld, wrote later. And he added: "Though it would be difficult to exaggerate the importance of this paper, Einstein's original ideas as viewed from the perspective of the present day are antiquated, if not wrong." Since Einstein had worked out his cosmology through the effort of pure thought, and had been able to materialize, as he said, the dreams of the ancients, the methods of observation of the universe had been multiplied, and had acquired a power that he himself did not suspect at the time. The human eye penetrated beyond the nebulae into this universe, which Einstein pictured as a finite universe, with curved space and populated with matter, and not as an infinite universe, a void around an island of matter.

Certain phenomena observed by these powerful means, such as the tendency of the spectrum of the nebulae to displace itself towards the red, the red shift of the nebulae, as it was called, seemed to other scientists, like Infeld, a breach in Einstein's theory. It never ceased, however, to occupy the minds of astronomers, physicists and mathematicians. A new spatial notion was born. The structure of the universe, more or less modified, will never be what it was before him.

The effect was immediate. Einstein's suggestion to the astronomers to direct their attention to the theory of relativity was accepted. In March 1917, the official organ of the Royal Astronomical Society, announced that on March 29, 1919, a total eclipse of the sun was taking place and that one would therefore have particularly favorable conditions for submitting Einstein's theory to a decisive experiment.

The mind of one man had spanned the abyss of hatred and sorrow which existed between nation and nation: it was a span, however, invisible to those who fought and died in the vain hope that there would be no more wars.

It was like a great epic or adventure into the unknown—an escape beyond the barriers of human misery which war had erected. Two great expeditions were organized in February 1919. One was to Sobral, in the north of Brazil; the other to the island of Principe, in the Gulf of Guinea. These two modern expeditions left because of their faith in one man, who had elaborated a bold theory, guided only by his scientific logic. Never before was there an adventure of such daring, so totally confined to the domains of pure thought.

Sir Arthur Eddington, the great British astronomer, has described the expedition to Principe, which he had insisted on joining himself. Nothing was left to the last minute. The expedition arrived a month before the eclipse. On the day itself dawn broke in a clouded, misty sky. When the eclipse became total, the dark disc of the moon, surrounded with its halo, appeared among the clouds as one often sees it at night when the stars are invisible. "There was nothing to do but carry out the arranged programme and hope for the best." A strange, ghostly half-light covered the earth, accompanied by deep silence, broken only by whispered conversations, the click of the observers changing their plates, and the inexorable ticking of the metronome squandering the precious seconds. Suddenly a flame shone out above the invisible sun and remained floating in space hundreds of millions of miles above the surface of the sun. The team at Principe had no time to take in this strange sight: they were too anxious about the success of the experiment.

The sky clouded over more and more: it seemed determined to thwart the efforts of man and baffle his curiosity. On the first photograph there was no sign of a star. About sixteen photographs were taken, however, with an exposure of 2–20 seconds. Towards the end of the eclipse, the clouds vanished and the last photographs gave a clear picture. In many of them one or other of the essential stars was missing. But one plate eventually succeeded in capturing the light of five stars, and this was good enough to be used for an examination of the Einstein theory.

Months passed by, devoted to the careful examination of the results obtained, and to a comparison of the photographs brought from Sobral with those of the Greenwich Observatory. After repeatedly verified calculations, the deviation of light of 1.64 seconds was established: the deviation that Einstein, from his writing-desk, had fixed at 1.75 seconds.

It was at the beginning of November 1919, at a solemn meeting of the Royal Astronomical Society, that the results achieved were made public, amid considerable tension. A philosopher who was present compared it later to that of a Greek drama in which the chorus awaits the verdict of destiny. The President of the Society opened the meeting by describing Einstein's theory as one of the greatest achievements in the history of human thought. "It is not the discovery of an outlying island, but of a whole continent of new scientific ideas. It is the greatest discovery in connection with gravitation that has been made since Newton first enunciated its principles."

The hall in which the meeting took place was dominated by a large portrait of Newton. The shadows of a great man of the past and of the absent stranger loomed over the audience. A stranger, under suspicion because of the very language in which he had enunciated his theory, had dared to challenge one of the most glorious names in English history.

The first contact between the enthusiastic public and Einstein occurred the day after the memorable meeting. *The Times* had sent a correspondent to Berlin to ask him for a few words of explanation of his theory. Einstein was not at all reluctant—it was the first time he had been approached. He even expressed his gratification at being able to say a few words about "relativity" (he still put the word in inverted commas) and made use of the occasion, after the lamentable collapse of international relations among scientists, to express his gratitude to the English physicists and astronomers. "It is quite in keeping with the great and proud traditions of scientific work in your country that eminent men of research should devote a lot of time and a lot of effort, and your scientific institutions a lot of money, to examine the results of a theory that was elaborated and published during the war in the country of your enemies."

4.2 The Special Theory of Relativity
Albert Einstein

In your schooldays most of you who read this book made acquaintance with the noble building of Euclid's geometry, and you remember—perhaps with more respect than love—the magnificent structure, on the lofty staircase of which you were chased about for uncounted hours by conscientious teachers. By reason of your past experience, you would certainly regard everyone with disdain who should pronounce even the most out-of-the-way proposition of this science to be untrue. But perhaps this feeling of proud certainty would leave you immediately if someone to ask you: "What, then, do you mean by the assertion that these propositions are true?" Let us proceed to give this question a little consideration.

Geometry sets out from certain conceptions such as "plane," "point," and "straight line," with which we are able to associate more or less definite ideas, and from certain simple propositions (axioms) which, in virtue of these ideas, we are inclined to accept as "true." Then, on the basis of a logical process, the justification of which we feel ourselves compelled to admit, all remaining propositions are shown to follow from these axioms, i.e., they are proven.

A proposition is then correct ("true") when it has been derived in the recognized manner from the axioms. The question of the "truth" of the individual geometrical propositions is thus reduced to one of the "truth" of the axioms. Now it has long been known that the last question is not only unanswerable by the methods of geometry, but that it is in itself entirely without meaning. We cannot ask whether it is true that

SOURCE. From *Relativity: The Special and General Theories*, translated by R. W. Lawson (London: Methuen, 1931); copyright 1961 by the Estate of Albert Einstein, reprinted by permission of Crown Publishers, Inc.

only one straight line goes through two points. We can only say that Euclidean geometry deals with things called "straight lines," to each of which is ascribed the property of being uniquely determined by two points situated on it. The concept "true" does not tally with the assertions of pure geometry, because by the word "true" we are eventually in the habit of designating always the correspondence with a "real" object; geometry, however, is not concerned with the relation of the ideas involved in it to objects of experience, but only with the logical connection of these ideas among themselves.

Every description of the scene of an event or of the position of an object in space is based on the specification of the point on a rigid body (body of reference) with which that event or object coincides. This applies not only to scientific description, but also to everyday life. If I analyze the place specification "Times Square, New York,"[1] I arrive at the following result. The earth is the rigid body to which the specification of place refers; "Times Square, New York," is a well-defined point, to which a name has been assigned, and with which the event coincides in space.

This primitive method of place specification deals only with places on the surface of rigid bodies, and is dependent on the existence of points on this surface which are distinguishable from each other. But we can free ourselves from both of these limitations without altering the nature of our specification of position. If, for instance, a cloud is hovering over Times Square, then we can determine its position relative to the surface of the earth by erecting a pole perpendicularly on the Square, so that it reaches the cloud. The length of the pole measured with the standard measuring rod, combined with the specification of the position of the foot of the pole, supplies us with a complete place specification. On the basis of this illustration,

we are able to see the manner in which a refinement of the conception of position has been developed.

We thus obtain the following result: Every description of events in space involves the use of a rigid body to which such events have to be referred. The resulting relationship takes for granted that the laws of Euclidean geometry hold for "distances," the "distance" being represented physically by means of the convention of two marks on a rigid body.

"The purpose of mechanics is to describe how bodies change their position in space with time." I should load my conscience with grave sins against the sacred spirit of lucidity were I to formulate the aims of mechanics in this way, without serious reflection and detailed explanations. Let us proceed to disclose these sins.

It is not clear what is to be understood here by "position" and "space." I stand at the window of a railway carriage which is traveling uniformly, and drop a stone on the embankment, without throwing it. Then, disregarding the influence of the air resistance, I see the stone descend in a straight line. A pedestrian who observes the misdeed from the footpath notices that the stone falls to earth in a parabolic curve. I now ask: Do the "positions" traversed by the stone lie "in reality" on a straight line or on a parabola? Moreover, what is meant here by motion "in space"? From the considerations of the previous section the answer is self-evident. In the first place, we entirely shun the vague word "space," of which, we must honestly acknowledge, we cannot form the slightest conception, and we replace it by "motion relative to a practically rigid body of reference." The positions relative to the body of reference (railway carriage or embankment) have already been defined in detail in the preceding section. If instead of "body of reference" we insert "system of coordinates," which is a useful idea for mathe-

---

[1] "Times Square, New York" has been used here because it is more familiar than the "Potsdamer Platz, Berlin," which is referred to in the original. [Ed.]

matical description, we are in a position to say: The stone traverses a straight line relative to a system of coordinates rigidly attached to the carriage, but relative to a system of coordinates rigidly attached to the ground (embankment) it describes a parabola. With the aid of this example it is clearly seen that there is no such thing as an independently existing trajectory (literally, "path curve"), but only a trajectory relative to a particular body of reference.

In order to have a complete description of the motion we must specify how the body alters its position *with time*; i.e., for every point on the trajectory it must be stated at what time the body is situated there. These data must be supplemented by such a definition of time that, in virtue of this definition, these time values can be regarded essentially as magnitudes (results of measurements) capable of observation. If we take our stand on the ground of classical mechanics, we can satisfy this requirement for our illustration in the following manner. We imagine two clocks of identical construction; the man at the railway carriage window is holding one of them, and the man on the footpath the other. Each of the observers determines the position on his own reference body occupied by the stone at each tick of the clock he is holding in his hand. In this connection we have not taken account of the inaccuracy involved by the finiteness of the velocity of propagation of light. With this and with a second difficulty prevailing here we shall have to deal in detail later.

As long as one was convinced that all natural phenomena were capable of representation with the help of classical mechanics, there was no need to doubt the validity of this principle of relativity. But in view of the more recent development of electrodynamics and optics, it became more and more evident that classical mechanics affords an insufficient foundation for the physical description of all natural phenomena. At this juncture the question of the validity of the principle of relativity became ripe for discussion, and it did not appear impossible that the answer to this question might be in the negative.

Now in virtue of its motion in an orbit round the sun, our earth is comparable with a railway carriage traveling with a velocity of about 30 kilometers per second. If the principle of relativity were not valid we should therefore expect that the direction of motion of the earth at any moment would enter into the laws of nature, and also that physical systems in their behavior would be dependent on the orientation in space with respect to the earth.

However, the most careful observations have never revealed such anisotropic properties in terrestrial physical space, i.e., a physical non-equivalence of different directions.[3] This is very powerful argument in favor of the principle of relativity.

There is hardly a simpler law in physics than that according to which light is propagated in empty space. Every child at school knows, or believes he knows, that this propagation takes place in straight lines with a velocity $c = 300,000$ km sec. At all events we know with great exactness that this velocity is the same for all colors, because if this were not the case, the minimum of emission would not be observed simultaneously for different colors during the eclipse of a fixed star by its dark neighbor. By means of similar considerations based on observations of double stars, the Dutch astronomer De Sitter was also able to show that the velocity of propagation of light cannot depend on the velocity of motion of the body emitting the light. The assumption that this velocity of propagation is dependent on the direction "in space" is in itself improbable.

At this juncture the theory of relativity entered the arena. As a result of an analysis of the physical conceptions of time and space, it became evident that *in reality there is not the*

---

[2]That is, a curve along which a body moves.
[3]The reference here is to the negative results of the Michelson-Morley experiment. [Ed.]

least incompatibility between the principle of relativity and the law of propagation of light, and that by systematically holding fast to both these laws a logically rigid theory could be arrived at. This theory has been called the *special theory of relativity* to distinguish it from the extended theory, with which we shall deal later.

Lightning has struck the rails on our railway embankment at two places *A* and *B* far distant from each other. I make the additional assertion that these two lightning flashes occurred simultaneously. If I ask you whether there is sense in this statement, you will answer my question with a decided "Yes." But if I now approach you with the request to explain to me the sense of the statement more precisely, you find after some consideration that the answer to this question is not so easy as it appears at first sight.

Are two events (e.g., the two strokes of lightning *A* and *B*) which are simultaneous with reference to the *railway embankment* also simultaneous *relatively to the train*? We shall show directly that the answer must be in the negative [Figure 4.2–1].

When we say that the lightning strokes *A* and *B* are simultaneous with respect to the embankment, we mean: the rays of light emitted at the places *A* and *B* where the lightning occurs, meet each other at the mid-point *M* of the length *A* → *B* of the embankment. But the events *A* and *B* also correspond to positions *A* and *B* on the train. Let *M'* be the mid-point of the distance *A* → *B* on the traveling train. Just when the flashes[4] of lightning occur, this point *M'* naturally coincides with the point *M*, but it moves

toward the right, with the velocity *v* of the train. If an observer sitting in the position *M'* in the train did not possess this velocity, then he would remain permanently at *M*, and the light rays emitted by the flashes of lightning *A* and *B* would reach him simultaneously, i.e., they would meet just where he is situated. Now in reality (considered with reference to the railway embankment) he is hastening toward the beam of light coming from *B*, whilst he is riding on ahead of the beam of light coming from *A*. Hence the observer will see the beam of light emitted from *B* earlier than he will see that emitted from *A*. Observers who take the railway train as their reference body must therefore come to the conclusion that the lightning flash *B* took place earlier than the lightning flash *A*. We thus arrive at the important result:

Events which are simultaneous with reference to the embankment are not simultaneous with respect to the train, and vice versa (relativity of simultaneity).

Our train of thought in the foregoing pages can be epitomized in the following manner. Experience has led to the conviction that, on the one hand, the principle of relativity holds true, and that on the other hand the velocity of transmission of light *in vacuo* has to be considered equal to a constant *c*. By uniting these two postulates we obtained the law of transformation for the rectangular coordinates *x, y, z,* and the time *t* of the events which constitute the processes of nature. In this connection we did not obtain the Galilei transformation, but, differing from classical mechanics, the *Lorentz transformation.* General laws of nature are covar-

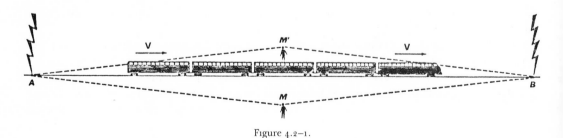

Figure 4.2–1.

iant with respect to Lorentz transformations.[5]

It is clear from our previous considerations that the (special) theory of relativity has grown out of electrodynamics and optics. In these fields it has not appreciably altered the predictions of theory, but it has considerably simplified the theoretical structure, i.e., the derivation of laws, and—what is incomparably more important—it has considerably reduced the number of independent hypotheses forming the basis of theory. The special theory of relativity has rendered the Maxwell-Lorentz theory so plausible that the latter would have been generally accepted by physicists even if experiment had decided less unequivocally in its favor.

Classical mechanics required to be modified before it could come into line with the demands of the special theory of relativity. For the main part, however, this modification affects only the laws for rapid motions, in which the velocities of matter v are not very small as compared with the velocity of light. We have experience of such rapid motions only in the case of electrons and ions; for other motions the variations from the laws of classical mechanics are too small to make themselves evident in practice. We shall not consider the motion of stars until we come to speak of the general theory of relativity.

Let me add a final remark of a fundamental nature. The success of the Faraday-Maxwell interpretation of electromagnetic action at a distance resulted in physicists becoming convinced that there are no such things as instantaneous actions at a distance (not involving an intermediary medium) of the type of Newton's law of gravitation. According to the theory of relativity, action at a distance with the velocity of light always takes the place of instantaneous action at a distance or of action at a distance with an infinite velocity of transmission.

4.3   City Speed Limit

George Gamow

It was a bank holiday, and Mr. Tompkins, the little clerk of a big city bank, slept late and had a leisurely breakfast. Trying to plan his day, he first thought about going to some afternoon movie and, opening the morning paper, turned to the entertainment page. But none of the films looked attractive to him. He detested all this Hollywood stuff, with infinite romances between popular stars.

If only there were at least one film with some real adventure, with something unusual and maybe even fantastic about it. But there was none. Unexpectedly, his eye fell on a little notice in the corner of the page. The local university was announcing a series of lectures on the problems of modern physics, and this afternoon's lecture was to be about Einstein's Theory of Relativity. Well, that might be something! He had often heard the statement that only a dozen people in the world really understood Einstein's theory. Maybe he could become the thirteenth! Surely he would go to the lecture; it might be just what he needed.

He arrived at the big university auditorium after the lecture had begun. The room was full of students, mostly young, listening with keen attention to the tall, white-bearded man near the blackboard who was trying to explain to his audience the basic ideas of the Theory of Relativity. But Mr. Tompkins got only as far as understanding that the whole point of Einstein's theory is that there is a maximum velocity, the velocity of light, which cannot be surpassed by any moving material

SOURCE. From *Mr. Tompkins in Paperback* (Cambridge Univ. Press, 1965).
[5]That is, all laws of Nature have the same *form* in all inertial reference frames. [Ed.]

body, and that this fact leads to very strange and unusual consequences. The professor stated, however, that as the velocity of light is 186,000 miles per second, the relativity effects could hardly be observed for events of ordinary life. But the nature of these unusual effects was really much more difficult to understand, and it seemed to Mr. Tompkins that all this was contradictory to common sense. He was trying to imagine the contraction of measuring rods and the odd behavior of clocks—effects which should be expected if they move with a velocity close to that of light—when his head slowly dropped on his shoulder.

When he opened his eyes again, he found himself sitting not on a lecture room bench but on one of the benches installed by the city for the convenience of passengers waiting for a bus. It was a beautiful old city with medieval college buildings lining the street. He suspected that he must be dreaming but to his surprise there was nothing unusual happening around him; even a policeman standing on the opposite corner looked as policemen usually do. The hands of the big clock on the tower down the street were pointing to five o'clock and the streets were nearly empty. A single cyclist was coming slowly down the street and, as he approached, Mr. Tompkin's eyes opened wide with astonishment. For the bicycle and the young man on it were unbelievably shortened in the direction of the motion, as if seen through a cylindrical lens. The clock on the tower struck five, and the cyclist, evidently in a hurry, stepped harder on the pedals. Mr. Tompkins did not notice that he gained much in speed, but, as the result of his effort, he shortened still more and went down the street looking exactly like a picture cut out of cardboard. Then Mr. Tompkins felt very proud because he could understand what was happening to the cyclist—it was simply the contraction of moving bodies, about which he had just heard. "Evidently nature's speed limit is lower here," he concluded, "that is why the bobby on the corner looks so lazy, he need not watch for speeders." In fact, a taxi moving along the street at the moment and making all the noise in the world could not do much better than the cyclist, and was just crawling along. Mr. Tompkins decided to overtake the cyclist, who looked a good sort of fellow, and ask him all about it. Making sure that the policeman was looking the other way, he borrowed somebody's bicycle standing near the kerb and sped down the street. He expected that he would be immediately shortened, and was very happy about it as his increasing figure had lately caused him some anxiety. To his great surprise, however, nothing happened to him or to his cycle. On the other hand, the picture around him completely changed. The streets grew shorter, the windows of the shops began to look like narrow slits, and the policeman on the corner became the thinnest man he had ever seen.

"By Jove!" exclaimed Mr. Tompkins excitedly, "I see the trick now. This is where the word *relativity* comes in. Everything that moves relative to me looks shorter for me, whoever works the pedals!" He was a good cyclist and was doing his best to overtake the young man. But he found that it was not at all easy to get up speed on this bicycle. Although he was working on the pedals as hard as he possibly could, the increase in speed was almost negligible. His legs already began to ache, but still he could not manage to pass a lamp-post on the corner much faster than when he had just started. It looked as if all his efforts to move faster were leading to no result. He understood now very well why the cyclist and the cab he had just met could not do any better, and he remembered the words of the professor about the impossibility of surpassing the limiting velocity of light. He noticed, however, that the city blocks became still shorter and the cyclist riding ahead of him did not now look so far away. He overtook the cyclist at the second turning, and when they had been riding side by side for a moment, was surprised to see the cyclist was actually quite a normal, sporting-looking young man. "Oh, that must be because we do not move relative to each other," he concluded; and he addressed the young man.

"Excuse me, sir!" he said, "Don't you find

it inconvenient to live in a city with such a slow speed limit?"

"Speed limit?" returned the other in surprise, "we don't have any speed limit here. I can get anywhere as fast as I wish, or at least I could if I had a motor-cycle instead of this nothing-to-be-done-with old bike!"

"But you were moving very slowly when you passed me a moment ago," said Mr. Tompkins. "I noticed you particularly."

"Oh you did, did you?" said the young man, evidently offended. "I suppose you haven't noticed that since you first addressed me we have passed five blocks. Isn't that fast enough for you?"

"But the streets became so short," argued Mr. Tompkins.

"What difference does it make anyway, whether we move faster or whether the street becomes shorter? I have to go ten blocks to get to the post office, and if I step harder on the pedals the blocks become shorter and I get there quicker. In fact, here we are," said the young man getting off his bike.

Mr. Tompkins looked at the post office clock, which showed half-past five. "Well!" he remarked triumphantly, "it took you half an hour to go this ten blocks, anyhow—when I saw you first it was exactly five!"

"And did you *notice* this half hour?" asked his companion. Mr. Tompkins had to agree that it had really seemed to him only a few minutes. Moreover, looking at his wrist watch he saw it was showing only five minutes past five. "Oh!" he said, "is the post office clock fast?" "Of course it is, or your watch is too slow, just because you have been going too fast. What's the matter with you, anyway? Did you fall down from the moon?" and the young man went into the post office.

After this conversation, Mr. Tompkins realized how unfortunate it was that the old professor was not at hand to explain all these strange events to him. The young man was evidently a native, and had been accustomed to this state of things even before he had learned to walk. So Mr. Tompkins was forced to explore this strange world by himself. He put his watch right by the post office clock, and to make sure that it went all right waited

for ten minutes. His watch did not lose. Continuing his journey down the street he finally saw the railway station and decided to check his watch again. To his surprise it was again quite a bit slow. "Well, this must be some relativity effect, too," concluded Mr. Tompkins; and decided to ask about it from somebody more intelligent than the young cyclist.

The opportunity came very soon. A gentleman obviously in his forties got out of the train and began to move towards the exit. He was met by a very old lady, who, to Mr. Tompkins's great surprise, addressed him as "dear Grandfather." This was too much for Mr. Tompkins. Under the excuse of helping with the luggage, he started a conversation.

'Excuse me, if I am intruding into your family affairs," said he, "but are you really the grandfather of this nice old lady? You see, I am a stranger here, and I never . . . ."

"Oh, I see," said the gentleman, smiling with his moustache. "I suppose you are taking me for the Wandering Jew or something. But the thing is really quite simple. My business requires me to travel quite a lot, and, as I spend most of my life in the train, I naturally grow old much more slowly than my relatives living in the city. I am so glad that I came back in time to see my dear little grand-daughter still alive! But excuse me, please, I have to attend to her in the taxi," and he hurried away leaving Mr. Tompkins alone again with his problems. A couple of sandwiches from the station buffet somewhat strengthened his mental ability, and he even went so far as to claim that he had found the contradiction in the famous principle of relativity.

"Yes, of course," thought he, sipping his coffee, "if all were relative, the traveller would appear to his relatives as a very old man, and they would appear very old to him, although both sides might in fact be fairly young. But what I am saying now is definitely nonsense: One could not have relative grey hair!" So he decided to make a last attempt to find out how things really are, and turned to a solitary man in railway uniform sitting in the buffet.

"Will you be so kind, sir," he began, "will you be good enough to tell me who is responsible for the fact that the passengers in the train grow old so much more slowly than the people staying at one place?"

"I am responsible for it," said the man, very simply.

"Oh!" exclaimed Mr. Tompkins. "So you have solved the problem of the Philosopher's Stone of the ancient alchemists. You should be quite a famous man in the medical world. Do you occupy the chair of medicine here?"

"No," answered the man, being quite taken aback by this, "I am just a brakeman on this railway."

"Brakeman! You mean a brakeman . . . ," exclaimed Mr. Tompkins, losing all the ground under him. "You mean you—just put the brakes on when the train comes to the station?"

"Yes, that's what I do: and every time the train gets slowed down, the passengers gain in their age relative to other people. Of course," he added modestly, "the engine driver who accelerates the train also does his part in the job."

"But what has it to do with staying young?" asked Mr. Tompkins in great surprise.

"Well, I don't know exactly," said the brakeman, "but it is so. When I asked a university professor traveling in my train once, how it comes about, he started a very long and incomprehensible speech about it, and finally said that it is something similar to 'gravitation redshift'—I think he called it— on the sun. Have you heard anything about such things as redshifts?"

"No-o," said Mr. Tompkins, a little doubtfully; and the brakeman went away shaking his head.

Suddenly a heavy hand shook his shoulder, and Mr. Tompkins found himself sitting not in the station café but in the chair of the auditorium in which he had been listening to the professor's lecture. The lights were dimmed and the room was empty. The janitor who wakened him said: "We are closing up, Sir; if you want to sleep, better go home." Mr. Tompkins got to his feet and started toward the exit.

4.4   *Albert Einstein and the Cosmic World Order*

Cornelius Lanczos

The twenty years between 1905 and 1925 saw a revolution in our traditional way of thinking, unprecedented in the entire history of science. And this revolution was almost singlehandedly the work of one man: Albert Einstein.

The beginnings were not particularly startling. A paper published in the *Annalen der Physik* had the title: "On the electrodynamics of moving bodies." On the surface it dealt with some specific properties of optical and electromagnetic phenomena which were now presented in novel fashion. In actual fact it presented an entirely new approach to our most ingrained fundamental ideas

concerning the structure of the physical world: the concepts of space and time. The idea of an absolute space and an absolute time, which was at the foundation of Newtonian physics, was shown to be untenable in the face of certain undeniable experimantal evidence. A new foundation of physics was demanded which harmonized with the results of the experiments.

The novel feature in Einstein's approach was that where other people saw an isolated event, he recognized the operation of fundamental principles. This feature of his early papers of 1905 remained a characteristic feature of his entire scientific

SOURCE. From *Albert Einstein and the Cosmic World Order* (New York: Wiley (Interscience), 1965) Chapter 6.

career. Other physicists explained the same phenomena that Einstein explained, but he saw deeper than the others because for him the universe was an integrated whole in which all-pervasive principles were at work. And thus every one of his discoveries was of principal significance. This put him more in the category of a philosopher than of a specialist who sees things with professional eyes.

The principle of the equal admissibility of all reference systems in uniform motion relative to each other (special relativity); the principle of the equivalence of mass and energy; the principle of the equivalence of a gravitational field with an accelerated frame of reference; the principle of the equal admissibility of arbitrary reference systems (general relativity)—these are landmarks in the history of physics, which before had recognized the operation of *laws* but now recognized the operation of interconnecting *principles*. Planck explained his radiation law by assuming that energy is not radiated out continuously but in definite discrete bundles, called "quanta." No, said Einstein, this is not enough. Something much deeper is here at work. The difference between particle and wave breaks down. Certain phenomena of light emission can be understood only if we conceive of light as a particle which moves with light velocity. This was the concept of the "light quanta" (now called "photons")—a radical departure from our earlier concepts.

This kind of thinking was completely new in the annals of science, and the early antagonism of the physicists toward Einstein's way of seeing things is easily understandable. Where Einstein saw that the concepts of space and time were involved and demanded modification, they wanted an explanation in terms of physical concepts. And thus they complained that Einstein did not solve the problem but hid it in a maze of mathematical formulae. For them Einstein's special relativity was not a physical theory but a clever conjuring trick of mathematical magic. But then Minkowski came and demonstrated that Einstein's theory in fact allows no

strictly physical interpretation because it is equivalent to a new foundation of geometry in which space and time are no longer independent categories, but merely two different aspects of a unified space-time world in which a unified geometry was at work. Newton's absolute space and absolute time were abandoned in favor of an absolute four-dimensional world, the world of space-time.

To get used to this much more abstract way of thinking was not easy. In an early meeting on relativity one of the participants walked out in anger with the remark: "I am a physicist, not a mathematician." Somewhat later another participant walked out in anger, saying: "I am a physicist, not a philosopher." The idea that there are watertight compartments into which we put our concepts was repugnant to Einstein, who saw the world as an integrated whole, without sharp boundaries between the various branches of science. In the earlier phases of physics, one thought in predominantly physical pictures rather than in abstract mathematical formulations. Mathematics was tolerated as a kind of necessary evil, but the demands on the advanced tools of mathematics were rather modest. With the development of both relativity and quantum theory the shift toward abstract thinking became very pronounced. We no longer believe that physical action is explainable in terms of little hard balls which are the centers of all kinds of forces. A much more abstract kind of concept is demanded in which the particle is no longer something that can be considered in isolation but rather as a part of an all-embracing entity, called a "field." In this development Einstein's thought constructions played a vital role.

When Minkowski in 1908 demonstrated that Einstein's new theory was equivalent to a new geometrical approach uniting space and time in one single entity, the voices for a "physical" explanation of relativity fell silent, since it now became obvious that an explanation in physical terms cannot be given. Special relativity was a geometrical, not a physical, phenomenon and thus the explanation had to be given in the language of geometry and not in the language of physics. But this was only the beginning. In a few years time the mathematization of physics became even more pronounced, when Einstein began to speculate on the nature of gravitation. Here the heavy apparatus of tensor calculus had to be invoked and finally a theory evolved which in speculative boldness surpassed everything that ever existed in the realm of physical thought. The problem of geometry now came into the center of discussion and a new foundation of geometry was the result, a geometry in which the geometry realized in nature became amalgamated with the physical properties of matter. If before, time was absorbed by space, now matter became absorbed by space, giving us a new physical world picture in which matter appeared as a certain curvature property of the space-time world.

This gradual abstractization of our primitive concepts may appear on the surface as a loss. We can regret the fact that so many of our cherished ideas concerning space, time and matter had to be thrown overboard. How simple was the picture of the physical world before Einstein's relativity and quantum theory appeared on the scene. We could go a long way with the simple concepts of force, work, energy, and a few similar ideas. Today the picture of the physical world is infinitely more complex. Where Newton succeeded with one single quantity, Einstein introduced *ten* quantities and the relations between these quantities is by far more complex than the simple equation found by Newton. Should we not deplore then the passing of the naive phase of physics, comparable to the golden age of mankind, spent in a paradisiac innocence of fairy tales? Today, from the perspective of history, the evolution from Newton to Einstein appears to us in a different light. We admit the loss of simplicity, but we are willing to pay the price for the sake of the tremendous advance in *unity*. It is the wisdom of cosmic events which came so eminently in the foreground through the work of Einstein. In the early years of relativity many people com-

plained about the peculiar and apparently absurd conclusions deduced from the theory which seemed to contradict the commands of common sense. "Look at this wise guy," they said, "he wants to show how smart he is by making fools of us." In fact, Einstein had not the least intention to extol his own smartness, rather he had the astonishing higher wisdom which comes to light in the structure of the physical world. In deep humility he mirrored in his speculations the cosmic spirit which manifests itself in all phases of physical events. The penetrating insight into the inner structure of the universe was the aim and the reward of this unique genius. Thus it happened that something that appeared as a loss in the beginning turned into a blessing when the full implications of the new world picture were properly understood.

This probing into the depths was something new in the annals of science. It never happened before that the relentless pursuit of one single idea—the relativity of reference systems—led to such sweeping results. In fact, it is to be wondered how man with the limited tools at his disposal can even claim to penetrate beyond the "veil of Maya," the deception of the external appearance. There was a time in the history of mankind when it was considered blasphemy to make any kind of inquiry into the working of the natural law. God made the universe in his infinite wisdom and it is not up to us to question his work. Any kind of experimentation with physical instruments are discouraged and looked upon with suspicion. Roger Bacon, the English monk of the thirteenth century, the excellent physicist who discovered the fundamental laws of optics by his ingenious physical experiments, was suspected of black magic and of being a disciple of the devil. He spent years of his life in prison and would have been burnt at the stake had he not had the good fortune to enjoy the protection of a few powerful friends among the higher clergy. Man has no right to question the handiwork of God. He should accept it in all humility and discard any second thoughts about it.

But the inquisitive mind of man could not be put to sleep. The dogmatic narrowness of a misunderstood religion gradually slackened and the time came when the scientific mind emancipated itself completely from the clutches of theological thinking. Now a new era began in which the emphasis was on experimentation. The theoretical approach came to be discredited because of the exaggerated emphasis of Aristotelian ideas during the Middle Ages, a period during which people frequently misunderstood the true intentions of this great philosopher who was also a great scientist. Now the pendulum swung too far in the opposite direction. The human mind became enslaved to the purely empirical and looked askance at Plato's world of ideas which tried to salvage the eternal substance from the fleeting world of sense impressions.

And now suddenly a man appeared on the platform who restored the equilibrium. He was a scientist of the highest order and nobody could suspect him of any kind of alchemistic tendencies. And yet he saw science in a new light. To him science did not mean the primacy of the experiment or the primacy of the theory, but the primacy of a deep reverence for the all-embracing lawfulness which manifests itself in the universe. He was more a visionary filled with rapturous admiration for the majesty of creation than a sober scientist. "The most incomprehensible thing about the world is that it is comprehensible," he said.

Marvellous are thy works: and that my soul knoweth right well.
How precious are thy thoughts unto me, O God! How great is the sum of them.
If I should count them, they are more in number than the sand: when I awake, I am still with thee.

Such words, spoken by the Psalmist, would sound strange in a scientific treatise. But reading Einstein's papers we feel that here is a scientist who is imbued with the prophetic spirit, and that the Psalmist merely stated his ideas in poetic language.

This was religion as much as science. This was poetry and music and philosophy as much as science. No wonder that some of Einstein's greatest contemporaries from entirely different walks of life were attracted to this new way of integrated thinking. The great Irish dramatist and social philosopher G. B. Shaw had a strong predilection for the cosmic way of thinking and a great admiration for Einstein, whom he hailed as the greatest mind of the century (next to himself). Einstein came to London in 1930 and, associated with some charity drive, a banquet was given in his honor in the Hotel Savoy, to which H. G. Wells and G. B. Shaw were also invited. Shaw usually improvised his speeches. But this occasion he fortunately considered so important that he prepared his address in advance, dictating it to his secretary, Miss Patch,[1] else this splendid toast would have been lost to posterity. He started out by comparing the two fundamental activities of the human intellect, Religion and Science:

Religion is always right. Religion solves every problem and thereby abolishes problems from the Universe. Religion gives us certainty, stability, peace and the absolute. It protects us against progress which we all dread. Science is the very opposite. Science is always wrong. It never solves a problem without raising ten more problems.

In order to prove his thesis, G. B. S. now conjured up a wonderful scheme. In his opinion there were only eight great men of science, all the others were only tinkers who chiselled away on the ideas of the eight great leaders. (How he came to the number eight and according to what scheme he chose his eight great men of science is a secret that probably not even G. B. Shaw would have been able to tell.) These eight great prophets of science are: Pythagoras, Aristotle, Ptolemy, Copernicus, Galileo, Kepler, Newton, and Einstein. But even among these

[1]B. Patch, *Thirty Years with G. B. S.* London: (V. Gollancz, 1951), pp. 193–94.

great natural philosophers there were only three who built complete universes— Ptolemy, Newton, and Einstein. And now G. B. S. continued as follows:

Copernicus proved that Ptolemy was wrong. Kepler proved that Copernicus was wrong. Galileo proved that Aristotle was wrong. But at that point the sequence broke down, because science then came up for the first time against that incalculable natural phenomenon, an Englishman. As an Englishman, Newton was able to combine prodigious mental faculty with credulities and delusions that would disgrace a rabbit. As an Englishman, he postulated a rectilinear universe because the English always used the word "square" to denote honesty, truthfulness, in short: rectitude. Newton knew that the universe consisted of bodies in motion, and that none of them moved in straight lines, nor ever could. But an Englishman was not daunted by facts. To explain why all the lines in his rectilinear universe were bent, he invented a force called gravitation and thus erected a complete British universe and established it as a religion which was devoutly believed in for 300 years. The book of this Newtonian religion was not that oriental magic thing, the Bible. It was that British and matter-of-fact thing, a Bradshaw.[2] It gives the stations of all the heavenly bodies, their distances, the rates at which they are travelling, and the hour at which they reach eclipsing points or crash into the earth like Sirius. Every item is precise, ascertained, absolute and English.

Three hundred years after its establishment a young professor rises calmly in the middle of Europe and says to our astronomers: "Gentlemen: if you will observe the next eclipse of the sun carefully, you will be able to explain what is wrong with the perihelion of Mercury." The civilized Newtonian world replies that, if the dreadful thing is true, if the eclipse makes good the blasphemy, the next thing the young professor will do is to question the existence of gravitation. The young professor smiles and says that gravitation is a very useful hypothesis and gives fairly close results in most cases, but that personally he can do without it. He is asked to explain how, if there is no gravitation, the heavenly bodies do not move in straight lines and run clear out of the universe. He replies that

no explanation is needed because the universe is not rectilinear and exclusively British; it is curvilinear. The Newtonian universe thereupon drops dead and is supplanted by an Einsteinian universe. Einstein has not challenged the facts of science but the axioms of science, and science has surrendered to the challenge."

After this magnificent caricature which no one but Shaw could have drawn with such charm and boldness, he concluded:

In London great men are six-a-penny and are a very mixed lot. When we drink their health and make speeches about them, we have to be guilty of scandalous suppressions and disgraceful hypocrisies. Suppose I had to rise to propose a toast to Napoleon. The one thing which I should not possibly be able to say would be perhaps the most important—that it would have been better for the human race if he had never been born. Tonight, at least, we have no need to be guilty of suppression. There are great men who are great among small men. There are great men who are great among great men, and that is the sort of man that we are honoring tonight. Napoleon and other great men of his type were makers of Empire. But there is an order of man who gets beyond that. They are makers of universes and as makers of universes their hands are unstained by the blood of any human being.

Needless to say, the mischievous remarks about Newton were in no way backed by Einstein who always had the highest respect for his predecessors and held Newton particularly in greatest esteem. In Newton's own time his theory of gravitation was a scientific deed of the first magnitude, particularly since Newton had to create also the mathematical tools of his discovery which did not exist before his time. His merits are by no means diminished by the fact that two hundred years later another genius came along who recognized the deeper implications of the Newtonian theory.

But what happened after 1925, when Einstein had still thirty years of his life ahead of him and his intellectual powers were still in full bloom? It was around that time that

[2]A *Bradshaw* is a British railway timetable.

Einstein voluntarily abdicated from his role as the leading physicist of his generation and went into hiding. Many of his colleagues regretted this voluntary exile and felt that for many years to come Einstein could still have given great impetus to the new evolution which took place around that time. And yet, if we are aware of all the circumstances, we can only admire the intellectual honesty which prompted Einstein to withdraw and continue his speculations in the stillness of his study-room, undisturbed by the gathering clouds which increasingly obscured the vistas.

A new generation of physicists was on the march which threw some of the most ingrained ideas of our customary physical thinking to the winds. But why should Einstein resent that when he himself introduced so many new revolutionary concepts and demonstrated the falsity of so many commonly accepted ideas? Here, however, something was on the march that he was unable to accept. In all his strivings he was driven by the philosophical desire to understand. His general relativity brought this desire to full fruition. In this he could demonstrate to what unexpectedly great results a philosophically oriented speculation can ascend if it is driven by a central idea which demands clarification. The new quantum theory was of a different kind, involving the manipulation of concepts whose philosophical meaning remained obscure. To assume that everything in nature is only statistical, that all our predictions in physics can only be based on the law of large numbers because the elementary processes of nature are governed by nothing but chance, was something that he could not reconcile with his own way of thinking. "The Lord does not throw dice" was the way he expressed it. From time to time he came out from his isolation and raised his voice in favor of that strict determinism without which he could not envisage rational science. But in every case he was the loser and soon he realized that he and his antagonists spoke two fundamentally different languages. Thus he gave up any further bickering, without modifying his own unalterable convictions.

The tremendous horizontal expansion of scientific knowledge frightened him more than it filled him with confidence. "Who would have thought around 1900," he wrote in a letter, "that in fifty years time we will know so much more and understand so much less"—a typical Einstein saying. The same letter contains another characteristic utterance: "Nowadays one can be happy if one is not trampled down by the stampede of the buffalos." It was clear that in a world in which science became a corporate endeavor and teamwork took the place of the flights of the lonely genius, an Einstein could no longer strive. He continued in practically complete isolation to work on the great project which fascinated his mind ever since the success of his gravitational theory of 1916: the "unified field theory." If geometry is the great sink which swallows more and more of physics, why should we come to an end with gravitation? Perhaps, if we look very closely, we will find a place for electricity and maybe even the quantum phenomena. In fact, what else could be assumed, believing in the fundamental one-ness of nature?

But the old days of delirious victories did not return. There was no central idea which guided him through the dark with hypnotic power, as of old. Again and again he came up with possible solutions, but they did not have the convincing power of his earlier work. Riemann's geometry was somehow antagonistic to the type of symmetry structure which seemed to characterize the electromagnetic field. Shall we then shelve this geometry in favor of something still more general? But a generalization which is not carried by inner necessity is fraught with dangers. The geometry of Riemann was so natural and so self-consistent that no inherently valid motivation for its abandonment could be found, except the desire to go beyond the gravitational phenomena. The formal structure of equations now took the place of that marvelous physical intuition that characterized Einstein's earlier work. Here are Maxwell's equations, let us find some geometrical scheme which will lead to their interpretation. Such schemes could be found but they did not lead to anything fruitful.

The addition of a fifth dimension, the enrichment of Riemannian geometry by absolute parallelism, the change from metrical geometry to projective geometry—these are some of the schemes which were tried and abandoned again. Some relativists suggested another generalization in which even the existence of a line element is sacrificed in favor of an "affine geometry" in which forty, or even more formidably sixty-four, quantities took the place of the ten components of the metrical tensor of Riemann—a tremendous formal complication for which no physical justification could be found. Einstein experimented for a while with this idea but found it too brittle and hopelessly out of touch with physical reality.

Finally, during the last years of his life, he settled on the theory of the "nonsymmetric line element," in which the metrical tensor is retained but its symmetry abandoned, which means that instead of ten we have sixteen fundamental quantities. But then the geometrical significance of the theory is sacrificed and a formal edifice erected whose inherent necessity cannot be perceived. The mathematical complexity of this theory is such that we can take refuge in the hope that new results will follow after the mathematical difficulties are out of the way. Yet we have little reason to assume that this theory will solve the outstanding problems of physics, particularly the existence of atomism and the structure of elementary particles which became such a burning problem in the physics of our day.

In the meantime "modern physics" continues to grow and advance without taking account of Einstein's unifying attempts and, in fact, denying even the possibility of such an attempt being successful. Science is not out for the ultimates but continues to play the game along the well-beaten path of positivism, which denies all transcendental yearning toward the ultimates and confines itself to the "description and prediction of experimental results." But how refreshing to know that in our sober and drab times there lived a man who was full of inspiration and reverence for the magnificence of creation, who dared to speculate and produce marvelous results. How poor would our physical world picture be if this man had not broken away from the sober confines of contemporary science and erected a new edifice, singing in the wilderness and with laurel in his hair.

We have come to the close of this discussion of the place of Albert Einstein in the history of physics. It is not within the bounds of our subject to talk about Einstein as a human personality. And yet it is difficult to take leave of this great man without mentioning the extraordinary human qualities which singled him out among other great men of science to the same extent as did his scientific achievements. He was a man of destiny and he lived out his life as a man of destiny. He was fully aware of the extraordinary pedestal to which his unique mental faculties raised him, and this knowledge filled him with humbleness and humility. In a historical era of the greatest aberrations of human nature and the most absurd barbaric outburst which has ever swept the world, he did not despair of the ultimate victory of sanity and the old-testamental principles of justice and mercy. He did not retire into the cloistered silence of the scientific study, but raised his voice vigorously again and again against injustice and the suppression of individual liberties, unmindful of the most outrageous slanders to which he thus exposed himself and of the audacity of his adversaries who tried to identify him with the camp of the liars and murderers. With singular consistency he recognized the danger of annihilation of the human race caused by the blind nationalism which separates the people of the world and the unholy armament race to which they are driven. He saw the establishment of a world court and a world government as the only possible solution of the outstanding issues— tragically out of date in a world which recognizes the principle of "might is right" as the only practical basis of world politics.

In taking leave I could hardly do so more

adequately than by quoting the inspired words in which the Protestant theologian Schleiermacher praised the memory of Spinoza, the great Dutch philosopher of the seventeenth century, who shows so many analogies to Einstein and whose cosmic philosophy was so near to Einstein's heart—words which equally apply to Einstein himself.

Him pervaded the Cosmic Spirit, the Infinity was his beginning and his end, the Universe his only and everlasting love. In holy innocence and deep humility he beheld himself mirrored in the eternal world, and perceived how himself was its most amiable mirror. Full of religion was he and full of Holy Spirit. Wherefore he stands there, alone and unequalled, a master of his art, but sublime above the profane rabble, a peerless beacon forever.

4.5   *Gravity*

David Park

... Of all the fields of force, gravity is the most familiar. It is by far the strongest of the various fields that we see acting in daily life, and we have a tendency to think of it as playing a dominating role in physics; indeed people thought of it this way up until the nineteenth century. But ... gravity is a weak force. The only reason that we experience so much of it is that we live in the neighborhood of a very large and heavy piece of rock, toward which everything is attracted. If, instead, we lived on an asteroid out in otherwise empty space, we would know very little about gravity, and it would be a laboratory curiosity like the electrical tricks that can be done with a comb and a piece of paper on a cold day. We know that our hands attract each other by the force of gravity, but we can never feel so weak a force.

If we compare the strength of the gravitational field with that of the electric field in a situation where both of them are present, we get a surprising result. The force of electrostatic attraction between an electron and a proton (for example, in a hydrogen atom) is about 10^{28} times as great as the force of gravitational attraction between them. Gravity still has a special relationship to all the other fields of force, but it belongs at the end of the list. We have to think of it as being quite

distinct from electric and magnetic forces because it is vastly weaker than they are.

.   .   .   .   .

The gravitational field is crucial to Newton's theory of planetary motion. Newton's first law of motion states that an object with no force on it at all moves in a straight line at constant speed. The effect of the gravitational forces is to bend the planetary paths away from these straight lines into closed orbits. But a question arises: A straight line at constant speed with respect to what? From a practical point of view, one cannot describe any kind of motion without saying with respect to what the motion is taking place. Newton had an answer to this to which he was compelled by logic; it was that there is such a thing as absolute space. The properties of this absolute space were not very clear—for Newton they were at least partly theological—but absolute space did provide a standard of straightness by which a particle could regulate its course. Many philosophers in Newton's time, particularly Leibniz and Huygens, vigorously disputed Newton's contention. They said that geometry as we know it, which is the study of space, consists entirely of the spatial relationships between objects

SOURCE. From *Contemporary Physics* (Harcourt, Brace, and World, New York, 1964); reprinted by permission.

and that nothing in our experience requires the existance of absolute space.

Leibniz and Huygens were unable, however, to answer Newton's argument: Consider a pail of water hanging at the end of a long rope. If the rope is twisted and then allowed to untwist, the pail will spin and the water will climb part way up the sides of the pail, making the surface of water concave. If the vessel is not turning, the surface is essentially flat. Therefore, Newton argued, there is an absolute difference between a vessel of water (or anything else) that is rotating and the same vessel when it is not rotating. This absolute difference he referred to as absolute space, saying that the pail in the one case was rotating with respect to absolute space and in the other case was not.

A more quantitative example of the same argument is found if we consider the solar system. The planets, moving according to Newton's laws, travel in ellipses. The long axis of each of these ellipses is almost exactly fixed in space. (The extent to which it is not exactly fixed in space is also given almost exactly by Newton's laws and can be taken into account.) Let us imagine that the solar system is alone and that no other stars at all are visible to the naked eye. Suppose then that one wanted to know what absolute space was. By observing relative positions of the planets and finding out which way the axes of the various ellipses pointed, one could make a diagram in space with certain fixed directions, and these would be taken as directions in absolute space. Now let us suppose that the stars are lighted up and we see them for the first time. We would then be amazed to find that the set of lines in space that were deduced from a purely mathematical argument turned out to point always to the same stars in the sky, that is to say, as if Newton's absolute space were in some way attached to star. This is exactly the situation that exists, and one has the unpleasant sensation of two scientific facts running along on independent parallel tracks because one cannot see the connection between them.

Suppose we think of a little particle moving through space without any force acting on it. It moves in a straight line at constant speed, but we no longer have to invoke absolute space in order to explain what we mean by this statement, for we may say just as well that the motion is in a straight line with respect to the fixed stars. (The fixed stars are not really as fixed as they look. All of them are moving to some extent in one direction or another, but the apparent motion is very slow and can be ignored or averaged out.)

This circumstance excited the interest of one of the great speculative physicists of the nineteenth century, Ernst Mach of Vienna. He reasoned that if a free particle moves in a straight line at constant speed with respect to the fixed stars, as is true from an observational point of view, and if any deviation from such a motion requires a force to be exerted, as is also true, then in a certain sense the fixed stars must cause the force. This idea is known as Mach's principle. Mach further proposed that if one were to imagine all the fixed stars taken away, leaving a single particle alone in the middle of a completely empty universe, there would be nothing in all of physics to tell us what laws of motion this particle would obey. He remarked that if the stars in some way make their presence felt by a particle moving among them, in the sense that the particle requires a force to make it depart from uniform motion with respect to them, then some sort of influence must reach from the stars to the particle. It is repugnant, and anti-Newtonian, to assume that one can have an action without an equal opposite reaction, and we might suppose that the particle itself reacts back on the stars and exerts a tiny force on them. This suggests the existence of a field of force in space, a force exerted by the stars on the traveling particle and by the traveling particle on the stars.

THE PRINCIPLE OF EQUIVALENCE

It occurred to Einstein in the years following the publication of the special theory of relativity in 1905 that Mach's field of force was probably very closely connected with gravity and might be either Newtonian gravity itself or some extension of it. We know, after all, that gravity is a universal property of all objects, as are mass and the tendency to move uniformly, and perhaps these properties are all related. In particular, if the gravitational influences of the stars have something to do with the inertial properties of a small mass, then it must also be true that the inertial properties of the mass have something to do with its own gravitational properties.

A remarkable fact about gravity, which falls in with this line of thought, is that if one drops a heavy weight and a light weight from the same height, they will both reach the ground at the same time. This is the experiment that somebody (not Galileo, but possibly Simon Stevin of Bruges) performed from a tower in Pisa. Of course, we all know they will not reach the ground at exactly the same time because air resistance differs on different objects. But there is good evidence that in the absence of air resistance the weights would fall at very nearly or exactly the same rate.

Einstein proposed to examine the consequences of assuming that the weights fall at exactly the same rate and to attempt a physical picture of gravity with this fact in the foreground. He imagined an elevator in the middle of space, with no appreciable gravitation acting upon it. It has a string attached to the top of it, and there is some agent by which the string can be pulled. In the elevator are an observer standing on the floor and two different-sized masses floating freely near the ceiling [Figure 4.5–1]. What happens if the string is pulled? The observer feels the elevator pushing on his feet, and the masses, not being attached to the elevator at all, remain at rest in space. The elevator floor accelerates up toward them and a moment later hits them. Note how this would be interpreted by the man in the elevator, if he did not know that the elevator was free to move. He might report like this: "Suddenly a field of force began to act. I could feel the force on my own feet as I was pushed against the floor of the elevator, and I had independent ocular evidence of it because two objects that I thought were perfectly safe at the top of the room suddenly started to

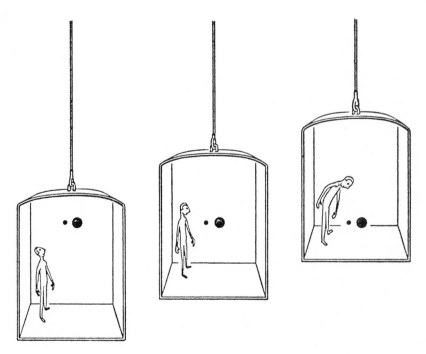

Figure 4.5–1.   To an observer in an elevator accelarating upward in force-free space, two unlike balls appear to fall to the floor at exactly the same rate. To an outside observer, they do not move at all.

fall toward the floor." He is asked, "Did you notice anything peculiar about the way these objects fell?" He answers, "One thing was very peculiar. Although one of the objects is much larger and more massive than the other, they both fell at precisely the same rate."

The physics of Newton is quite able to deal with this situation. According to Newton's second law, an object of mass m is given an acceleration a by the application of a force F equal to the product of mass and acceleration,

$$F = ma.$$

If all objects fall with the same acceleration, then a is a constant quantity in this relationship, and the force is proportional to the mass. But we have already seen that this proportionality of the force acting upon a body to the mass of the body is an essential feature of Newton's theory of gravity; we have thus come back to that famous hypothesis, and in fact by an argument quite similar to that used by Newton in the first place. Nevertheless, it is valuable to look at a gravitational field in such a way that this property is built into it at the very beginning and is essential to the description, quite apart from any of the algebra the theory may contain.

Einstein now went further. He asked, "What if *every* experiment that the man is able to do with moving weights in his little accelerating laboratory were to give exactly the same results as if it were done in a laboratory at rest on the surface of the earth and subject to its gravity?" If this were so, then one would be able to say that the force of inertia, as encountered in Newton's second law of motion, is the same as a gravitational force. The hypothesis that gravitational and inertial forces are basically equivalent is known as the *principle of equivalence*. Several experiments can at least be imagined with a view to testing its truth. In the first place, suppose that we have a set of objects—a piece of iron, a book, a stone, and an onion, all of exactly the same weight. (Remember that the weight of a thing is the force with which gravity pulls it toward the earth.) Then we

measure their inertial masses as defined by Newton's second law, by seeing whether a given force will produce the same acceleration in all of them. The question is, will the inertial masses be exactly the same? Or to put it more succinctly, is weight always precisely proportional to inertial mass? If there is a little difference from one object to another, then the principle of equivalence cannot possibly hold.

In the late seventeenth century, Newton carried out experiments with pendulums of wood and metal in his chambers at Trinity College in Cambridge, which allowed him to conclude that mass and weight were quite accurately proportional to each other. In the early years of this century, Baron Roland Eötvös of Hungary carried out essentially the same experiments, but to an accuracy of one part in 10^7, and recent experiments conducted by Professor Robert H. Dicke at Princeton University have increased the precision to one part in 10^{10}. So far, the results support Newton and Einstein.

The principle of equivalence has been introduced here as an alternative way of explaining phenomena that are already clearly explained by Newton's theory of gravity. But the test of a scientific idea is to see whether it leads to anything new, and to this end we look for some phenomenon that falls outside the domain of mechanics governed by Newton's theory. Let us, for example, think about the effect of gravity on a beam of light. To get a notion of what it might be, let us consider Einstein's famous elevator, with a beam of light projected across from one side to the other, forming a spot on the opposite wall [Figure 4.5-2]. Now suppose that a force is applied to the string. Light is emitted from the light source and starts across the elevator. While it is in motion, the elevator picks up speed, and the light reaches the opposite wall of the elavator farther down than it would if the elevator were not accelerating. An observer in the elevator would deduce that the same downward force that affects everything else in the elavator is also bending the light downward. One would, therefore, conclude that light has weight and that it is affected by

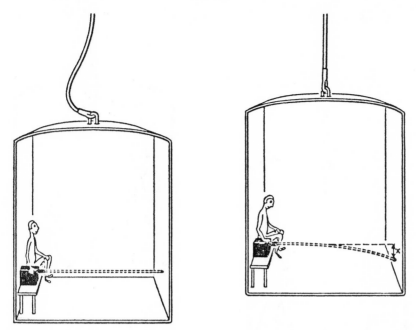

Figure 4.5–2. A beam of light is directed across an elevator at rest. Then, while the light is in transit, the elevator rises a little, and the light hits a lower spot on the wall than it did before.

the "gravitational" field in the usual way. Now let us estimate how big this effect would be.

The normal rate of acceleration of objects falling to the earth is about 32 feet per second in each second. The apparent deflection of the beam, given by x in the figure, is the distance the elevator rises in the time available. Since light travels at about 10^9 feet per second, it takes about $10/10^9$, or 10^{-8}, second to cross the elevator. During this time, the elevator's speed increases by 32×10^{-8}, or 3.2×10^{-7}, feet per second, and its average increase in velocity is half this, or 1.6×10^{-7} feet per second. During the interval of 10^{-8} second, it therefore rises $1.6 \times 10^{-7} \times 10^{-8}$, or 1.6×10^{-15}, feet more than it would have if it were unaccelerated—a little less than the diameter of an atomic nucleus. The measurement would be impossible to make.

We accordingly must look for some other property of light to use. A suitable one is the light's frequency. Imagine that the source of light is located at the top of the elevator and

shining toward a receiver at the floor [Figure 4.5–3]. At a certain moment a little pulse of light is emitted from the ceiling and starts toward the floor. Then, as the figure shows, the elevator accelerates, the floor moves up toward the ceiling, and by the time the light arrives at the floor, the floor is moving upward toward the source with a velocity that it did not have at the time the light was emitted. Effectively, therefore, the receiver of light is moving with respect to the source. We now need a way of detecting this fact.

The Doppler effect, quite a familiar phenomenon, does precisely this. Everybody is familiar with the effect of riding on a train past a ringing signal bell and hearing the pitch drop as one passes it; an automobile with its horn blowing as it passes us on a road gives us the same effect. If the source of the sound is moving with respect to the receiver (it does not matter which is moving), then there is a change of frequency at the receiver. Now let us estimate the change of frequency in this case. Suppose that the room is a generous 100 feet high. Traveling at 10^9 feet per

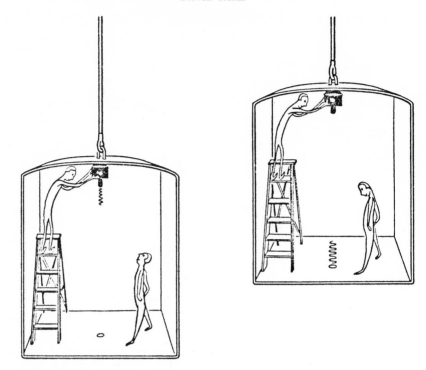

Figure 4.5–3. When a pulse of light arrives at the floor of an accelerating elevator, the floor has acquired an upward velocity with respect to the velocity of the light source when the light was emitted.

second, a flash of light would take $100/10^9$, 10^{-7}, second to travel from the ceiling to the floor. Accelerating at 32 feet per second each second, the floor would acquire a speed of 32×10^{-7}, or 3.2×10^{-6}, feet per second by the time the flash arrived there. The rule governing the Doppler effect is that the fractional change in the light's frequency or its wavelength is the velocity of the source or receiver divided by the velocity of light. This fractional change is $3.2 \times 10^{-6}/10^9 \cong 3.2 \times 10^{-15}$, or about three parts in a million billion. But although this fraction is amost unimaginably small (it corresponds to approximately 10^{-5} inch in the distance from the earth to the moon), there is no essential physical principle that makes it undetectable. As a matter of fact, in certain situations it is just barely possible to pick it up.

Here is a new fact whose existence is predicted by the principle of equivalence. Unfortunately, it can also be explained from the mechanical viewpoint if we assume that light is mechanical and allow ourselves a few more liberties. If the experiment works, we can say that it is in accordance with the principle of equivalence but not that it proves the principle to be correct. On the other hand, if the experiment does not come out as expected, the principle must be wrong.

INERTIA

Now let us return to Mach's principle— that inertial forces in some way reflect a field property of all the mass in the universe— and consider its present status. The principle is very possibly true. We do not know because we do not know enough about how the universe is put together. Einstein was never able to form a complete theory corresponding to Mach's ideas, and in order to do this properly, one may have to know all about the large-scale structure of the universe.

The expression "all about the...universe" refers to its past history as well as its present condition. Physical influences seem never to act instantaneously. The inertial properties of a particle are not really determined by the present positions of all the stars but by their positions in the past when their supposed influence started. According to Einstein's theory, which is not necessarily true, the influence would travel at the speed of light—so fast that over all ordinary distances the delay would not matter at all. But some of the objects of the universe seen through a telescope are so distant that the light started from them billions of years ago, and therefore what is seen through the telescope is not the present state of the universe at all but its state at progressively earlier epochs as one looks farther out. However, this situation may be just what is appropriate to the question at hand, because the influence that produces inertia may be traveling at the same speed. The question is complicated, and there is no very simple answer until we know very much more than we do now.

We must also admit that we are not entirely sure what field we are talking about. Still, there is one property that seems to characterize all fields and, in addition, seems eminently reasonable: the farther one is from the object that is creating the field, the weaker the field is. The gravitational field falls off inversely as the square of the distance from the object that causes it, and so does the electric field; the magnetic field is somewhat more complex, but it too becomes weaker as one goes farther away. If this property applies to the field responsible for inertial effects, then we might expect that a distribution of mass relatively near the place where an inertial experiment is being carried out would have more effect on the result of this experiment than a similar distribution of mass at a great distance away and that, if there is a large distribution of mass somewhere in the vicinity of the experiment, the direction toward this mass could somehow be distinguished in its inertial properties from any direction perpendicular to it.

Living on the earth as we do, we are about halfway out toward the edge of a galaxy consisting of some billions of stars. To one side of us is the galactic center with its tremendous mass. In other directions space is comparatively empty. There are of course a vast number of other galaxies surrounding ours, but they are all relatively far away. Let us suppose, therefore, that the effect of the galactic center on one side of us is larger in producing inertia than the effect of similar concentrations of mass at greater distances. Then is it possible that this center could produce some special effect?

In Newton's second law, the mass m has no directional properties at all, and the law can be taken to imply that it takes exactly as much force to give the same object a given acceleration toward the east as toward the north. The question we ask is, is this statement exactly true, or is it only very nearly true? (There is no question of its being seriously inaccurate; if that were so, the inaccuracy would have been discovered by astronomers long ago.) What interests us here is the possibility of a very slight anisotropy (that is, directional character) of inertia that could be detected by delicate measurements. Let us suppose, for example, that a pendulum is swinging above the surface of the earth at the latitude of New York City in a north-south direction. If we note its period of swing and then wait for some 12 hours, the earth will have turned halfway around, and we will easily see that the pendulum is swinging along a different line, roughly 90 degrees away from the first. Presumably the force of the earth's gravity on the pendulum will be unchanged. On the other hand, the east-west line, if previously it pointed toward the galactic center, will now point about 90 degrees away. If the inertial mass were, for example, a little greater as a result of this, then we would expect the pendulum to swing a little more slowly. It is very difficult to perform the experiment with a pendulum because it must be done with great accuracy (and besides, a pendulum works by gravity, which we do not fully understand) . . . . [Experiments have shown that if there is an anisotropy in mass] the effect must be less than one part in 10^{20}. This result is disappointing. We dislike concluding that there is no effect whatever because life would be so much more interesting if there were one. Hence we remain hopeful and say that it is perhaps just barely too small to observe. But to have brought the matter to this degree of accuracy within a few years after the publication of the first paper on the subject is a most remarkable achievement.

Although mathematics stands in the way of our knowing many of the consequences of Einstein's general theory, the equations seem to predict that inertia will be isotropic no matter how the mass of the universe is distributed. Thus the most important result of these experiments may be a check on the theory itself.

4.6   *Relativity and the Philosophers*

Richard P. Feynman

Poincaré made the following statement of the principle of relativity: "According to the principle of relativity, the laws of physical phenomena must be the same for a fixed observer as for an observer who has a uniform motion of translation relative to him, so that we have not, nor can we possibly have, any means of discerning whether or not we are carried along in such a motion."

When this idea descended upon the world, it caused a great stir among philosophers, particularly the "cocktail-party philosophers," who say, "Oh, it is very simple: Einstein's theory says all is relative!" In fact, a surprisingly large number of philosophers, not only those found at cocktail parties (but rather than embarrass them, we shall just call them "cocktail-party philosophers"), will say, "That all is relative is a conequence of Einstein, and it has profound influences on our ideas." In addition, they say "It has been demonstrated in physics that phenomena depend upon your frame of reference." We hear that a great deal, but it is difficult to find out what it means. Probably the frames of reference that were originally rered to were the coordinate systems which we use in the analysis of the theory of relativity. So the fact that "things depend upon your frame of reference" is supposed to have had a profound effect on modern

thought. One might well wonder why, because, after all, that things depend upon one's point of view is so simple an idea that it certainly cannot have been necessary to go to all the trouble of the physical relativity theory in order to discover it. That what one sees depends upon his frame of reference is certainly known to anybody who walks around, because he sees an approaching pedestrian first from the front and then from the back; there is nothing deeper in most of the philosophy which is said to have come from the theory of relativity than the remark that "A person looks different from the front than from the back." The old story about the elephant that several blind men describe in different ways is another example, perhaps, of the theory of relativity from the philosopher's point of view.

But certainly there must be deeper things in the theory of relativity than just this simple remark that "A person looks different from the front than from the back." Of course relativity is deeper than this, because *we can make definite predictions with it*. It certainly would be rather remarkable if we could predict the behavior of nature from such a simple observation alone.

There is another school of philosophers who feel very uncomfortable about the theory of relativity, which asserts that we cannot determine our absolute velocity without

SOURCE. From Feynman, Leighton, and Sands, *The Feynman Lectures on Physics* (Reading, Mass.: Addison-Wesley, 1963), Chapter 16.

looking at something outside, and who would say, "It is obvious that one cannot measure his velocity without looking outside. It is self-evident that it is *meaningless* to talk about the velocity of a thing without looking outside; the physicists are rather stupid for having thought otherwise, but it has just dawned on them that this is the case. If only we philosophers had realized what the problems were that the physicists had, we could have decided immediately by brainwork that it is impossible to tell how fast one is moving without looking outside, and we could have made an enormous contribution to physics." These philosophers are always with us, struggling in the periphery to try to tell us something, but they never really understand the subtleties and depths of the problem.

Our inability to detect absolute motion is a result of *experiment* and not a result of plain thought, as we can easily illustrate. In the first place, Newton believed that it was true that one could not tell how fast he is going if he is moving with uniform velocity in a straight line. In fact, Newton first stated the principle of relativity . . . Why then did the philosophers not make all this fuss about "all is relative," or whatever, in Newton's time? Because it was not until Maxwell's theory of electrodynamics was developed that there were physical laws that suggested that one *could* measure his velocity without looking outside; soon it was found *experimentally* that one could *not*.

Now, *is* it absolutely, definitely, philosophically *necessary* that one should not be able to tell how fast he is moving without looking outside? One of the consequences of relativity was the development of a philosophy which said, "You can only define what you can measure! Since it is self-evident that one cannot measure a velocity without seeing what he is measuring it relative to, therefore it is clear that there is no *meaning* to absolute velocity. The physicists should have realized that they can talk only about what they can measure." But *that is the whole problem*: whether or not one *can define* absolute velocity is the same as the problem of whether or not one *can detect in an experiment*,

without looking outside, whether he is moving. In other words, whether or not a thing is measurable is not something to be decided *a priori* by thought alone, but something that can be decided only by experiment. Given the fact that the velocity of light is 186,000 mi/sec, one will find few philosophers who will calmly state that it is self-evident that if light goes 186,000 mi/sec inside a car, and the car is going 100,000 mi/sec, that the light also goes 186,000 mi/sec past an observer on the ground. That is a shocking fact to them; the very ones who claim it is obvious find, when you give them a specific fact, that it is not obvious.

Finally, there is even a philosophy which says that one cannot detect *any* motion except by looking outside. It is simply not true in physics. True, one cannot perceive a *uniform* motion in a *straight line*, but if the whole room were *rotating* we would certainly know it, for everybody would be thrown to the wall—there would be all kinds of "centrifugal" effects. That the earth is turning on its axis can be determined without looking at the stars, by means of the so-called Foucault pendulum, for example. Therefore it is not true that "all is relative"; it is only *uniform velocity* that cannot be detected without looking outside. Uniform *rotation* about a fixed axis *can* be. When this is told to a philosopher, he is very upset that he did not really understand it, because to him it seems impossible that one should be able to determine rotation about an axis without looking outside. If the philosopher is good enough, after some time he may come back and say, "I understand. We really do not have such a thing as absolute rotation; we are really rotating *relative to the stars*, you see. And so some influence exerted by the stars on the object must cause the centrifugal force."

Now, for all we know, that is true; we have no way, at the present time, of telling whether there would have been centrifugal force if there were no stars and nebulae around. We have not been able to do the experiment of removing all the nebulae and then measuring our rotation, so we simply do not know. We must admit that the philo-

sopher may be right. He comes back, therefore, in delight and says, "It is absolutely
necessary that the world ultimately turn out
to be this way: *absolute* rotation means
nothing; it is only *relative* to the nebulae."
Then we say to him, "*Now*, my friend, is it
or is it not obvious that uniform velocity in
a straight line, *relative to the nebulae* should
produce no effects inside a car?" Now that
the motion is no longer absolute, but is a
motion *relative to the nebulae*, it becomes a
mysterious question, and a question that can
be answered only by experiment.

What, then, *are* the philosophic influences
of the theory of relativity? If we limit ourselves to influences in the sense of *what kind
of new ideas and suggestions* are made to the
physicist by the principle of relativity, we
could describe some of them as follows. The
first discovery is, essentially, that even those
ideas which have been held for a very long
time and which have been very accurately
verified might be wrong. It was a shocking

discovery, of course, that Newton's laws are
wrong, after all the years in which they
seemed to be accurate. Of course it is clear,
not that the experiments were wrong, but
that they were done over only a limited range
of velocities, so small that the relativistic
effects would not have been evident. But
nevertheless, we now have a much more
humble point of view of our physical laws—
everything *can* be wrong!

Secondly, if we have a set of "strange"
ideas, such as that time goes slower when
one moves, and so forth, whether we *like* them
or do *not* like them is an irrelevant question.
The only relevant question is whether the
ideas are consistent with what is found experimentally. In other words, the "strange
ideas" need only agree with *experiment*, and
the only reason that we have to discuss the
behavior of clocks and so forth is to demonstrate that although the notion of the time
dilation is strange, it is *consistent* with the way
we measure time.

CHAPTER 5

ELECTRONS AND THE ORIGIN
OF QUANTUM THEORY

5.1

Cathode Rays

J. J. Thomson

The experiments discussed in this paper were undertaken in the hope of gaining some information as to the nature of the Cathode Rays. The most diverse opinions are held as to these rays; according to the almost unanimous opinion of German physicists they are due to some process in the æther to which—inasmuch as in a uniform magnetic field their course is circular and not rectilinear—no phenomenon hitherto observed is analogous: another view of these rays is that, so far from being wholly ætherial, they are in fact wholly material, and that they mark the paths of particles of matter charged with negative electricity. It would seem at first sight that it ought not to be difficult to discriminate between views so different, yet experience shows that this is not the case, as amongst the physicists who have most deeply studied the subject can be found supporters of either theory.

The electrified-particle theory has for purposes of research a great advantage over the ætherial theory, since it is definite and its consequences can be predicted; with the ætherial theory it is impossible to predict what will happen under any given circumstances, as on this theory we are dealing with hitherto unobserved phenomena in the æther, of whose laws we are ignorant.

The following experiments were made to test some of the consequences of the electrified-particle theory.

CHARGE CARRIED BY THE CATHODE RAYS

If these rays are negatively electrified particles, then when they enter an enclosure they ought to carry into it a charge of negative electricity. This has been proved to be the case by Perrin, who placed in front of a plane cathode two coaxial metallic cylinders which were insulated from each other: the outer of these cylinders was connected with the earth, the inner with a gold-leaf electroscope. These cylinders were closed except for two small holes, one in each cylinder, placed so that the cathode rays could pass through them into the inside of the inner cylinder. Perrin found that when the rays passed into the inner cylinder the electroscope received a charge of negative electricity, while no charge went to the electroscope when the rays were deflected by a magnet so as no longer to pass through the hole.

SOURCE. From the *Philosophical Magazine* **44**: 293 (1897).

This experiment proves that something charged with negative electricity is shot off from the cathode, traveling at right angles to it, and that this something is deflected by a magnet; it is open, however, to the objection that it does not prove that the cause of the electrification in the electroscope has anything to do with the cathode rays. Now the supporters of the ætherial theory do not deny that electrified particles are shot off from the cathode; they deny, however, that these charged particles have any more to do with the cathode rays than a rifle-ball has with the flash when a rifle is fired. I have therefore repeated Perrin's experiment in a form which is not open to this objection. The arrangement used was as follows:—Two coaxial cylinders [Figure 5.1–1] with slits in them are placed in a bulb connected with the discharge-tube; the cathode rays from the cathode A pass into the bulb through a slit in a metal plug fitted into the neck of the tube; this plug is connected with the anode and is put to earth. The cathode rays thus do not fall upon the cylinders unless they are deflected by a magnet. The outer cylinder is connected with the earth, the inner with the electrometer. When the cathode rays (whose path was traced by the phosphorescence on the glass) did not fall on the slit,

the electrical charge sent to the electrometer when the induction-coil producing the rays was set in action was small and irregular; when, however, the rays were bent by a magnet so as to fall on the slit there was a large charge of negative electricity sent to the electrometer. I was surprised at the magnitude of the charge; on some occasions enough negative electricity went through the narrow slit into the inner cylinder in one second to alter the potential of a capacity of 1.5 microfarads by 20 volts. If the rays were so much bent by the magnet that they overshot the slits in the cylinder, the charge passing into the cylinder fell again to a very small fraction of its value when the aim was true. Thus this experiment shows that however we twist and deflect the cathode rays by magnetic forces, the negative electrification follows the same path as the rays, and that this negative electrification is indissolubly connected with the cathode rays.

When the rays are turned by the magnet so as to pass through the slit into the inner cylinder, the deflection of the electrometer connected with this cylinder increases up to a certain value, and then remains stationary although the rays continue to pour into the cylinder. This is due to the fact that the gas in the bulb becomes a conductor of electricity when the cathode rays pass through it, and thus, though the inner cylinder is perfectly insulated when the rays are not passing, yet as soon as the rays pass through the bulb the air between the inner cylinder and the outer one becomes a conductor, and the electricity escapes from the inner cylinder to the earth. Thus the charge within the inner cylinder does not go on continually increasing; the cylinder settles down into a state of equilibrium in which the rate at which it gains negative electricity from the rays is equal to the rate at which it loses it by conduction through the air. If the inner cylinder has initially a positive charge it rapidly loses that charge and acquires a negative one; while if the initial charge is a negative one, the cylinder will leak if the initial negative potential is numerically greater than the equilibrium value.

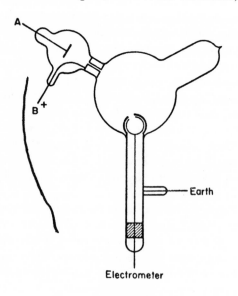

Figure 5.1–1.

DEFLECTION OF THE CATHODE RAYS BY AN ELECTROSTATIC FIELD

An objection very generally urged against the view that the cathode rays are negatively electrified particles, is that hitherto no deflection of the rays has been observed under a small electrostatic force, and though the rays are deflected when they pass near electrodes connected with sources of large differences of potential, such as induction-coils or electrical machines, the deflection in this case is regarded by the supporters of the ætherial theory as due to the discharge passing between the electrodes, and not primarily to the electrostatic field. Hertz made the rays travel between two parallel plates of metal placed inside the discharge-tube, but found that they were not deflected when the plates were connected with a battery of storage-cells; on repeating this experiment I at first got the same result, but subsequent experiments showed that the absence of deflection is due to the conductivity conferred on the rarefied gas by the cathode rays. On measuring this conductivity it was found that it diminished very rapidly as the exhaustion increased; it seemed then that on trying Hertz's experiment at very high exhaustions[1] there might be a chance of detecting the deflection of the cathode rays by an electrostatic force. The apparatus used is represented in [Figure 5.1–2].

The rays from the cathode C pass through a slit in the anode A, which is a metal plug fitting tightly into the tube and connected with the earth; after passing through a second slit in another earth-connected metal plug B, they travel between two parallel aluminium plates about 5 cm. long by 2 broad and at a distance of 1.5 cm. apart; they then fall on the end of the tube and produce a narrow well-defined phosphorescent patch. A scale pasted on the outside of the tube serves to measure the deflection of this patch. At high exhaustions the rays were deflected when the two aluminium plates were connected with the terminals of a battery of small storage-cells; the rays were depressed when the upper plate was connected with the negative pole of the battery, the lower with the positive, and raised when the upper plate was connected with the positive, the lower with the negative pole. The deflection was proportional to the difference of potential between the plates, and I could detect the deflection when the potential-difference was as small as two volts. It was only when the vacuum was a good one that the deflection took place, but that the absence of deflection is due to the conductivity of the medium is shown by what takes place when the vacuum has just arrived at the stage at which the deflection begins. At this stage there is a deflection of the rays when the plates are first connected with the terminals of the battery, but if this connection is maintained the patch of phosphorescence gradually creeps back to its undeflected position. This is just what would happen if the space between the plates were a conductor, though a very bad one, for then the positive and negative ions between the plates would slowly diffuse, until the positive plate became coated with negative ions, the negative plate with

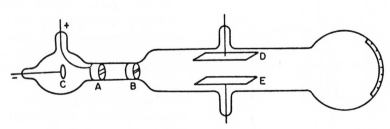

Figure 5.1–2.

---

[1]That is, under conditions of good vacuum. [Ed.]

positive ones; thus the electric intensity between the plates would vanish and the cathode rays be free from electrostatic force.....

[A description of the details of electrostatic deflection under various conditions
is omitted—Ed.]

MAGNETIC DEFLECTION OF THE CATHODE RAYS IN DIFFERENT GASES

The deflection of the cathode rays by the
magnetic field was studied with the aid of
the apparatus shown in [Figure 5.1–3]. The
cathode was placed in a side-tube fastened
on to a bell-jar; the opening between this
tube and the bell-jar was closed by a metallic plug with a slit in it; this plug was
connected with the earth and was used as
the anode. The cathode rays passed through
the slit in this plug into the bell-jar, passing
in front of a vertical plate of glass ruled into
small squares. The bell-jar was placed between two large parallel coils arranged as
a Helmholtz galvanometer. The course of
the rays was determined by taking photographs of the bell-jar when the cathode rays
were passing through it; the divisions on the
plate enabled the path of the rays to be
determined. Under the action of the magnetic
field the narrow beam of cathode rays spreads
out into a broad fan-shaped luminosity
in the gas. The luminosity in this fan is not
uniformly distributed, but is condensed
along certain lines. The phosphorescence on
the glass is also not uniformly distributed;
it is much spread out, showing that the
beam consists of rays which are not all deflected to the same extent by the magnet.
The luminosity on the glass is crossed by

bands along which the luminosity is very
much greater than in the adjacent parts.
These bright and dark bands are called by
Birkeland, who first observed them, the
magnetic spectrum. The brightest spots on
the glass are by no means always the termina-
tions of the brightest streaks of luminosity
in the gas; in fact, in some cases a very
bright spot on the glass is not connected
with the cathode by any appreciable luminosity, though there may be plenty of
lumonsity in other parts of the gas. One
very interesting point brought out by the
photographs is that in a given magnetic field,
and with a given mean potential-difference
between the terminals, the path of the rays is
independent of the nature of the gas. Photographs were taken of the discharge in
hydrogen, air, carbonic acid, methyl iodide,
i.e., in gases whose densities range from 1
to 70, and yet, not only were the paths of the
most deflected rays the same in all cases,
but even the details, such as the distribution
of the bright and dark spaces, were the same;
in fact, the photographs could hardly be
distinguished from each other. It is to be
noted that the pressures were not the same;
the pressures in the different gases were
adjusted so that the mean potential-dif-
ferences between the cathode and the anode
were the same in all the gases. When the
pressure of a gas is lowered, the potential-difference between the terminals increases, and
the deflection of the rays produced by a magnet diminishes, or at any rate the deflection
of the rays when the phosphorescence is a
maximum diminishes. If an air-break is inserted an effect of the same kind is produced.

In the experiments with different gases,
the pressures were as high as was consistent
with the appearance of the phosphorescence
on the glass, so as to ensure having as much
as possible of the gas under consideration
in the tube.

As the cathode rays carry a charge of
negative electricity, are deflected by an electrostatic force as if they were negatively
electrified, and are acted on by a magnetic
force in just the way in which this force
would act on a negatively electrified body

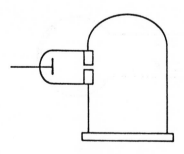

Figure 5.1–3.

moving along the path of these rays, I can see no escape from the conclusion that they are charges of negative electricity carried by particles of matter. The question next arises, What are these particles? are they atoms, or molecules, or matter in a still finer state of subdivision? To throw some light on this point, I have made a series of measurements of the ratio of the mass of these particles to the charge carried by it. To determine this quantity, I have used two independent methods. The first of these is as follows:—Suppose we consider a bundle of homogeneous cathode rays. Let m be the mass of each of the particles, e the charge carried by it. Let N be the number of particles passing across any section of the beam in a given time; then Q the quantity of electricity carried by these particles is given by the equation

$$Ne = Q.$$

We can measure Q if we receive the cathode rays in the inside of a vessel connected with an electrometer. When these rays strike against a solid body, the temperature of the body is raised; the kinetic energy of the moving particles being converted into heat; if we suppose that all this energy is converted into heat, then if we measure the increase in the temperature of a body of known thermal capacity caused by the impact of these rays, we can determine $\mathcal{E}$, the kinetic energy of the particles, and if v is the velocity of the particles,

$$\tfrac{1}{2}Nmv^2 = \mathcal{E}$$

If ρ is the radius of curvature of the path of these rays in a uniform magnetic field B, then[2]

$$\frac{mvc}{e} = B\rho,$$

from these equations we get

$$\tfrac{1}{2}\frac{mc^2}{e}v^2 = \frac{\mathcal{E}}{Q},$$

$$v = \frac{2\mathcal{E}}{Q(B\rho)},$$

$$\frac{m}{e} = \frac{(B\rho)^2\,Qc^2}{2\mathcal{E}}.$$

Thus, if we know the values of Q, c, $\mathcal{E}$, and $B\rho$, we can deduce the values of v and m/e.

. . . . . .

[Thomson used this method to determine v and m/e. The description of the apparatus and results are omitted because the alternative method which follows proved to be more accurate—Ed.]

I shall describe another method of measuring the quantities m/e and v of an entirely different kind from the preceding; this method is based upon the deflection of the cathode rays in an electrostatic field. If we measure the deflection experienced by the rays when traversing a given length under a uniform electric intensity, and the deflection of the rays when they traverse a given distance under a uniform magnetic field, we can find the values of m/e and v in the following way:

Let the space passed over by the rays under a uniform electric intensity E be l, the time taken for the rays to traverse this space is l/v, the velocity in the direction of E is therefore

$$\frac{Ee}{m}\frac{l}{v},$$

so that θ, the angle through which the rays are deflected when they leave the electric field and enter a region free from electric force, is given by the equation

$$\theta = \frac{Ee}{m}\frac{l}{v^2}$$

If, instead of the electric intensity, the rays are acted on by a magnetic force B at right angles to the rays, and extending across the distance l, the velocity at right angles to the original path of the rays is

$$\frac{Bev}{mc}\frac{l}{v},$$

---

[2]In order to conform to modern usage, some of the notation below has been altered and cgs (Gaussian) units are used throughout. [Ed.]

so that ϕ, the angle through which the rays are deflected when they leave the magnetic field, is given by the equation

$$\phi = \frac{Be}{mc}\frac{l}{v}.$$

From these equations we get

$$v = \frac{\phi}{\theta}\frac{cE}{B}$$

and

$$\frac{m}{e} = \frac{B^2\theta l}{c^2 E \phi^2}.$$

In the actual experiments B was adjusted so that $\phi = \theta$; in this case the equations become[3]

$$v = \frac{cE}{B},$$

$$\frac{m}{e} = \frac{B^2 l}{c^2 E \theta}$$

The apparatus used to measure v and m/e by this means is that represented in [Figure 5.1–2]. The electric field was produced by connecting the two aluminium plates to the terminals of a battery of storage-cells. The phosphorescent patch at the end of the tube was deflected, and the deflection measured by a scale pasted to the end of the tube. As it was necessary to darken the room to see the phosphorescent patch, a needle coated with luminous paint was placed so that by a screw it could be moved up and down the scale; this needle could be seen when the room was darkened, and it was moved until it coincided with the phosphorescent patch. Thus, when light was admitted, the deflection of the phosphorescent patch could be measured.

The magnetic field was produced by placing outside the tube two coils whose diameter was equal to the length of the plates; the coils were placed so that they covered the space occupied by the plates, the distance between the coils was equal to the radius of either. The mean value of the magnetic force over the length l was determined in the following way: a narrow coil C whose length was l, connected with a ballistic galvanometer, was placed between the coils; the plane of the windings of C was parallel to the planes of the coils; the cross section of the coil was a rectangle 5 cm by 1 cm. A given current was sent through the outer coils and the kick α of the galvanometer observed when this current was reversed. The coil C was then placed at the center of two very large coils, so as to be in a field of uniform magnetic force: the current through the large coils was reversed and the kick β of the galvanometer again observed; by comparing α and β we can get the mean value of the magnetic force over a length l; this was found to be $60 \times I$, where I is the current flowing through the coils.

A series of experiments was made to see if the electrostatic deflection was proportional to the electric intensity between the plates; this was found to be the case. In the following experiments the current through the coils was adjusted so that the electrostatic deflection was the same as the magnetic:

TABLE 5.1–1

Gas	θ (radians)	B (gauss)	E (statvolts/cm)	l (cm)	m/e* (g/statcoul)	v (cm/sec)
Air	0.080	5.5	0.5	5	4.3×10^{-18}	2.8×10^9
Air	0.095	5.4	0.5	5	3.7×10^{-18}	2.8×10^9
Air	0.118	6.6	0.5	5	4.0×10^{-18}	2.3×10^9
Hydrogen	0.082	6.3	0.5	5	5.0×10^{-18}	2.5×10^9
Carbonic acid	0.100	6.9	0.5	5	5.0×10^{-18}	2.2×10^9
Air	0.055	5.0	0.6	5	4.3×10^{-18}	3.6×10^9
Air	0.064	3.6	0.33	5	3.7×10^{-18}	2.8×10^9

*The value for m/e currently accepted is 1.90×10^{-18} g/statcoul. [Ed.]

[3]A device that operates under these conditions is called a "crossed-field analyzer" and selects particles with a particular *velocity*. [Ed.]

The cathode in the first five experiments was aluminium, in the last two experiments it was made of platinum; in the last experiment Sir William Crookes's method of getting rid of the mercury vapour by inserting tubes of pounded sulphur, sulphur iodide, and copper filings between the bulb and the pump was adopted. In the calculation of m/e and v no allowance has been made for the magnetic force due to the coil in the region outside the plates; in this region the magnetic force will be in the opposite direction to that between the plates, and will tend to bend the cathode rays in the opposite direction: thus the effective value of B will be smaller than the value used in the equations, so that the values of m/e are larger and those of v less than they would be if this correction were applied. The method of determining the values of m/e and v is much less laborious and probably more accurate than the former method; it cannot, however, be used over so wide a range of pressures.

From these determinations we see that the value of m/e is independent of the nature of the gas, and that its value 10^{-18} is very small compared with the value 10^{-14}, which is the smallest value of this quantity previously known, and which is the value for the hydrogen ion in electrolysis.

Thus for the carriers of the electricity in the cathode rays m/e is very small compared with its value in electrolysis. The smallness of m/e may be due to the smallness of m or the largeness of e, or to a combination of these two. That the carriers of the charges in the cathode rays are small compared with ordinary molecules is shown, I think, by Lenard's results as to the rate at which the brightness of the phosphorescence produced by these rays diminishes with the length of path traveled by the ray. If we regard this phosphorescence as due to the impact of the charged particles, the distance through which the rays must travel before the phosphorescence fades to a given fraction (say $1/e$, where $e = 2.71$) of its original intensity, will be some moderate multiple of the mean free path. Now Lenard found that this distance depends solely upon the density of the medium, and not upon its chemical nature or physical state. In air at atmospheric pressure the distance was about half a centimeter, and this must be comparable with the mean free path of the carriers through air at atmospheric pressure. But the mean free path of the molecules of air is a quantity of quite a different order. The carrier, then, must be small compared with ordinary molecules.

The two fundamental points about these carriers seem to me to be (1) that these carriers are the same whatever the gas through which the discharge passes, (2) that the mean free paths depend upon nothing but the density of the medium traversed by these rays.

It might be supposed that the independence of the mass of the carriers of the gas through which the discharge passes was due to the mass concerned being the quasi mass which a charged body possesses in virtue of the electric field set up in its neighborhood; moving the body involves the production of a varying electric field, and, therefore, of a certain amount of energy which is proportional to the square of the velocity. This causes the charged body to behave as if its mass were increased by a quantity, which for a charged sphere is $\frac{1}{5}e^2/\mu a$ ("Recent Researches in Electricity and Magnetism"), where e is the charge and a the radius of the sphere. If we assume that it is this mass which we are concerned with in the cathode rays, since m/e would vary as e/a, it affords no clue to the explanation of either of the properties (1 and 2) of these rays. This is not by any means the only objection to this hypothesis, which I only mention to show that it has not been overlooked.

The explanation which seems to me to account in the most simple and straightforward manner for the facts is founded on a view of the constitution of the chemical elements which has been favorably entertained by many chemists: this view is that the atoms of the different chemical elements are different aggregations of atoms of the same kind. In the form in which this hypothesis was enunciated by Prout, the atoms of the different elements were hydrogen atoms; in this precise form the hypothesis is not tenable, but if we substitute for hydrogen some unknown primordial substance X, there is nothing known which is inconsistent with this hypothesis.

.   .   .   .   .

If, in the very intense electric field in the neighborhood of the cathode, the molecules of the gas are dissociated and are split up, not into the ordinary chemical atoms, but into these primordial atoms, which we shall for brevity call corpuscles; and if these corpuscles are charged with electricity and projected from the cathode by the electric field, they would behave exactly like the cathode

rays. They would evidently give a value of m/e which is independent of the nature of the gas and its pressure, for the carriers are the same whatever the gas may be.

.   .   .   .   .

Thus we have in the cathode rays matter in a new state, a state in which the subdivision of matter is carried very much further than in the ordinary gaseous state: a state in which all matter—that is, matter derived from different sources such as hydrogen, oxygen, etc.—is of one and the same kind; this matter being the substance from which all the chemical elements are built up.

5.2

The Septuagenarian Electron

Sir George Thomson[1]

The year 1897 has the best claim to be the birth year of the electron; for in it the main ideas, which for the next 30 years at least made up this concept, were first clearly stated and supported by strong evidence—though a little short of proof.

I have only space to pick out certain points in this long and intricate history; so I shall choose those that seem best to show how the many able searchers for truth in this matter have been helped or hindered by experimental and psychological causes.

The electron is a concept that has developed, and changed greatly in the process. It originated as the name of a unit of charge. Johnstone Stoney, an Irish physicist who favored the idea that Michael Faraday's laws of electrolysis imply a natural unit of charge, introduced the word in 1891 as follows: "In electrolysis a definite quantity of electricity, the same in all cases, passes for each chemical bond which is ruptured . . . a charge of this amount is associated in the chemical atom with each bond . . . . These charges, which it

will be convenient to call 'electrons' cannot be removed from the atom but they become disguised when atoms chemically unite." Stoney was to be proved partly wrong; the name he gave is now applied to a detachable part of the atom and implies a group of properties of which charge is only one.

In the latter part of the nineteenth century James Clerk Maxwell, following Faraday, had stressed the importance of the medium much above that of the electrical charges it separates. Thus, "According to this theory all charge is the residual effect of the polarization of the dielectric," and again, "It is extremely improbable, however, that when we come to understand the true nature of electrolysis we shall retain in any form the theory of electrical charges."

Faraday himself had been by no means convinced of what we now consider the obvious explanation of his laws. However, by about 1895 the evidence for a natural unit charge, which had been stressed by Hermann Helmholtz in the 1880's was very strong, but

SOURCE. From *Physics Today*, May, 1967, p. 55.
[1]Son of Sir J. J. Thomson and also a Noble Prize winner in physics.

there was little evidence for its material character. It might have turned out to be just a unit governing the transfer of electricity, more like a quantum of radiation than the electron we know. The decisive evidence that changed Stoney's concept to one that lasted for the next 30 years came from the study of cathode rays.

THE NATURE OF CATHODE RAYS

Cathode rays had long been a problem. Were they charged particles or some phenomenon in the ether, perhaps the often surmised longitudinal waves? On January 1, 1897 the relevant evidence was as follows:

They were deflected by a magnetic field as would be a wire carrying negative electricity from the cathode (known since their discovery).

They could go through thin solid films, for example gold foil, without making holes in them. This fact, discovered by Heinrich Hertz in 1892, was much used by Philip Lenard, who studied them in the atmosphere and in a vessel filled with gas. Lenard proved that they were absorbed by matter and that equal weights per square centimeter of many widely different substances absorbed them nearly equally.

Lenard also had proved that the absorption coefficient of cathode rays depended on the voltage of the discharge that made them and that, if this was kept constant, the magnetic deflection was independent of the gas through which the rays were passed. He had not tried varying the gas in the discharge.

Jean Perrin had proved that when the rays were caught in a hollow container facing the cathode, they gave it a negative charge; this replaced erroneous results of William Crookes and Hertz.

Finally, according to Hertz's paper of 1883 the rays were not deflected by an electric field.

This last was a serious hurdle for the supporters of charged particles, as was the observation that the rays could penetrate thin solids without leaving holes (more so than the modern physicist can easily realize).

In favor of the charged-particle explanation were the magnetic deflection and Perrin's experiment, then more convincing than it might be now with our wide experience of secondary radiations.

On January 7, 1897 E. Wiechert read a paper to the Königsberg Society for Science and Economics that deserves more credit than it usually receives. He believed in particles and tried to determine their m/e by combining mv/e found by magnetic deflection with a direct measure of v, using a Hertzian oscillator of known frequency to operate a shutter by its magnetic effect. This last was due to T. des Coudres.

Unfortunately in Wiechert's first paper the oscillation was not rapid enough actually to measure v but only gave a lower limit, with some indication that the true value was about twice as large, giving m/e about a two-thousandth part of the corresponding ratio M/e for the hydrogen ion in electrolysis.

A great merit of the paper was that it spoke boldly of the cathode rays as "electric atoms," and it stressed their universality, though not considering them as composing chemical atoms. The weakness of this early experiment is that it forms no real argument against those who believed in the ether-wave theory, for which one would expect an even larger velocity than Wiechert's limit, perhaps even the speed of light. In a later paper Wiechert was able to show that the speed varied with magnetic deflection about as it should for particles, but this was after Thomson's work. Kaufmann in a paper sent to the *Annalen der Physik und Chemie* in May 1897 studied the magnetic deflections of rays formed in discharges of different potentials. He believed in waves, but calculated what m/e would have to be if the rays were particles, assuming that these rays acquired the full energy of the discharge potential—an assumption open to criticism but in fact nearly true under his conditions. He found m/e about $(\frac{1}{1000}) M/e$ but rejected this as much too small and with it the idea of particles, because he supposed they should vary with the gas, the cathode material or both.

CATHODE RAYS CHARGED PARTICLES

On the last day of April, J. J. Thomson gave a lecture at Faraday's old home, the Royal Institution. He put forward evidence which had made him conclude that the cathode rays are particles 1000 times lighter than a hydrogen atom and a universal constituent of matter. It is interesting that among his experiments were some closely resembling those of Kaufmann with which they agree, but from which Thomson deduced precisely the opposite conclusions to Kaufmann's!

In this lecture, which was printed in extenso in the *Electrician* of May 21, (and which contains some work that had appeared in *Nature*, March 15) J. J. makes an unanswerable case for the particle theory—unanswerable at least at the time.

He improves Perrin's experiment by showing that the rays when deflected carry the charge with them; therefore deflection is not a secondary phenomenon. He improves Lenard's experiment by showing that the deflection of the rays for equal potential is the same for any gas in the discharge; he measures m/e by measuring the magnetic deflection and also the heat given to a thermocouple per unit of charge carried by the rays, in other words mv/e and $(\frac{1}{2}) mv^2/e$ giving m/e of the order $\frac{1}{1000} M/e$. Most important of all he explains the negative experiment of Hertz on electric deflection. He was guided here by an old experiment of Eugen Goldstein, who had shown that if a discharge tube is provided with two cathodes, the rays from each are deflected when the other is connected to it. Thomson thought this occured because the electric field from the second cathode deflected the rays. He improved on the experiment by putting two parallel plates in the dark space of the discharge from a single cathode and earthing each in turn. The rays passing between the plates were deflected.

This experiment was shown at the Royal Institution and is the basis of the e/m experiment illustrated in all the textbooks, itself the origin of the cathode-ray oscilloscope. Because this is the vital point in the matter, it is worth going back to Hertz's experiment of 1883 to see what went wrong there. In these experiments Hertz first made his discharge in a flat box formed by square glass plates held a few centimeters apart by a brass frame. The purpose of this shape was to explore the distribution of current in the flat discharge by measuring the magnetic field close to the glass plates. The visible cathode rays did not show any connection with the lines of current flow thus found. He then worked with a more conventional cylindrical tube, its cathode near one end being surrounded by an anode of wire gauze through which the cathode rays escaped. These he caught in a vessel further down the cylinder connected to an electrometer, but he could detect no charge.

Fairly convinced by now that the cathode rays were merely secondary (he compared them to the visible light of the discharge), he tried the crucial experiment of electric deflection, applying an electric field to the ray by connecting the poles of a battery to two metal plates between which the rays passed and found no deflection of the phosphorescence on the end of the tube. At first the plates were outside the tube, but guessing correctly that the field might cause electrification of the glass, which would cancel its effect inside, Hertz put the plates themselves inside. Again no effect. It is strange that he did not realize the possibility of ions from his discharge polarizing his electrodes. Thomson did so, helped I suspect by his study in the previous year of the conductivity caused in air by rays. He was working on the space-charge effects, as we now call them, of the ions.

A UNIVERSAL CONSTITUENT

Though these experiments of Thomson proved that cathode rays were electrified particles, the further step of claiming them as a universal constituent of matter depended on other evidence. In doing so J. J. was influenced by the dependence of their absorption on density alone, which contrasted so strongly with the dependence on chemistry of the absorption of light, visible and ultra-violet. Even X-rays, whose absorption is atomic, do not show the simple proportionality to density that Lenard found. In his definitive paper, sent to the *Philosophical Magazine* (published October), J. J. not only claimed that the particles of cathode rays are a universal constituent of matter, in number roughly proportional to the density, so that collisions with them account for the absorption of cathode rays, but he indicated a way in which they might be arranged. He used an experiment of the American physicist, Alfred Mayer, in which floating magnets arrange themselves in concentric rings on the surface of water under the influence of a powerful magnet held vertically above them.

The average charge on the ions produced by X-rays was, as is well known, measured by Thomson in 1898, using C. T. R. Wilson's discovery of ions as centers of condensation. The result was equal to that on a monovalent ion in solution within the very wide limits to which this last was known.

A more accurate proof of this equality, though without determining either quantity, was made by John Townsend, who compared the motions of these ions moving under electric fields and diffusing through a gas under their own partial pressure.

There was, however, no proof that the charge on a cathode ray was equal to this "electron." J. J. then showed that two other processes produced particles with the same m/e as cathode rays, namely emission from a hot wire and the photoelectric effect.

This paper appeared in 1899 and contained also a measurement of "e" for the photoelectric ions by a slight modification of the method used for X-ray ions, getting the same result. In this way the connection between Stoney's definition of his term and the new concept was established. But in this paper, as in the Royal Institution lecture, J. J. used the word "corpuscle" for these elementary particles and continued to use it for some years. The word "electron" was still not in common use. Hendrik Lorentz, for example, spoke of "ions." No doubt electron is much the neater and more distinctive name, but its use for "corpuscle," even without the complication introduced by discovery of positrons, leaves one with no single word for Stoney's unit charge as a purely electrical quantity.

J. J. stated that the charge on the corpuscle was that on a monovalent ion, and that ionization involves breaking an atom by detaching a corpuscle.

Pieter Zeeman's discovery of the effect that carries his name and its explanation in terms of Lorentz's theory, was submitted to the Amsterdam Academy in October 1896. It appeared in Dutch in May of the next year but had already been published in English in the *Philosophical Magazine* of March, dated January with a postscript dated February. There is a curious error in this paper, the sign of the charge of the moving particle being given as positive.

Thomson referred to this paper in his Royal Institution lecture and especially the agreement in the value found for m/e, namely 10^{-7}.[2]

Though Thomson's ideas were not accepted immediately, even after the Dover meeting of the British Association in 1899, the evidence after the 1899 paper was adequate except on one important point. All the measurements of the supposed unit charge from Faraday's law were statistical; they gave average values for e. This was not so for m/e; even Thomson's measurement using electrical deflection gave some resolution, as did Kaufmann's and though poor at first the measurements rapidly improved. Most people assumed that the charge was also a true constant of nature, not just a mean. But it needed proof. Conceivably m/e could be more fundamental than either m or e. Thus in a gas such as argon containing isotopes, the mean value of $(\frac{1}{2}) mv^2$ is the same for every atom (over a period of time), but that of m can vary. As you all know, this matter was brilliantly cleared up by Robert A. Millikan in 1909 and with him this chapter ends.

---

[2]In cgs (Gaussian) units, the value of m/e is approximately 10^{-18} g/statcoul—see the preceding paper. [Ed.]

FIRST-RATE MINDS BLINDED

In looking back at it, one is impressed by the extent to which a theory long held can blind even first-rate minds to new ideas and by how easy it is to explain almost anything in terms of a favorite theory. Also, see how important technique can be. A slight reduction in the very small amount of gas present in a discharge changes the whole appearance of the problem. How cautious one should be in accepting the experiments even of able workers at their face value!

The electron thus established served as the basis of atomic physics for nearly 30 years with comparatively little modification. In the form of beta rays, electrons were used by Kaufmann, Bucherer and others to examine experimentally the variation of mass with velocity and to establish the truth of the formula given for this by the theory of relativity. Less successfully a number of physicists, including Thomson, used it to form theories of metallic conduction, but ran up against the difficulty that the necessary number of free electrons, regarded as Maxwellian particles, subject to equipartition of energy, would increase the specific heats of a metal much beyond the observed values.

Niels Bohr was the first to emancipate the electron from full obedience to the laws of Maxwellian electrodynamics by assuming that electrons could move forever round a nucleus without radiating energy, provided that their angular momentum was an integral multiple of $h/2\pi$. This made possible theories of atomic structure capable of explaining many of the facts of spectroscopy, but otherwise left the idea of a small spherical particle unchanged. In the middle twenties it was becoming clear that in spite of its many successes there was something wrong with the Bohr-Sommerfeld theory. I remember vividly hearing Arthur H. Compton give a paper to the Cavendish Laboratory Colloquium on the "ring electron," showing how it could deal with some of these difficulties; however, this idea never fully caught on and was superseded in 1926 by George Uhlenbeck and Samuel Goudsmit's idea of the spinning electron. Even this was not too satisfactory. The

point electron failed because it had not enough degrees of freedom to account for the variety of spectral lines—why, for example, the sodium lines are double. The spinning electron had too many.

WAVES OR PARTICLES?

But of course the real difficulty of physics in the 1920's was the radiation problem, waves or particles? Solutions came nearly simultaneously from Werner Heisenberg and Louis de Broglie. The former did not immediately affect one's ideas on the nature of electrons, but de Broglie stated from the first that an electron as well as a photon has waves associated with it. In his first paper in English he used the idea of the electronic wavelength to explain Bohr's condition for the permitted orbits. I had been partly prepared for this by a remark Lawrence Bragg once made to me that he thought the electron was not as simple as people then supposed. Anyhow I wrote a brief attempt to amplify de Broglie's ideas in the *Philosophical Magazine* of July 1925.

Meanwhile Clinton Davisson and C. H. Kunsman had been working on production of secondary electrons from metals by the impact of electrons of energy around 200 volts. The first paper appeared in 1921. This was by no means a virgin field, and Davisson's research is an outstanding example of what can sometimes be achieved by greatly superior technique in a field that might have been supposed unpromising.

Davisson's earlier papers used poly-crystalline targets and showed "peaks," that is, directions of preferred scattering that varied with the speed of the primary electron. The theoretical physicist Walter Elsasser in a letter sent to *Naturwissenshaften* in July 1925 suggested that the effects were analogous to diffraction by an optical grating and that in de Broglie's theory deviations from normal mechanics were to be expected in "an accessible range of velocities."

Davisson read Elsasser's paper, but states that he was not influenced by it as he did not think Elsasser's theory valid. He was inclined at first to explain these experiments by shells of electrons in the atoms and later by "transparent directions." However, by a fortunate accident (April 1925) a nickel target was subjected to a prolonged heat treatment that converted it into a few large crystals. The scattering pattern was now completely different. Davisson decided to work in future with a single crystal.

In 1926 he attended the British Association meeting at Oxford where he showed some curves relating to single crystals to Max Born, to D. R. Hartree and perhaps to P. M. Blackett and James Franck. On his return he spent his time, as he says, "trying to understand Schrödinger's papers, as I had then an inkling (probably derived from the Oxford discussions) that the explanation might reside in them." On return he instituted a search for the beams and after some failures found on January 6, 1927 strong beams due to the line gratings of the surface atoms. I was at the same meeting, where I heard de Broglie's theory discussed informally. I do not remember Elsasser's name being mentioned, but some of those who were discussing the theory may well have read his work. I think I did not meet Davisson and certainly did not discuss his experiments with him. The experiments of Davison and Lester Germer, first published in a note to *Nature* in April 1927, and later more fully in the *Physical Review* of December of that year, completely established that electrons are diffracted by the surface atoms of a crystal of nickel cut with a (111) face in the surface, as would be waves of $\lambda = h/mv$—the de Broglie formula; the atoms below the surface modify the effects in a way that can be explained as due to the electrons acquiring energy on entering the metal. The electrons Davisson used had energies in the region 50—400 volts.

Meanwhile it had occurred to me that one ought to be able to get an electron analog of Thomas Young's optical "eriometer," in which grains or fibers oriented at random on a plate of glass give halos whose diameters depend on the size of the fibers. Celluloid was known to be composed of long molecules and could easily be made in very thin films. As it happened a pupil of mine at

Aberdeen, A. Reid (killed the next year in a motor accident), had an apparatus that could readily be used for this experiment, and I asked him to try it.

He sent cathode rays of about 30 kV through a thin film of celluloid onto a photographic plate, which, when developed, showed fuzzy rings.

As it happened we had had trouble with some optical illusions on photographic plates, and I was chary of trusting these fuzzy rings though they were about the right size assuming $\lambda = h/mv$ and a reasonable guess for the effective thickness of the celluloid molecules. This delayed us a bit, but eventually there was no doubt and we published a note in *Nature*, June 1927.

However, it was essential to get patterns from some substance whose crystal pattern was well established. Thanks to the dexterous manipulation of C. G. Fraser, I was able to use films of gold, aluminum and platinum, which gave patterns on the photographic plate that reproduced in all respects those to be expected on the de Broglie theory.

A note was published in *Nature* in December 1927, and the paper appeared in *Proceedings of the Royal Society* in February 1928. Because of the higher energy of the electrons there was no measurable correction for the inner potential of the metal.

In reply to a criticism I was able to show by the use of a magnetic deflection that the patterns were actually caused by the photographic action of electrons not by some kind of secondary X-rays formed in the target. In a later paper I showed that the relative intensity of the rings formed in thin gold by Bragg reflection from the different crystal planes agreed with the predictions of a wave-mechanical theory due to Nevil Mott.

In comparing these experiments I must stress the far greater difficulty of those of Davisson and Germer, which are among the greatest experimental triumphs of physics, even now hard to repeat, while mine, once we learned how to make thin films, were among the easiest. This is mainly due to the difference in energy of the rays used. In addition, reflection experiments, though not very difficult with fast electrons, are naturally more complicated than transmission ones. I set out to test a theory; Davisson to explain an experiment.

The original simple form of de Broglie's theory is the wave-mechanical analog of the point electron with three degrees of freedom only. P. A. M. Dirac in 1928 put forward a more complicated theory that removes this objection and at the same time staisfies the requirements of relativity. It also led him to predict existence of electronic states of negative energy, an infinite number of which are supposed occupied. A vacancy in this sea of

states then behaves like an electron of normal mass but positive charge. With the discovery of the positron by Carl Anderson in 1932 and the creation of pairs by the materialization of gamma rays, discovered by Seth Neddermeyer and Anderson the next year, this stage in the history of the electron came to an end.

No doubt there is much more to come. The relation of mu mesons to electrons is still a mystery 20 years after the discovery of muon decay. Perhaps electrons will lose their unique position and become merely senior partners in a firm of leptons; but 70 years is a good life.

5.3   The Particle-like Properties of Light
Norman Feather

In 1887 Hertz found that small sparks could be obtained across a minute gap in an isolated circuit when an induction coil was operated so as to produce sparks across a "primary" gap in its neighborhood. From this observation, he proceeded, by successive stages, to build up the experimental case for the validity of Maxwell's inference that light is propagated as an electromagnetic wave process. In the same series of "preliminary" researches, he also observed that when two spark gaps are fairly close together, and sparks are passing freely across one, the breakdown voltage of the other gap is less than normal. He followed up this observation, also, and identified the incidence of ultra-violet light from the sparks of the first gap on the metal electrodes of the second as the immediate cause of the phenomenon. Unwittingly, he had made the first observation in the long series of experiments by which the particle-like properties of light have since been demonstrated and explored. In the whole history of physics, in retrospect, antithesis is nowhere so strikingly exemplified as in these preliminary researches of Hertz in 1887.

In 1888 W. L. F. Hallwachs showed that the action of the ultra-violet light in Hertz's experiments was to cause the leakage of negative electricity from the metal electrode on which it was incident. Over a number of years from 1889 J. P. L. T. Elster and H. F. K. Geitel investigated this "photo-electric" action, showing that the material of the electrode was an important factor in the process: for the same irradiation, the more electro-positive the metal the greater the effect. A. G. Stoletow (1890) was the first to investigate the phenonenon systematically with electrodes "in vacuum." Under these conditions he found that the leakage of negative electricity was independent of the strength of the collecting field, within wide limits. In 1897 Joseph John Thomson (1856–1940), Rayleigh's successor as Cavendish professor at Cambridge—and others independently, though not so conclusively—discovered the negative electron as the invariable carrier of the cathode rays in the discharge tube—whatever the residual gas in the tube, and in 1899 he showed that the leakage of negative electricity in the photoelectric effect was by virtue of the emission of these same

SOURCE. From *Vibrations and Waves* (Edinburgh: Edinburgh Univ. Press, 1961) p. 276; copyright by Edinburgh University Press.

particles from the irradiated surface of the metal. But the culminating research was that of P. E. A. Lenard (1862–1947), published in 1902. Lenard, a former pupil of Hertz at Bonn, found that, though the magnitude of the leakage current was proportional to the intensity of the light incident on the surface, the kinetic energies of the individual electrons constituting that current were independent of the intensity of the light. Moreover, for the same light, the photoelectrons had in general the greater energy the more electro-positive the metal from which they were emitted. When this stage had been reached, it had clearly become evident that the photoelectric effect posed very serious difficulties for any interpretation based on the classical electromagnetic theory of light.

According to the classical theory, the rate of transport of energy, across any surface on which radiation is incident normally, is proportional to the area of the surface and to the square of the amplitude of the electric field intensity at the surface. In the photoelectric effect, according to observation, the energy incident on a metal surface, far from being uniformly effective over the surface, becomes concentrated on a very few of the very many electrons lying in the immediate surface layers of the metal. In the case of zinc, for example, there are some 6.6×10^{22} atoms per cm^3, and if we allow only one electron per atom, as a conservative estimate, the first layer of atoms provides some 1.6×10^{15} electrons per cm^2 of surface available for photoelectric emission—or some $6.1 \times 10^{-16} cm^2$ of surface as "catchment area" for each such electron, if all were to compete equally for the incident radiation.

In Lenard's experiments the average energy carried away by a single electron was found to be of the order of 3×10^{-12} erg. Ten years previously Stoletow had shown, in a similar arrangement, that the photoelectric current reaches its full value in less than 10^{-3} sec from the instant of first illumination. According to a crude classical picture, therefore, 3×10^{-12} erg of energy must have been intercepted by an area of $6 \times 10^{-16} cm^2$ of surface in 10^{-3} sec—and the incident flux

must have been at least as great as 5×10^6 erg $cm^{-2} sec^{-1}$. In actual fact it was probably of the order of 1 erg $cm^{-2} sec^{-1}$ in Lenard's experiments.

This one illustrative example shows clearly that the electromagnetic theory was in real difficulties over the photoelectric effect. It is little wonder, then, that, as first Silliman Memorial lecturer at Yale University in 1903, Thomson should say "on the view we have taken of a wave of light the wave itself must have a structure, and the front of the wave, instead of being, as it were, uniformly illuminated, will be represented by a series of bright specks on a dark ground." Thomson, indeed, had already recognized a similar difficulty in the ionization produced by X-rays in passing through a gas. Ionization of gas molecules implies the liberation of electrons at the expense of the energy of the radiation, so Thomson said "we suppose that the front of the Röntgen ray consists of specks of great intensity separated by considerable intervals where the intensity is very small."

Thomson saw the difficulties clearly, but his attempt to circumvent them by use of classical analogies provided no real basis of solution. It was Einstein in 1905, who, breaking with the old theory altogether, introduced an entirely different point of view. Five years previously, Max Planck (1858–1947), confronted with the difficulties which had then become critical for any classical theory of the equilibrium between radiant (heat) energy and matter in a constant-temperature enclosure, had originated the "quantum" hypothesis. He had shown that, formally at least, these difficulties could be removed if it were supposed that interchange of energy between radiation and matter can occur only in finite amounts proportional to the (classically specified) frequency of the radiation. In this way a new "universal" constant, Planck's constant h, was introduced into physics—albeit speculatively. Planck supposed that, for radiation of frequency v (v is now conventionally used, rather than f, for frequency in quantum physics), the basic unit or quantum of energy is hv. By 1905 the new hypothesis had not in fact been accorded wide acceptance, but Einstein recognized in the phenomena of the photoelectric effect just the type of behavior that Planck had imagined—and, with the single-mindedness, and, as it later appeared, the blind wisdom, of youth (he was then twenty-six), he postulated that light is indeed not only absorbed and emitted in energy quanta but also propagated as "photons" (a term not used by Einstein at the time, but appropriated later). In this thorough-going revision of viewpoint, the energy-content of the photon became its prime characteristic, the associated frequency no more than a convenient alternative parameter of description. The measure of the associated frequency was given, in appropriate units, by dividing the measure of the energy by the measure of h, but the classical electromagnetic connotation of frequency was tacitly abandoned.

Thus Einstein's photons replaced Thomson's "bright specks"—and the "dark ground" became absolutely dark. When it was first put forward, this whole idea appeared wildly paradoxical—except in the context of the photoelectric effect and the theory of "black-body" radiation—but as time passed it was seen to have more and more relevance to experiment. On the other hand, there was no escaping the paradoxes: they had simply to be endured. They showed up most acutely, perhaps, in Taylor's experiment on interference under weak illumination. This could be described in the statement that the normal interference pattern is obtained even though the time spent by a single quantum in traversing the apparatus is considerably less than the average time between the entry of one quantum and the next. Classical electromagnetic theory, as we have seen, was already in some difficulty here, though it might provide a reasonable formal basis for understanding: Einstein's hypothesis, in its original crude form, certainly provided none. Yet more and more of the discovered properties of ultra-violet light and of X-rays appeared to require a photon-type explanation. In particular, the detailed experiments of O. W.

Richardson (1879–1959) and K. T. Compton (1887–1954) in 1912, and of R. A. Millikan (1868–1953) in 1916, demonstrated this quite clearly. The photoelectric effect with light occurs in precise conformity with the quantitative predictions of Einstein's hypothesis: when the frequency of the incident light, classically determined, is increased from v to v', the maximum energy of the photoelectrons from a given metal increases by an amount $h(v' - v)$. Finally, in 1923, A. H. Compton (b. 1892) elucidated experimentally a mode of action of X-rays on matter which previous observers had failed to understand. In the "Compton effect," now so called, it is a very close approximation to the truth to say that the basic action is a particle-like collision between an X-ray photon and an electron, free and at rest. In this action the electron acquires energy at the expense of the photon; after the collision the photon appears as of diminished frequency (or increased wavelength), for its energy has decreased.

Reviewing these developments in brief, we see, first, that there are certain phenomena, which attain greater importance as the classical wavelength of the radiation decreases, for which no classical electromagnetic interpretation is possible, and, secondly, that, though these phenomena may be brought within the framework of a non-classical (quantum) scheme of interpretation, yet this scheme still retains formal contact with the classical theory through the, now ambivalent, concepts of frequency and wavelength. And we add the other side of the picture—that quite obviously the phenomena of diffraction and interference (and these are in evidence at all wavelengths) are quite incomprehensible within an interpretive scheme in which wholly individualized photons are the only structural units.

Here is the essence of the duality which the observed facts force upon the theorist who surveys the whole field of the physics of "radiation" in an attempt to formulate its laws.

5.4   *The Wave-Particle Dualism*
                              Wilhelm H. Westphal

The fact that both light and the elementary particles of matter behave—according to the kind of the observations made on them—either like particles or like waves is now perfectly confirmed by experience. This wave-particle dualism was bound to throw physics into hopeless confusion so long as it clung to the belief that the wave model or the particle model constituted something like a statement about the *true nature of light and matter*. It can be shown that these two models are in no way compatible with one another—as, for instance, by the invention of some new model which would combine the properties of both existing models. In order to solve this difficulty physics had to take a very close look at the very foundations of our physical cognition.

Our physical thinking has evolved from our everyday thinking through refinement and specialization, but we do not possess for it any other intellectual equipment than for our ordinary everyday reasoning. And this has the peculiarity that, whenever we think about anything, we cannot manage without some visual concept; this is true even for our physical thinking. We feel compelled to form a picture of the objects of our thinking, and we shape these pictures in such a way that they correspond somehow to the physical behavior of the objects pictured.

Until well into the twentieth century it had been universally believed that we should be able in principle to form a mental picture of any physical object—a picture which not only corresponds to the object's behavior but which would moreover be *correct* in the true sense of the word. An electron, for instance, would have to be something one could make a model of, a model which—apart from size—would be a faithful replica of an electron, with the same modes of behavior and capable of correct description in words. The same way of thinking was applied to light, which was why recourse had to be taken to the ether so that light could be visualized as the vibrations of a material medium. It was believed, in short, that all natural phenomena were susceptible of correct mechanical-visual comprehension, and indeed the attainment of such understanding was regarded as the proper task of physics. This belief appears to meet with full confirmation in our everyday experiences, in the sphere directly accessible to our senses.

However, physics has advanced a long way beyond that sphere of everyday experience and in doing so has come into insoluble conflict with this belief in visual comprehension—first with light and then also with matter. Neither of these proved susceptible of comprehension in terms of a single mental picture; both needed two totally different visual concepts according to circumstances. Consequently, neither mental picture could be truly "correct"—and that is why we have always been careful to use the term *model*.

The necessary re-examination of the foundations of physical thought at last led to the realization that the belief that all natural objects should be understood in mechanical-visual terms and that all natural processes were changes of state of such mechanical-visual objects, was not an *intellectual necessity* but merely an *intellectual habit*. This habit is fully confirmed by our everyday experience with the macroscopic bodies directly perceived by us, and indeed a good way beyond that sphere, but it ceases to apply if we go too far beyond its lower or upper limit. Our intellectual habit fails us completely once we approach the order of magnitude of the

SOURCE. From *A Short Textbook of Physics* (New York: Springer-Verlag. 1968) pp. 297–298.

atoms and elementary particles, and it similarly fails us if we try to apply it to the immense dimensions of the universe. It is by no means an intellectual necessity that everything existing in nature should be capable of visualization. Elementary particles or light are certainly not susceptible to such visual comprehension, and any question posed on these lines is, in principle, incapable of being answered.

The quantum model and the wave model are therefore no more than *analogies* with a limited field of application. *Elementary particles and light are in reality neither particles nor waves*—they merely, according to circumstances, i.e., according to the kind of measurement applied to them, exhibit the kind of behavior which a particle or a wave model would exhibit in analogous conditions. Although these models therefore are by no means "correct," we still cannot do without them, either for theoretical or experimental research into the objects under consideration; and indeed they are enormously useful for that purpose. What matters is that we should know which model is applicable to the task in hand—and that is a question which can be clearly answered. We have already seen in the case of light that for all processes concerned with the propagation of light the wave model meets our requirements, whereas all interactions with matter require the application of the corpuscular quantum model. Much the same applies when we are dealing with elementary particles of matter.

Admittedly, this realization means resigning ourselves to the fact that we can never gain any insight into the "true nature" of matter and of light. But in fact this resignation is an extremely important step forward, a step which rids physics of an insoluble problem and opens the way to new thought and research.

5.5   *The Uncertainty Principle*

Richard P. Feynman

This is the way Heisenberg stated the uncertainty principle originally: If you make the measurement on any object, and you can determine the x-component of its momentum with an uncertainty Δp, you cannot, at the same time, know its x-position more accurately than $\Delta x = h/\Delta p$. The uncertainties in the position and momentum at any instant must have their product greater than Planck's constant. . . . The more general statement of the uncertainty principle is that one cannot design equipment in any way to determine which of two alternatives is taken, without, at the same time, destroying the pattern of interference.

. . . . .

The uncertainty principle "protects" quantum mechanics. Heisenberg recognized that if it were possible to measure the momentum and the position simultaneously with a greater accuracy, the quantum mechanics would collapse. So he proposed that it must be impossible. Then people sat down and tried to figure out ways of doing it, and nobody could figure out a way to measure the position and the momentum of anything—a screen, an electron, a billiard ball, anything—with any greater accuracy. Quantum mechanics maintains its perilous but accurate existence.

PHILOSOPHICAL IMPLICATIONS

Let us consider briefly some philosophical implications of quantum mechanics. As al-

SOURCE. From Feynman, Leighton, and Sands, The *Feynman Lectures on Physics* (Reading, Mass.: Addison-Wesley, 1963) Chapter 37.

ways, there are two aspects of the problem: one is the philosophical implication for physics, and the other is the extrapolation of philosophical matters to other fields. When philosophical ideas associated with science are dragged into another field, they are usually completely distorted. Therefore we shall confine our remarks as much as possible to physics itself.

First of all, the most interesting aspect is the idea of the uncertainty principle; making an observation affects the phenomenon. It has always been known that making observations affects a phenomenon, but the point is that the effect cannot be disregarded or minimized or decreased arbitrarily by rearranging the apparatus. When we look for a certain phenomenon we cannot help but disturb it in a certain minimum way, and *the disturbance is necessary for the consistency of the viewpoint*. The observer was sometimes important in prequantum physics, but only in a rather trivial sense. The problem has been raised: if a tree falls in a forest and there is nobody there to hear it, does it make a noise? A *real* tree falling in a *real* forest makes a sound, of course, even if nobody is there. Even if no one is present to hear it, there are other traces left. The sound will shake some leaves, and if we were careful enough we might find somewhere that some thorn had rubbed against a leaf and made a tiny scratch that could not be explained unless we assumed the leaf were vibrating. So in a certain sense we would have to admit that there is a sound made. We might ask: was there a *sensation* of sound? No, sensations have to do, presumably, with consciousness. And whether ants are conscious and whether there were ants in the forests, or whether the tree was conscious, we do not know. Let us leave the problem in that form.

Another thing that people have emphasized since quantum mechanics was developed is the idea that we should not speak about those things which we cannot measure. (Actually relativity theory also said this.) Unless a thing can be defined by measurement, it has no place in a theory. And since an accurate value of the momentum of a localized particle cannot be defined by measurement it therefore has no place in the theory. The idea that this is what was the matter with classical theory *is a false position*. It is a careless analysis of the situation. Just because we cannot *measure* position and momentum precisely does not *a priori* mean that we *cannot* talk about them. It only means that we *need* not talk about them. The situation in the sciences is this: A concept or an idea which cannot be measured or cannot be referred directly to experiment may or may not be useful. It need not exist in a theory. In other words, suppose we compare the classical theory of the world with the quantum theory of the world, and suppose that it is true experimentally that we can measure position and momentum only imprecisely. The question is whether the *ideas* of the exact position of a particle and the exact momentum of a particle are valid or not. The classical theory admits the ideas; the quantum theory does not. This does not in itself mean that classical physics is wrong. When the new quantum mechanics was discovered, the classical people—which included everybody except Heisenberg, Schrödinger, and Born—saids "Look, your theory is not any good because you cannot answer certain questions like: what is the exact position of a particle?, which hole does it go through?, and some others." Heisenberg's answer was: "I do not need to answer such questions because you cannot ask such a question experimentally." It is that we do not *have* to. Consider two theories (a) and (b); (a) contains an idea that cannot be checked directly but which is used in the analysis, and the other, (b), does not contain the idea. If they disagree in their predictions, one could not claim that (b) is false because it cannot explain this idea that is in (a), because that idea is one of the things that cannot be checked directly. It is always good to know which ideas cannot be checked directly, but it is not necessary to remove them all. It is not true that we can pursue science completely by using only those concepts which are directly subject to experiment.

In quantum mechanics itself there is a wave function amplitude, there is a potential, and there are many constructs that we cannot measure directly. The basis of a science is its ability to *predict*. To predict means to tell what will happen in an experiment that has never been done. How can we do that? By assuming that we know what is there, independent of the experiment. We must extrapolate the experiments to a region where they have not been done. We must take our concepts and extend them to places where they have not yet been checked. If we do not do that, we have no prediction. So it was perfectly sensible for the classical physicists to go happily along and suppose that the position—which obviously means something for a baseball—meant something also for an electron. It was not stupidity. It was a sensible procedure. Today we say that the law of relativity is supposed to be true at all energies, but someday somebody may come along and say how stupid we were. We do not know where we are "stupid" until we "stick our neck out," and so that the whole idea is to put our neck out. And the only way to find out that we are wrong is to find out *what* our predictions are. It is absolutely necessary to make constructs.

We have already made a few remarks about the indeterminacy of quantum mechanics. That is, that we are unable now to predict what will happen in physics in a given physical circumstance which is arranged as carefully as possible. If we have an atom that is in an excited state and so is going to emit a photon, we cannot say *when* it will emit the photon. It has a certain amplitude to emit the photon at any time, and we can predict only a probability for emission; we cannot predict the future exactly. This has given rise to all kinds of nonsense and questions on the meaning of freedom of will, and of the idea that the world is uncertain.

Of course we must emphasize that classical physics is also indeterminate, in a sense. It is usually thought that this indeterminacy, that we cannot predict the future, is an important quantum-mechanical thing, and this is said to explain the behavior of the mind, feelings of free will, etc. But if the world *were* classical—if the laws of mechanics

were classical—it is not quite obvious that the mind would not feel more or less the same. It is true classically that if we knew the position and the velocity of every particle in the world, or in a box of gas, we could predict exactly what would happen. And therefore the classical world is deterministic. Suppose, however, that we have a finite accuracy and do not know *exactly* where just one atom is, say to one part in a billion. Then as it goes along it hits another atom, and because we did not know the position better than to one part in a billion, we find an even larger error in the position after the collision. And that is amplified, of course, in the next collision, so that if we start with only a tiny error it rapidly magnifies to a very great uncertainty. To give an example: if water falls over a dam, it splashes. If we stand nearby, every now and then a drop will land on our nose. This appears to be completely random, yet such a behavior would be predicted by purely classical laws. The exact position of all the drops depends upon the precise wigglings of the water before it goes over the dam. How? The tiniest irregularities are magnified in falling, so that we get complete randomness. Obviously, we cannot really predict the position of the drops unless we know the motion of the water *absolutely exactly*.

Speaking more precisely, given an arbitrary accuarcy, no matter how precise, one can find a time long enough that we cannot make predictions valid for that long a time. Now the point is that this length of time is not very large. It is not that the time is millions of years if the accuracy is one part in a billion. The time goes, in fact, only logarithmically with the error, and it turns out that in only a very, very tiny time we lose all our information. If the accuracy is taken to be one part in billions and billions and billions—no matter how many billions we wish, provided we do stop somewhere—then we can find a time less than the time it took to state the accuracy—after which we can no longer predict what is going to happen! It is therefore not fair to say that from the apparent freedom and indeterminacy of the human mind, we should have realized that classical "deterministic" physics could not ever hope to understand it, and to welcome quantum mechanics as a release from a "completely mechanistic" universe. For already in classical mechanics there was indeterminability from a practical point of view.

5.6   *Two Worlds Are one*

Eugene Rabinowitch

There is another aspect of modern science which also should affect man's idea of the world he lives in. Since science first arose, man has lived in two separate worlds. One is the world of physical existence, populated by material objects—from stars to atoms—which follow certain discoverable causal or statistical laws. The other is the world of mind and emotion. The first is dominated by concepts such as heavy or light, fast or slow, positively or negatively charged; the second by concepts such as good or bad, beautiful or hideous, virtuous or evil. Our growing understanding of the physical world—in which observations are reproducible and future behavior can be predicted on the basis of past history—has gradually narrowed the field of human experience dominated by emotional and spiritual concepts. Arbitrary manipulation of material objects by gods, once thought to explain many events, is used in today's religious systems only to explain exceptional cases, "miracles" interrupting the natural course of events. This narrowing of the realm of the spiritual and broadening of the realm of the material, susceptible to scientific analysis, has led many to believe that ultimately everything in human experience will be interpreted in material terms, and that concepts taken from man's emotional and spiritual experience are merely temporary tools, useful to describe material phenomena of such complexity that their interpretation by means of the laws of the material world is as yet impossible.

However, as justifiable as is the reaction against the past belief in the constant, crude intervention of spiritual forces into material events, this belief should not be replaced by the opposite—the extension of laws derived from observation of the material world to the spiritual aspects of human existence. Spiritual experience is as much an incontestable part of human existence as the experience of material phenomena. Everybody knows the experience of free will, of capacity to choose, of distinguishing between ethical or esthetic values. Some men can be brought to deny this experience, to say that it is based on self-deception, and to find satisfaction in considering themselves only very complex machines acting on the basis of the causal or statistical laws of science.

But does scientific analysis of the material world really require such denial of personal spiritual experience? Certain developments in theoretical physics suggest that man does not need to do violence to himself by denying his spiritual existence. The relevant theoretical principle, which has emerged from the extension of experimental physics into the world of elementary particles, is Bohr's principle of complementarity. It asserts the legitimacy and the inevitability of the parallel existence of two apparently mutually exclusive pictures in the description of physical events in microphysics. It further asserts that these two descriptions cannot contradict each other, because, as shown by Heisenberg, every time man attempts by any conceivable experiment whatsoever to obtain an increasingly precise definition of magnitudes belonging to one picture, he makes, by this

SOURCE. From *Bulletin of the Atomic Scientists*, September, 1963; copyright 1970 by the Educational Foundation for Nuclear Science, reprinted by permission of Science and Public Affairs.

very act, the magnitudes belonging to the other picture more and more vague. Together, the two sets of concepts give a full description of all we know now about the material universe; separately, each can represent only one aspect of it.

·   ·   ·   ·   ·

There are eminent scientists who are dissatisfied with this dual picture, which implies renunciation of strict causal interpretation of the behavior of elementary particles, replacing it by statistical laws of probability determined by the propagation of the correlated waves. But the majority believe that this duality is not a temporary weakness of theory, but represents the ultimate realization of the capacity to understand the material world possible for man as a sensing organism. They consider the dualistic picture created by Bohr's complementarity principle and Heisenberg's uncertainty relation as logically satisfactory. The apparent contradiction of the wave and the particle pictures is, in their eyes, only a weakness of traditional ways of thinking, based on observations of macroscopic objects.

I believe that acceptance of the legitimate existence of two physical pictures—incompatible on the surface but never contradictory in predicting future observations—implies a major revolution in man's concept of the world. It offers a glimpse of how the apparent contradiction between the material and the spiritual elements in human existence could be ultimately resolved.

There is obviously no simple analogy between the particle and the wave aspects of the material world and the spiritual and the material aspects of human existence, but it suggests that the modes of interaction between man and the world around him may, by their very nature, make it impossible to form a unified picture, and requires "coexistence" of two (or perhaps more) independent systems of concepts and relations. If the analogy holds, the two systems can become contradictory only if one of them is applied to the types of observations which rightfully belong to the other.

The spiritual aspect of the world is, then, not contradictory to and not separable from its material aspect; it is complementary to it. The existence of spiritual forces cannot reveal itself, as many believe, in occasional violations of the laws which govern the material world. Rather, the whole world confronted by man has its spiritual and its material aspects. The material aspects reveal themselves to man through his sense organs, refined by instruments. These organs or instruments could not reveal the existence and operation of the spiritual forces even if each event in the world has its physical as well as its spiritual aspects, because by the very process of observation of material parameters, the spiritual ones are made diffuse and escape observation.

The more we will learn about the physical aspects of the physiological processes which accompany man's thoughts and emotions, the vaguer will become their spiritual content. We may be able to induce emotions and visions by drugs; but this only illustrates the coupling of physicochemical and spiritual experience, and does not make the one "explain" the other.

ATOMS AND RADIATION

6.1

The Roots of Atomic Science
Werner Heisenberg

The concept of the atom goes back much further than the beginning of modern science in the seventeenth century; it has its origin in ancient Greek philosophy and was in that early period the central concept of materialism taught by Leucippus and Democritus. On the other hand, the modern interpretation of atomic events has very little resemblance to genuine materialistic philosophy; in fact, one may say that atomic physics has turned science away from the materialistic trend it had during the nineteenth century. It is therefore interesting to compare the development of Greek philosophy toward the concept of the atom with the present position of this concept in modern physics.

.   .   .   .   .

In the philosophy of Democritus the atoms are eternal and indestructible units of matter, they can never be transformed into each other. With regard to this question modern physics takes a definite stand against the materialism of Democritus and for Plato and the Pythagoreans. The elementary particles are certainly not eternal and indestructible units of matter, they can actually be transformed into each other. As a matter of fact, if two such particles, moving through space with a very high kinetic energy, collide, then many new elementary particles may be created from the available energy and the old particles may have disappeared in the collision. Such events have been frequently observed and offer the best proof that all particles are made of the same substance: energy. But the resemblance of the modern views to those of Plato and the Pythagoreans can be carried somewhat further. The elementary particles in Plato's *Timaeus* are finally not substance but mathematical forms. "All things are numbers" is a sentence attributed to Pythagoras. The only mathematical forms available at that time were such geometric forms as the regular solids or the triangles which form their surface.

In modern quantum theory there can be no doubt that the elementary particles will finally also be mathematical forms, but of a much more complicated nature. The Greek philosophers thought of static forms and found them in the regular solids. Modern science, however, has from its beginning in the sixteenth and seventeenth centuries started from the dynamic problem. The

SOURCE. From *Physics and Philosophy*, First ed. (New York: Harper and Row, 1958); copyright 1958 by Werner Heisenberg, reprinted by permission of Harper and Row Publishers, Inc.

constant element in physics since Newton is not a configuration or a geometrical form, but a dynamic law. The equation of motion holds at all times, it is in this sense eternal, whereas the geometrical forms, like the orbits, are changing. Therefore, the mathematical forms that represent the elementary particles will be solutions of some eternal law of motion for matter. Actually this is a problem which has not yet been solved. The fundamental law of motion for matter is not yet known and therefore it is not yet possible to derive mathematically the properties of the elementary particles from such a law. But theoretical physics in its present state seems to be not very far from this goal . . .

If we follow the Pythagorean line of thought we may hope that the fundamental law of motion will turn out as a mathematically simple law, even if its evaluation . . . may be very complicated. It is difficult to give any good argument for this hope for simplicity—except the fact that it has hitherto always been possible to write the fundamental equations in physics in simple mathematical forms. This fact fits in with the Pythagorean religion, and many physicists share their belief in this respect, but no convincing argument has yet been given to show that it must be so.

.   .   .   .   .

It may seem at first sight that the Greek philosophers have by some kind of ingenious intuition come to the same or very similar conclusions as we have in modern times only after several centuries of hard labor with experiments and mathematics. This interpretation of our comparison would, however, be a complete misunderstanding. There is an enormous difference between modern science and Greek philosophy, and that is just the empiristic attitude of modern science. Since the time of Galileo and Newton, modern science has been based upon a detailed study of nature and upon the postulate that only such statements should be made, as have been verified or at least can be verified by experiment. The idea that one could single out some events from nature by an experiment, in order to study the details and to find out what is the constant law in the continuous change, did not occur to the Greek philosophers. Therefore, modern science has from its beginning stood upon a much more modest, but at the same time much firmer, basis than ancient philosophy. Therefore, the statements of modern physics are in some way meant much more seriously than the statements of Greek philosophy. When Plato says, for instance, that the smallest particles of fire are tetrahedrons, it is not quite easy to see what he really means. Is the form of the tetrahedron only symbolically attached to the element fire, or do the smallest particles of fire mechanically act as rigid tetrahedrons or as elastic tetrahedrons, and by what force could they be separated into the equilateral triangles, etc.? Modern science would finally always ask: How can one decide experimentally that the atoms of fire are tetrahedrons and not perhaps cubes? Therefore, when modern science states that the proton is a certain solution of a fundamental equation of matter it means that we can from this solution deduce mathematically all possible properties of the proton and can check the correctness of the solution by experiments in every detail. This possibility of checking the correctness of a statement experimentally with very high precision and in any number of details gives an enormous weight to the statement that could not be attached to the statements of early Greek philosophy.

All the same, some statements of ancient philosophy are rather near to those of modern science. This simply shows how far one can get by combining the ordinary experience of nature that we have without doing experiments with the untiring effort to get some logical order into this experience to understand it from general principles.

6.2   *Note on the Spectral Lines of Hydrogen*
                              Johann Jacob Balmer

Using measurements by H. W. Vogel and by Huggins of the ultraviolet lines of the hydrogen spectrum I have tried to derive a formula which will represent the wavelengths of the different lines in a satisfactory manner. I was encouraged to take up this work by Professor E. Hagenbach. Ångström's very exact measurements of the four hydrogen lines enable one to determine a common factor for their wavelengths which is in as simple a numerical relation as possible to these wavelengths. I gradually arrived at a formula which, at least for these four lines, expresses a law by which their wavelengths can be represented with striking precision. The common factor in this formula, as it has been deduced from Ångström's measurements, is $b = 3645.6$ cm/10^8).

We may call this number the fundamental number of hydrogen; and if corresponding fundamental numbers can be found for the spectral lines of other elements, we may accept the hypothesis that relations which can be expressed by some function exist be-

do not form a regular series; but if we multiply the numerators in the second and the fourth terms by 4 a consistent regularity is evident and the coefficients have for numerators the numbers 3^2, 4^2, 5^2, 6^2 and for denominators a number that is less by 4.

For several reasons it seems to me probable that the four coefficients which have just been given belong to two series, so that the second series includes the terms of the first series; hence I have finally arrived at the present formula for the coefficients in the more general form: $m^2 / (m^2 - n^2)$ in which m and n are whole numbers.

For $n = 1$ we obtain the series $4/3$, $9/8$, $16/15$, $25/24$, and so on for $n = 2$ the series $9/5$, $16/12$, $25/21$, $36/32$, $49/45$, $64/60$, $81/77$, $100/96$, and so on. In this second series the second term is already in the first series but in a reduced form.

If we carry out the calculation of the wavelengths with these coefficients and the fundamental number 3645.6, we obtain the following numbers in 10^{-8}cm.

TABLE 6.2–1

According to the formula			Ångström gives	Difference
$H\alpha$ (C-line) $= \frac{9}{5}b$	$= 6562.08$		6562.10	$+0.02$
$H\beta$ (F-line) $= \frac{4}{3}b$	$= 4860.8$		4860.74	-0.06
$H\gamma$ (near G) $= \frac{25}{21}b$	$= 4340$		4340.1	$+0.1$
$H\delta$ (h-line) $= \frac{9}{8}b$	$= 4101.3$		4101.2	-0.1

tween these fundamental numbers and the corresponding atomic weights.

The wavelengths of the first four hydrogen lines are obtained by multiplying the fundamental number $b = 3645.6$ in succession by the coefficients $9/5$; $4/3$; $25/21$; and $9/8$. At first it appears that these four coefficients

The deviations of the formula from Ångström's measurements amount in the most unfavorable case to not more than $1/40,000$ of a wavelength, a deviation which very likely is within the limits of the possible errors of observation and is really striking evidence for the great scientific skill and

SOURCE. From *Annalen de Physik and Chemie* **25**: 80 (1885). Translated by H. A. Boorse and L. Motz, Ed. of *The World of the Atom* (New York: Basic Book; 1966), p. 365.

care with which Ångström must have worked.

From the formula we obtained for a fifth
hydrogen line $49/45 \times 3645.6 = 3969.65 \times 10^{-8}$ cm. I knew nothing of such a fifth line,
which must lie within the visible part of the
spectrum just before H_i (which according to
Ångström has a wavelength 3968.1); and I
had to assume that either the temperature
relations were not favorable for the emission
of this line or that the formula was not
generally applicable.

On communicating this to Professor
Hagenbach he informed me that many more
hydrogen lines are known, which have been
measured by Vogel and by Huggins in the
violet and the ultraviolet parts of the
hydrogen spectrum and in the spectrum of
the white stars; he was kind enough himself
to compare the wavelengths thus determined
with my formula and to send me the result.

While the formula in general gives somewhat larger numbers than those contained
in the published lists of Vogel and of
Huggins, the difference between the calculated and the observed wavelengths is so
small that the agreement is striking in the
highest degree. Comparisons of wavelengths
measured by different investigators show
in general no exact agreement; and yet the
observations of one man may be made to
agree with those of another by a slight
reduction in an entirely satisfactory way.
[Here Balmer compares the wavelengths
calculated from his formula with the measurements of several observers—Ed.]

These comparisons show that the formula
holds also for the fifth hydrogen line, which
lies just before the first Fraunhofer H-line
(which belongs to calcium). It also appears
that Vogel's hydrogen lines and the corresponding Huggins lines of the white stars
can be represented by the formula very satis-
factorily. We may almost certainly assume
that the other lines of the white stars which
Huggins found farther on in the ultraviolet
part of the spectrum will be expressed by the
formula. I lack knowledge of the wavelengths. Using the fundamental number
3645.6, we obtain according to the formula

for the ninth and the following hydrogen
lines up to the fifteenth:

$$\tfrac{121}{117}b = 3770.24 \qquad \tfrac{225}{221}b = 3711.58$$
$$\tfrac{36}{35}b = 3749.76 \qquad \tfrac{64}{63}b = 3703.46$$
$$\tfrac{169}{165}b = 3733.98 \qquad \tfrac{289}{285}b = 3696.76$$
$$\tfrac{49}{48}b = 3721.55$$

Whether the hydrogen lines of the white
stars agree with the formula to this point or
whether other numerical relations gradually
replace it can only be determined by observation.

I add to what I have said a few questions
and conclusions.

Does the above formula hold only for the
single chemical element hydrogen, and will
not other fundamental numbers in the
spectral lines of other elements be found
which are peculiar to those elements? If not,
we may perhaps assume that the formula
that holds for hydrogen is a special case of
a more general formula which under certain
conditions goes over into the formula for
the hydrogen lines.

None of the hydrogen lines which correspond to the formula when $n = 3, 4$, and
so on, and which may be called lines of the
third or fourth order, is found in any
spectrum as yet known; they must be emitted
under entirely new relations of temperature
and pressure if they are to become perceptible.

If the formula holds for all the principal
lines of the hydrogen spectrum with $n = 2$, it
follows that these spectral lines on the
ultraviolet side approach the wavelength
3645.6 in a more closely packed series, but
they can never pass this limiting value, while
the C-line also is the extreme line on the red
side. Only if lines of higher orders are present
can lines be found on the infrared side.

The formula has no relation, so far as can
be shown, with the very numerous lines of
the second hydrogen spectrum which Has-
selberg has published in the *Mémoires de
l'Academie des Sciences de St. Petersbourg*, 1882.
For certain values of pressure and temperature hydrogen may easily change in such a
way that the law of formation of its spectral
lines becomes entirely different.

There are great difficulties in the way of finding the fundamental numbers for other chemical elements, such as oxygen or carbon, by means of which their principal spectral lines can be determined from the formula. Only extremely exact determinations of wavelengths of the most prominent lines of an element can give a common base for these wavelengths, and without such a base it seems as if all trials and guesses will be in vain. Perhaps by using a different graphical construction of the spectrum a way will be found to make progress in such investigations.

6.3   *On the Constitution of Atoms and Molecules*
Niels Bohr

INTRODUCTION

In order to explain the results of experiments on scattering of α rays by matter Prof. Rutherford has given a theory of the structure of atoms. According to this theory, the atoms consist of a positively charged nucleus surrounded by a system of electrons kept together by attractive forces from the nucleus; the total negative charge of the electrons is equal to the positive charge of the nucleus. Further, the nucleus is assumed to be the seat of the essential part of the mass of the atom, and to have linear dimensions exceedingly small compared with the linear dimensions of the whole atom. The number of electrons in an atom is deduced to be approximately equal to half the atomic weight. Great interest is to be attributed to this atom-model; for, as Rutherford has shown, the assumption of the existence of nuclei, as those in question, seems to be necessary in order to account for the results of the experiments on large angle scattering of the α rays.

In an attempt to explain some of the properties of matter on the basis of this atom-model we meet, however, with difficulties of a serious nature arising from the apparent instability of the system of electrons: difficulties purposely avoided in atom-models previously considered, for instance, in the one proposed by Sir J. J. Thomson. According to the theory of the latter the atom consists of a sphere of uniform positive electrification, inside which the electrons move in circular orbits.

The principal difference between the atom-models proposed by Thomson and Rutherford consists in the circumstance that the forces acting on the electrons in the atom-model of Thomson allow of certain configurations and motions of the electrons for which the system is in a stable equilibrium; such configurations, however, apparently do not exist for the second atom-model. The nature of the difference in question will perhaps be most clearly seen by noticing that among the quantities characterizing the first atom a quantity appears—the radius of the positive sphere—of dimensions of a length and of the same order of magnitude as the linear extension of the atom, while such a length does not appear among the quantities characterizing the second atom, viz. the charges and masses of the electrons and the positive nucleus; nor can it be determined solely by help of the latter quantities.

The way of considering a problem of this kind has, however, undergone essential alterations in recent years owing to the development of the theory of the energy radiation, and the direct affirmation of the new assumptions introduced in this theory, found by experiments on very different phenomena such as specific heats, photoelectric effect, Röntgen-rays, etc. The result of the discussion of these questions seems to be a general

SOURCE. From the *Philosophical Magazine* **26**: 1 (1913).

acknowledgment of the inadequacy of the classical electrodynamics in describing the behavior of systems of atomic size. Whatever the alteration in the laws of motion of the electrons may be, it seems necessary to introduce in the laws in question a quantity foreign to the classical electrodynamics, *i.e.* Planck's constant, or as it often is called the elementary quantum of action. By the introduction of this quantity the question of the stable configuration of the electrons in the atoms is essentially changed, as this constant is of such dimensions and magnitude that it, together with the mass and charge of the particles, can determine a length of the order of magnitude required.

This paper is an attempt to show that the application of the above ideas to Rutherford's atom-model affords a basis for a theory of the constitution of atoms. It will further be shown that from this theory we are led to a theory of the constitution of molecules.

In the present first part of the paper the mechanism of the binding of electrons by a positive nucleus is discussed in relation to Planck's theory. It will be shown that it is possible from the point of view taken to account in a simple way for the law of the line spectrum of hydrogen. Further, reasons are given for a principal hypothesis on which the considerations contained in the following parts are based.

I wish here to express my thanks to Prof. Rutherford for his kind and encouraging interest in this work.

BINDING OF ELECTRONS BY POSITIVE NUCLEI[1]

General Considerations. The inadequacy of the classical electrodynamics in accounting for the properties of atoms from an atom-model as Rutherford's, will appear very clearly if we consider a simple system consisting of a positively charged nucleus of very small dimensions and an electron describing closed orbits around it. For simplic-

ity, let us assume that the mass of the electron is negligibly small in comparison with that of the nucleus, and further, that the velocity of the electron is small compared with that of light.

Let us at first assume that there is no energy radiation. In this case the electron will describe stationary elliptical orbits. The frequency of revolution ω and the major-axis of the orbit $2a$ will depend on the amount of energy W which must be transferred to the system in order to remove the electron to an infinitely great distance apart from the nucleus. Denoting the charge of the electron and of the nucleus by $-e$ and Ze, respectively, and the mass of the electron by m, we thus get

$$\omega = \frac{\sqrt{2}\, W^{3/2}}{\pi\, Ze^2\, \sqrt{m}}, \quad 2a = \frac{Ze^2}{W} \tag{1}$$

Further, it can easily be shown that the mean value of the kinetic energy of the electron taken for a whole revolution is equal to W. We see that if the value of W is not given, there will be no values of ω and a characteristic for the system in question.

Let us now, however, take the effect of the energy radiation into account, calculated in the ordinary way from the acceleration of the electron. In this case the electron will no longer describe stationary orbits. "W" will continuously increase, and the electron will approach the nucleus describing orbits of smaller and smaller dimensions, and with greater and greater frequency; the electron on the average gaining in kinetic energy at the same time as the whole system loses energy. This process will go on until the dimensions of the orbit are of the same order of magnitude as the dimensions of the electron or those of the nucleus. A simple calculation shows that the energy radiated out during the process considered will be enormously great compared with that radiated out by ordinary molecular processes.

It is obvious that the behavior of such a system will be very different from that of

---

[1] Some of the notation in this section has been changed to conform with modern practice. [Ed.]

an atomic system occurring in nature. In the first place, the actual atoms in their permanent state seem to have absolutely fixed dimensions and frequencies. Further, if we consider any molecular process, the result seems always to be that after a certain amount of energy characteristic for the systems in question is radiated out, the systems will again settle down in a stable state of equilibrium, in which the distances apart of the partices are of the same order of magnitude as before the process.

Now the essential point in Planck's theory of radiation is that the energy radiation from an atomic system does not take place in the continuous way assumed in the ordinary electrodynamics, but that it, on the contrary, takes place in distinctly separated emissions, the amount of energy radiated out from an atomic vibrator of frequency v in a single emission being equal to nhv, where n is an integer number, and h is a universal constant.

Returning to the simple case of an electron and a positive nucleus considered above, let us assume that the electron at the beginning of the interaction with the nucleus was at a great distance apart from the nucleus, and had no sensible velocity relative to the latter. Let us further assume that the electron after the interaction has taken place has settled down in a stationary orbit around the nucleus. We shall, for reasons referred to later, assume that the orbit in question is circular; this assumption will, however, make no alteration in the calculations for systems containing only a single electron.

Let us now assume that, during the binding of the electron, a homogeneous radiation is emitted of a frequency v equal to half the frequency of revolution of the electron in its final orbit; then, from Planck's theory, we might expect that the amount of energy emitted by the process considered is equal to mhv, where h is Planck's constant and n an integer number. If we assume that the radiation emitted is homogeneous, the second assumption concerning the frequency of the radiation suggests itself, since the frequency of revolution of the electron at the beginning of the emission is o. . . . The question, how-

ever, of the rigorous validity of both assumptions, and also of the application made of Planck's theory, will be more closely discussed in [the section entitled "General Considerations Continued."]

Putting

$$W = nh\frac{\omega}{2}, \tag{2}$$

we get by help of the formula (1)

$$W = \frac{2\pi^2 m Z^2 e^4}{nh^2},$$

$$\omega = \frac{4\pi^2 m Z^2 e^4}{n^3 h^3}, \tag{3}$$

$$2a = \frac{n^2 h^2}{2\pi^2 m Z e^2}$$

If in these expressions we give n different values, we get a series of values for W, ω and a corresponding to a series of configurations of the system. According to the above considerations, we are led to assume that these configurations will correspond to states of the system in which there is no radiation of energy; states which consequently will be stationary as long as the system is not disturbed from outside. We see that the value of W is greatest if n has its smallest value 1. This case will therefore correspond to the most stable state of the system, i.e. will correspond to the binding of the electron for the breaking up of which the greatest amount of energy is required.

Putting in the above expressions $n = 1$ and $Z = 1$, and introducing the experimental values

$$e = 4.7 \times 10^{-10} \text{ statcoul,}$$

$$\frac{e}{m} = 5.31 \times 10^{17} \text{ statcoul/g,}$$

$$h = 6.5 \times 10^{-27} \text{ erg-sec,}$$

we get

$$2a = 1.1 \times 10^{-8} \text{ cm,}$$

$$\omega = 6.2 \times 10^{15} \text{ sec}^{-1},$$

$$W = 13 \text{ eV.}$$

We see that these values are of the same

order of magnitude as the linear dimensions of the atoms, the optical frequencies, and the ionization energies.

The general importance of Planck's theory for the discussion of the behavior of atomic systems was originally pointed out by Einstein. The considerations of Einstein have been developed and applied on a number of different phenomena, especially by Stark, Nernst, and Sommerfeld. The agreement as to the order of magnitude between values observed for the frequencies and dimensions of the atoms, and values for these quantities calculated by considerations similar to those given above, has been the subject of much discussion. It was first pointed out by Haas, in an attempt to explain the meaning and the value of Planck's constant on the basis of J. J. Thomson's atom-model, by help of the linear dimensions and frequency of an hydrogen atom.

Systems of the kind considered in this paper, in which the forces between the particles vary inversely as the square of the distance, are discussed in relation to Planck's theory by J. W. Nicholson. In a series of papers this author has shown that it seems to be possible to account for lines of hitherto unknown origin in the spectra of the stellar nebulae and that of the solar corona, by assuming the presence in these bodies of certain hypothetical elements of exactly indicated constitution. The atoms of these elements are supposed to consist simply of a ring of a few electrons surrounding a positive nucleus of negligibly small dimensions. The ratios between the frequencies corresponding to the lines in question are compared with the ratios between the frequencies corresponding to different modes of vibration of the ring of electrons. Nicholson has obtained a relation to Planck's theory showing that the ratios between the wave length of different sets of lines of the coronal spectrum can be accounted for with great accuracy by assuming that the ratio between the energy of the system and the frequency of rotation of the ring is equal to an integer multiple of Planck's constant. The quantity Nicholson refers to as the

energy is equal to twice the quantity which we have denoted above by W. In the latest paper cited Nicholson has found it necessary to give the theory a more complicated form, still however, representing the ratio of energy to frequency by a simple function of whole numbers.

The excellent agreement between the calculated and observed values of the ratios between the wave lengths in question seems a strong argument in favor of the validity of the foundation of Nicholson's calculations. Serious objections, however, may be raised against the theory. These objections are intimately connected with the problem of the homogeneity of the radiation emitted. In Nicholson's calculations the frequency of lines in a line-spectrum is identified with the frequency of vibration of a mechanical system in a distinctly indicated state of equilibrium. As a relation from Planck's theory is used, we might expect that the radiation is sent out in quanta; but systems like those considered, in which the frequency is a function of the energy, cannot emit a finite amount of a homogeneous radiation; for, as soon as the emission of radiation is started, the energy and also the frequency of the system are altered. Further, according to the calculation of Nicholson, the systems are unstable for some modes of vibration. Apart from such objections—which may be only formal—it must be remarked, that the theory in the form given does not seem to be able to account for the well-known laws of Balmer and Rydberg connecting the frequencies of the lines in the line-spectra of the ordinary elements.

It will now be attempted to show that the difficulties in question disappear if we consider the problems from the point of view taken in this paper. Before proceeding it may be useful to restate briefly the ideas characterizing the calculations [following formulas (2) and (3)]. The principal assumptions used are:

(1) That the dynamical equilibrium of the systems in the stationary states can be discussed by help of the ordinary mechanics,

while the passing of the systems between different stationary states cannot be treated on that basis.

(2) That the latter process is followed by the emission of a *homogeneous* radiation, for which the relation between the frequency and the amount of energy emitted is the one given by Planck's theory.

The first assumption seems to present itself; for it is known that the ordinary mechanics cannot have an absolute validity, but will only hold in calculations of certain mean values of the motion of the electrons. On the other hand, in the calculations of the dynamical equilibrium in a stationary state in which there is no relative displacement of the particles, we need not distinguish between the actual motions and their mean values. The second assumption is in obvious contrast to the ordinary ideas of electrodynamics, but appears to be necessary in order to account for experimental facts.

In the calculations [just referred to] we have further made use of the more special assumptions, viz. that the different stationary states correspond to the emission of a different number of Planck's energy quanta, and that the frequency of the radiation emitted during the passing of the system from a state in which no energy is yet radiated out to one of the stationary states, is equal to half the frequency of revolution of the electron in the latter state. We can, however, also arrive at the expressions (3) for the stationary states by using assumptions of somewhat different form. We shall, therefore, postpone the discussion of the special assumptions, and first show how by the help of the above principal assumptions, and of the expressions (3) for the stationary states, we can account for the line-spectrum of hydrogen.

Emission of Line-Spectra. SPECTRUM OF HYDROGEN. General evidence indicates that an atom of hydrogen consists simply of a single electron rotating round a positive nucleus of charge e. The re-formation of a hydrogen atom, when the electron has been removed to great distances away from the nucleus—*e.g.* by the effect of electrical dis-

charge in a vacuum tube—will accordingly correspond to the binding of an electron by a positive nucleus considered [earlier]. If in (3) we put $Z = 1$, we get for the total amount of energy radiated out by the formation of one of the stationary states,

$$W_n = \frac{2\pi^2 m e^4}{h^2 n^2}$$

The amount of energy emitted by the passing of a system from a state corresponding to $n = n_1$ to one corresponding to $n = n_2$, is consequently

$$W_{n_2} - W_{n_1} = \frac{2\pi^2 m e^4}{h^2}\left(\frac{1}{n_2{}^2} - \frac{1}{n_1{}^2}\right).$$

If now we suppose that the radiation in question is homogeneous, and that the amount of energy emitted is equal to $h v$, where v is the frequency of the radiation, we get

$$W_{n_2} - W_{n_1} = h v,$$

and from this

$$v = \frac{2\pi^2 m e^4}{h^3}\left(\frac{1}{n_2{}^2} - \frac{1}{n_1{}^2}\right). \tag{4}$$

We see that this expression accounts for the law connecting the lines in the spectrum of hydrogen. If we put $n_2 = 2$ and let n_1 vary, we get the ordinary Balmer series. If we put $n_2 = 3$, we get the series in the infra-red observed by Paschen and previously suspected by Ritz. If we put $n_2 = 1$ and $n_1 = 4, 5, \cdots$, we get series respectively in the extreme ultra-violet and the extreme infra-red, which are not observed, but the existence of which may be expected.

The agreement in question is quantitative as well as qualitative. Putting

$$e = 4.7 \times 10^{-10} \text{ statcoul,}$$

$$\frac{e}{m} = 5.31 \times 10^{17} \text{ statcoul/g,}$$

and $h = 6.5 \times 10^{-27}$ erg-sec,

we get

$$\frac{2\pi^2 m e^4}{h^3} = 3.1 \times 10^{15} \text{ sec}^{-1}.$$

The observer value for the factor outside

the bracket in the formula (4) is

$$3.290 \times 10^{15} \text{ sec.}^{-1}.$$

The agreement between the theoretical and observed values is inside the uncertainty due to experimental errors in the constants entering in the expression for the theoretical value. We shall return to consider the possible importance of the agreement in question.

It may be remarked that the fact, that it has not been possible to observe more than 12 lines of the Balmer series in experiments with vacuum tubes, while 33 lines are observed in the spectra of some celestial bodies, is just what we should expect from the above theory. According to the equation (3) the diameter of the orbit of the electron in the different stationary states is proportional to n^2. For $n = 12$ the diameter is equal to 1.6×10^{-6} cm, or equal to the mean distance between the molecules in a gas at a pressure of about 7 mm mercury; for $n = 33$ the diameter is equal to 1.2×10^{-3} cm, corresponding to the mean distance of the molecules at a pressure of about 0.02 mm mercury. According to the theory the necessary condition for the appearance of a great number of lines is therefore a very small density of the gas; for simultaneously to obtain an intensity sufficient for observation the space filled with the gas must be very great. If the theory is right, we may therefore never expect to be able in experiments with vacuum tubes to observe the lines corresponding to high numbers of the Balmer series of the emission spectrum of hydrogen; it might, however, be possible to observe the lines by investigation of the absorption spectrum of this gas [see section entitled "Absorption of Radiation"].

It will be observed that we in the above way do not obtain other series of lines, generally ascribed to hydrogen; for instance, the series first observed by Pickering in the spectrum of the star ζ Puppis, and the set of series recently found by Fowler by experiments with vacuum tubes containing a mixture of hydrogen and helium. We shall, however, see that, by help of the above theory, we can account naturally for these series of lines if we ascribe them to helium.

A neutral atom of the latter element consists, according to Rutherford's theory, of a positive nucleus of charge $2e$ and two electrons. Now considering the binding of a single electron by a helium nucleus, we get, putting $Z = 2$ in the expressions (3), and proceeding in exactly the same way as above,

$$\nu = \frac{8\pi^2 m e^4}{h^3}\left(\frac{1}{n_2^2} - \frac{1}{n_1^2}\right) = $$
$$\frac{2\pi^2 m e^4}{h^3}\left(\frac{1}{\left(\frac{n_2}{2}\right)^2} - \frac{1}{\left(\frac{n_1}{2}\right)^2}\right).$$

If we in this formula put $n_2 = 1$ or $n_2 = 2$, we get series of lines in the extreme ultra-violet. If we put $n_2 = 3$, and let n_1 vary, we get a series which includes 2 of the series observed by Fowler, and denoted by him as the first and second principal series of the hydrogen spectrum. If we put $n_2 = 4$, we get the series observed by Pickering in the spectrum of ζ Puppis. Every second of the lines in this series is identical with a line in the Balmer series of the hydrogen spectrum; the presence of hydrogen in the star in question may therefore account for the fact that these lines are of a greater intensity than the rest of the lines in the series. The series is also observed in the experiments of Fowler, and denoted in his paper as the Sharp series of the hydrogen spectrum. If we finally in the above formula put $n_2 = 5$, 6, ..., we get series, the strong lines of which are to be expected in the infra-red.

The reason why the spectrum considered is not observed in ordinary helium tubes may be that in such tubes the ionization of helium is not so complete as in the star considered or in the experiments of Fowler, where a strong discharge was sent through a mixture of hydrogen and helium. The condition for the appearance of the spectrum is, according to the above theory, that helium atoms are present in a state in which they have lost both their electrons. Now we must assume that the amount of energy to be used in removing the second electron from a helium atom is much greater than that to be used in removing the first. Further, it is known from experiments on positive rays, that hydrogen atoms can acquire a negative charge; therefore the presence of hydrogen in the experiments of Fowler may effect that more electrons are removed from some of the helium atoms than would be the case if only helium were present.

General Considerations Continued. . . . The possibility of an emission of a radiation of a frequency [$v = n\omega$] may also be interpreted from analogy with the ordinary electro-dynamics, as an electron rotating round a nucleus in an elliptical orbit will emit a radiation which according to Fourier's theorem can be resolved into homogeneous components, the frequencies of which are $n\omega$, if ω is the frequency of revolution of the electron.

We are thus led to assume that the interpretation of the equation (2) is not that the different stationary states correspond to an emission of different numbers of energy-quanta, but that the frequency of the energy emitted during the passing of the system from a state in which no energy is yet radiated out to one of the different stationary states, is equal to different multiples of $\omega/2$, where ω is the frequency of revolution of the electron in the state considered. From this assumption we get exactly the same expressions as before for the stationary states, and from these by help of the principal assumptions [given earlier] the same expression for the law of the hydrogen spectrum. Consequently we may regard our preliminary considerations only as a simple form of representing the results of the theory.

Before we leave the discussion of this question, we shall for a moment return to the question of the significance of the agreement between the observed and calculated values of the constant entering in the expressions (4) for the Balmer series of the hydrogen spectrum. From the above consideration it will follow that, taking the starting-point in the form of the law of the hydrogen spectrum and assuming that the different lines correspond to a homogeneous radiation emitted during the passing between different stationary states, we shall arrive at exactly

the same expression for the constant in question as that given by (4), if we only assume (1) that the radiation is sent out in quanta $h\nu$, and (2) that the frequency of the radiation emitted during the passing of the system between successive stationary states will coincide with the frequency of revolution of the electron in the region of slow vibrations.

As all the assumptions used in this latter way of representing the theory are of what we may call a qualitative character, we are justified in expecting—if the whole way of considering is a sound one—an absolute agreement between the values calculated and observed for the constant in question, and not only an approximate agreement. The formula (4) may therefore be of value in the discussion of the results of experimental determinations of the constants e, m, and h.

While there obviously can be no question of a mechanical foundation of the calculations given in this paper, it is, however, possible to give a very simple interpretation of the result of the calculation [following formulas (2) and (3)] by help of symbols taken from the ordinary mechanics. Denoting the angular momentum of the electron round the nucleus by M, we have immediately for a circular orbit $\pi M = T/\omega$, where ω is the frequency of revolution and T the kinetic energy of the electron; for a circular orbit we further have $T = W$ and from (2), we consequently get

$$M = nh/2\pi = n\hbar$$

where

$$\hbar = 1.04 \times 10^{-27} \text{ erg-sec.}$$

If we therefore assume that the orbit of the electron in the stationary states is circular, the result of the calculation can be expressed by the simple condition: that the angular momentum of the electron round the nucleus in a stationary state of the system is equal to an entire multiple of a universal value, independent of the charge on the nucleus. The possible importance of the angular momentum in the discussion of atomic systems in relation to Planck's theory is emphasized by Nicholson.

The great number of different stationary states we do not observe except by investigation of the emission and absorption of radiation. In most of the other physical phenomena, however, we only observe the atoms of the matter in a single distinct state, *i.e.* the state of the atoms at low temperature. From the preceding considerations we are immediately led to the assumption that the "permanent" state is the one among the stationary states during the formation of which the greatest amount of energy is emitted. According to the equation (3), this state is the one which corresponds to $n = 1$.

Absorption of Radiation. In order to account for Kirchhoff's law it is necessary to introduce assumptions on the mechanism of absorption of radiation which correspond to those we have used considering the emission. Thus we must assume that a system consisting of a nucleus and an electron rotating round it under certain circumstances can absorb a radiation of a frequency equal to the frequency of the homogeneous radiation emitted during the passing of the system between different stationary states. Let us consider the radiation emitted during the passing of the system between two stationary states A_1 and A_2 corresponding to values for n equal to n_1 and n_2, $n_1 > n_2$. As the necessary condition for an emission of the radiation in question was the presence of systems in the state A_1, we must assume that the necessary condition for an absorption of the radiation is the presence of systems in the state A_2.

These considerations seem to be in conformity with experiments on absorption in gases. In hydrogen gas at ordinary conditions for instance there is no absorption of a radiation of a frequency corresponding to the line-spectrum of this gas; such an absorption is only observed in hydrogen gas in a luminous state. This is what we should expect according to the above. We have assumed that the radiation in question was emitted during the passing of the systems between stationary states corresponding to $n \geq 2$. The state of the atoms in hydrogen gas at ordinary conditions

should, however, correspond to $n = 1$; furthermore, hydrogen atoms at ordinary conditions combine into molecules, *i.e.* into systems in which the electrons have frequencies different from those in the atoms. From the circumstance that certain substances in a non-luminous state, as, for instance, sodium vapor, absorb radiation corresponding to lines in the line-spectra of the substances, we may, on the other hand, conclude that the lines in question are emitted during the passing of the system between two states, one of which is the permanent state.

How much the above considerations differ from an interpretation based on the ordinary electrodynamics is perhaps most clearly shown by the fact that we have been forced to assume that a system of electrons will absorb a radiation of a frequency different from the frequency of vibration of the electrons calculated in the ordinary way. It may in this connection be of interest to mention a generalization of the considerations to which we are led by experiments on the photo-electric effect, and which may be able the throw some light on the problem in question. Let us consider a state of the system in which the electron is free, *i.e.* in which the electron possesses kinetic energy sufficient to remove to infinite distances from the nucleus. If we assume that the motion of the electron is governed by the ordinary mechanics and that there is no (sensible) energy radiation, the total energy of the system—as in the above considered stationary states—will be constant. Further, there will be perfect continuity between the two kinds of states, as the difference between frequency and dimensions of the systems in successive stationary states will diminish without limit if τ increases. In the following considerations we shall for the sake of brevity refer to the two kinds of states in question as "mechanical" states; by this notation only emphasizing the assumption that the motion of the electron in both cases can be accounted for by the ordinary mechanics.

Tracing the analogy between the two kinds of mechanical states, we might now expect the possibility of an absorption of radiation, not only corresponding to the passing of the system between two different stationary states, but also corresponding to the passing between one of the stationary states and a state in which the electron is free; and as above, we might expect that the frequency of this radiation was determined by the equation $E = h\nu$, where E is the difference between the total energy of the system in the two states. As it will be seen, such an absorption of radiation is just what is observed in experiments on ionization by ultra-violet light and by Röntgen rays. Obviously, we get in this way the same expression for the kinetic energy of an electron ejected from an atom by photo-electric effect as that deduced by Einstein, *i.e.* $T = h\nu - W$, where T is the kinetic energy of the electron ejected, and W the total amount of energy emitted during the original binding of the electron. . . .

Experiments on the phenomena of X-rays suggest that not only the emission and absorption of radiation cannot be treated by the help of the ordinary electrodynamics, but not even the result of a collision between two electrons of which the one is bound in an atom. This is perhaps most clearly shown by some very instructive calculations on the energy of β-particles emitted from radioactive substances recently published by Rutherford. These calculations strongly suggest that an electron of great velocity in passing through an atom and colliding with the electrons bound will lose energy in distinct finite quanta. As is immediately seen, this is very different from what we might expect if the result of the collisions was governed by

the usual mechanical laws. The failure of the classical mechanics in such a problem might also be expected beforehand from the absence of anything like equipartition of kinetic energy between free electrons and electrons bound in atoms. From the point of view of the "mechanical" states we see, however, that the following assumption—which is in accord with the above analogy—might be able to account for the result of Rutherford's calculation and for the absence of equipartition of kinetic energy: two colliding electrons, bound or free, will, after the collision as well as before, be in mechanical states. Obviously, the introduction of such an assumption would not make any alteration necessary in the classical treatment of a collision between two free particles. But, considering a collision between a free and a bound electron, it would follow that the bound electron by the collision could not acquire a less amount of energy than the difference in energy corresponding to successive stationary states, and consequently that the free electron which collides with it could not lose a less amount.

The preliminary and hypothetical character of the above considerations needs not to be emphasized. The intention, however, has been to show that the sketched generalization of the theory of the stationary states possibly may afford a simple basis of representing a number of experimental facts which cannot be explained by help of the ordinary electrodynamics, and that the assumptions used do not seem to be inconsistent with experiments on phenomena for which a satisfactory explanation has been given by the classical dynamics and the wave theory of light.

6.4   *The Chance Discovery of X-rays*

A. Sutcliffe and A. P. D. Sutcliffe

In the later part of the nineteenth century many scientists were studying the remarkable effects produced when electricity is discharged in a partial vacuum and were greatly

helped in their work by the invention of Crookes's tube in 1879. This is a long cylindrical tube of glass containing two terminals. One is connected through an induction coil

SOURCE. From *Stories from Science* (Cambridge Univ. Press, 1962) Vol. 2, Chapter 39.

to the positive pole of a battery and is called the anode. The other is similarly connected to the negative pole and is called the cathode. Practically all the air has been pumped out of the tube by a vacuum pump which has been connected to a small outlet tube which is then sealed off.

When the current is switched on the walls of the tube become a ghostly, apple-green, shimmering color, or, as scientists say, they become fluorescent. From their observations Sir William Crookes and others deduced that this fluorescence was caused by rays which come from the cathode and strike the inner walls of the tube.

A few years later, Professor Lenard discovered that these cathode rays, as they are called, whilst they were stopped by even a thin wall of glass, would pass through aluminium foil; and so he devized an improved kind of tube which had a window of aluminium inserted in its wall. Lenard found that the cathode rays passed out of the tube and into the air where they could be detected for a very short distance only.

A few other substances besides glass fluoresce when cathode rays fall on them, one being barium platino-cyanide; and towards the end of the century many scientists were using a screen made of a sheet of paper or cardboard coated with tiny crystals of this substance in their experiments on cathode rays.

One day towards the end of the year 1895 Professor Röntgen of Wurtzberg in Bavaria was experimenting with an improved Crookes's tube. He had darkened the laboratory by drawing the blinds and had covered the tube with a shield of black cardboard through which no light, however intense, could pass. It was thoroughly dark in the laboratory, therefore, when he switched on the coil. He then happened to glance round and saw that one of the fluorescent screens, which was standing on the table a few feet away, was glowing brightly. This sight puzzled him, for the tube was thoroughly blacked out and there were no signs of cathode rays escaping from it. Yet it seemed as if some kind of rays were coming, in a direct line, from the tube

to the screen; for, as he soon proved, they could not have come from anywhere else. He moved the screen nearer, and found that it continued to glow if he kept it pointing in the same direction.

Gradually he became convinced that a new kind of ray was being given out by the fluorescent tube; a ray which could pass through thick black paper. Perhaps, he reflected, it could pass through other things. He put a piece of wood directly between the tube and the screen. The screen glowed—the rays were passing through the wood. He replaced it with a piece of cloth, again the screen glowed—the rays were passing through the cloth. He then interposed a small piece of metal; but the metal cast a shadow of its own shape on the screen—evidently the mysterious rays were not passing through the metal.

Then he had a most brilliant yet simple idea. Rays of ordinary light, he reflected, affect a photographic plate; perhaps these mysterious rays would do so also. To test this idea he put a photographic plate in the path which the rays would take and persuaded his wife to place her hand between the tube and the plate. He switched on the coil. When the plate was developed he and his wife saw the bones appearing distinctly with the flesh dimly outlined around them. This was the first time a photograph of the skeleton of any living person had been taken and it must have been a staggering sight for the lady to see this photograph of part of her own skeleton.

.   .   .   .   .

Röntgen called the rays X-rays, because little was known about them and mathematicians always use the letter X to denote an unknown quantity. Later, an attempt was made to give them the name of "Röntgen rays" which would be more appropriate and would honor the discoverer. But the name was not used for long, in this country at any rate. As the editor of a scientific journal has written Professor Röntgen did not have the good fortune of having a name which sounded pleasant when spoken. (Its pronunciation is "*runtyen*," with the "u" as in ruck, and the "e" as in "peck.")

As with many other chance observations, it seems almost unbelievable that no one before Röntgen had observed what he had just seen, particularly because many astute scientists had been doing experiments with Crookes's tube for fifteen years or more before 1895.

After Röntgen's publication of the details of his discovery Sir William Crookes realized how near he himself had been to making the same discovery. Thus, as Lord Rayleigh, another brilliant physicist, wrote:

It was a source of great annoyance to Crookes that he missed the discovery of the X-rays. According to the account he gave in my hearing, he had definitely found previously unopened boxes of plates in his laboratory to be fogged for no assignable reason, and, acting I suppose in accordance with the usual human instinct of blaming someone else when things go wrong, he complained to the makers who naturally had no satisfactory explanation to offer. I believe that it was only after Röntgen's discovery that he connected this with the use of highly exhausted vacuum tubes in the neighborhood.

Röntgen first described his discovery to the Physico-Medical Society of Wurzburg in December, 1896, and shortly afterwards particulars were released to the press. The discovery created a great sensation in many countries. Early in January of the following year, a celebrated British professor of physics described the discovery in a journal which was read by most educated people. He began by stating that a very singular scientific discovery had just been made by Professor Röntgen of Wurzburg in Bavaria, who had succeeded in finding a means of photographing an object of metal which was completely enclosed in a wooden box with more ease than if it had been in a glass one. Röntgen, he added, could also photograph a man's skeleton through the skin, flesh and clothes, which under these rays were photographically transparent, while the bones were opaque, like metal.

"The discovery," he continued,

adds one more to the marvels of science. To photograph in total darkness seems inexplicable,

but that we should be able to photograph through walls of wood, or through solid opaque bodies, is little short of a miracle. We shall now be able to realize Dickens' fancy when he made Scrooge perceive through Marley's body the two brass buttons on the back of his coat. We shall now be able to discover photographically the position of a bullet in a man's body. Even stone walls will not a prison make to the revelations of the camera.

Lord Rayleigh also described this announcement when he wrote many years later as follows: "Röntgen's discovery was followed by a greater outburst of enthusiasm than any other experimental discovery before or since. Most physical laboratories had the means of taking X-ray photographs of hands and this was tried on all sides." For example, almost immediately after hearing about the discovery, Professor J. J. Thomson gave a lecture at the Cavendish Laboratory during the course of which a photograph of the hands of one of the ladies present was taken, developed and shown during the lecture.

It is not surprising, therefore, that the "man in the street" was led to think that Röntgen had invented a kind of camera which would produce a photograph showing the bones of the body, or that some newspapers called the discovery "a revolution in photography." Indeed, the editor of a scientific journal, commenting on this aspect, wrote: "There are very few persons who would care to sit for a portrait which would show 'only the bones and rings on the fingers."

Some people became alarmed at the thought that the new invention would enable, for example, a street photographer to take intimate photographs "which would be an insult to decency." Indeed it is said that an enterprising firm in London not only advertized underwear guaranteed to be X-ray proof but actually made a small fortune out of their sales of it.

Punch produced these lines:

> O, Röntgen, then the news is true
> And not a trick of idle rumour,
> That bids us each beware of you
> And your grim and graveyard humour.
> We do not want, like Dr. Swift
> To take our flesh off and pose in

Our bones, or show each little rift
And joint for you to poke your nose in.
We only crave to contemplate
Each other's usual full-dress photo,
Your worse than "altogether" state
 Of portraiture we bar in toto.
The fondest swain would scarcely prize
A picture of his lady's framework;
To gaze on this with yearning eyes
Would probably be voted tame work.

Meanwhile, serious students had begun to appreciate the tremendous boon which these rays might confer on mankind. Indeed the importance of X-rays in surgery was at once realized by doctors; and it is perhaps worthy of note that Röntgen's first announcement of his discovery of the rays was read before a Medical Society—that of Wurzbur. Thus "surgery was the first art to be closely identified with X-rays. On January 20, 1896, a doctor in Berlin detected a glass splinter in a finger; on February 7 a doctor at Liverpool X-rayed a bullet in a boy's head; in April a professor at Manchester took X-ray photographs through the head of a woman who had been shot."

Years later, Sir J. J. Thomson summarized the value of X-rays to surgery in these words. "Few have done more to relieve human suffering than Röntgen and those who by developing the application of X-rays, to surgery have supplied the surgeon with the most powerful means of diagnosis."

Doctors have other uses for X-rays, for example, to destroy cancer cells and such diseases as ringworm. Industry has also found them of value, especially in metallurgy, where they can detect flaws and cracks in iron structures which have been cast or molded.

6.5   *Lasers—The Light Fantastic*

          Max Gunther

The sign on the gray metal door warns in red: DANGER. LASER LIGHT. Within, a smaller sign quietly continues the warning: DANGER. INVISIBLE BEAMS. HIGH ENERGY.

There are laboratories such as this all over the country—indeed, all over the world. This one belongs to the Perkin-Elmer Corporation at Norwalk, Connecticut. It is a long, darkened room cluttered with equipment. Electrical cables lie tangled across the narrow floor space like spaghetti on a plate. The room is dominated by a single lab bench running down its middle, and on the bench is a glass pipe 30 feet long.

"This is a molecular laser," says the scientist who has been bending over the pipe, tinkering with some fixture of unguessable purpose. "It's quite new. It runs on a mix of carbon dioxide and other common gases."

He straightens up. Dr. Dane Rigden is his name; he is a British physicist who came to this country several years ago because, in his field, this is where the action is. "I'll show you how it works," he says. He turns to the 30-foot laser, then thinks of something and turns back. "By the way," he says, "keep your goggles on. And keep your hands by your sides."

He puts his own goggles on, flips some switches and turns some dials. A loud hum fills the dark room and a wave of heat comes from—where? You look at the gadgetry and frown, puzzled. The glass tube is glowing with a dim, cool, purplish light. Nothing else visible is happening.

"You've heard a lot of talk about laser death rays," says Dr. Rigden, still tinkering. "It's been mostly science fiction so far. But if there ever is such a weapon, this might be the laser used to make it." He points to one end of the long glass tube. "Coming out of there right now is a continuous thin beam of intense infrared light. You can't see it, but"—he looks to make sure your hands are

SOURCE. From *Playboy*, Vol. 15, No. 2 (February, 1968), p. 72; copyright by HMH Publishing Co., Inc.

down by your sides—"there's enough of it to chop your arm off."

He rummages in a grease-spotted paper bag, pulls out a thick meat bone and grips it with a pair of tongs. He holds the bone out, then lowers it into the invisible laser beam. There is a sudden blinding flash of light and a loud sput. Dr. Rigden withdraws the bone and holds its smoldering stump up for inspection. The bone has been chopped in half.

Dr. Rigden gazes at the bone stump thoughtfully. "As molecular lasers go," he says, "this one isn't unusually powerful. It delivers about two hundred and fifty watts in its beam. Raytheon and others have generated continuous beams of over 3000 watts. And last week I heard——Well." He stops, grinning. "Security regulations, you know. Come on, I'll show you something else."

A laser lab is strangely exciting. Little knots of men stand about in the corridors, talking. Many talk with foreign accents—for, like Dr. Rigden, they have been drawn from all over the world by the promise of action. There is a profound and mysterious feeling in the air, a Christmas-morning feeling of discoveries hidden just around the corner and large events about to happen. There are no bored or weary or disappointed faces here. Everyone in the business is a perpetually surprised newcomer. The lab equipment is equally new. The complex structures of metal and glass and rubber-sheathed cable stand untidily on tables and floors. Everything has an ephemeral look, as though it were put together in an eager hurry yesterday and will be rebuilt to try out some new idea tomorrow.

"Here's something else new," says Dr. Rigden. He ducks into another dark, cluttered laboratory room and puts practiced hands on a metallic tube about two feet long. He is obviously enjoying himself. "This is an argon-gas laser. Watch."

He switches on the device. A pencil-thin beam of intense blue light shoots out and makes a tiny brilliant spot on a target at the other end of a long metal table.

"Pretty, isn't it?" says the scientist

reflectively. He gazes at the target spot for a long time without moving. "You never get tired of looking at laser light."

The blue light is more than pretty. It is a blue of shining heartbreaking purity—bluer, unimaginably purer than any earthly light that shone until the 1960's. Its perfect jewellike brilliance hurts the eye, yet you can't look away. The blue target spot has a peculiar mobile quality; it seems to consist of a million tiny luminous specks that churn slowly about one another. The effect arises from the special qualities of laser light. It is hypnotic. You lean closer....

"Back off a little," says Dr. Rigden. "This beam can give you a nasty burn. Here—I'll show you something else."

He tinkers with a gray box mounted next to the laser. The box clicks. Long numbers appear in windows on its face. "You're looking at a new kind of yardstick," Dr. Rigden explains. "Laser light is so pure that we know exactly what its wave length is. By knowing the wave length and by doing a little arithmetic, we can measure distances in billionths of an inch. This gadget is now measuring the distance from the laser head to the target. Here: Put your hand next to the laser."

You rest your hand on the cold metal plate on which the laser is mounted. The counting device instantly starts to click.

"Know what just happened?" asks Dr. Rigden. "The warmth of your hand expanded the metal. The distance to the target increased. Not much. Less than a hairbreadth. But enough."

You begin to understand what all the excitement is about. You begin to see why industrial companies and governments are pouring millions of dollars a year into laser research. The U.S. military services and other Government agencies have been interested in lasers from the beginning. So have the Russians. So have the British, the French and the Italians. They're fascinated by the capacity of laser light to measure distances to targets. At the moment, they're particularly fascinated by the new carbon-dioxide laser, the Buck Rogers device that generates a steady beam of enormous energy—not just a brief burst but a continuous, miles-long invisible needle. And they're fascinated by what they see in the future.

Every once in a while, from behind some imperfectly closed security curtain in Rome or London or Washington, a few tantalizing words leak out. Ron Barker, energetic young associate publisher of *Laser Focus* magazine, spent much of 1966 touring laser labs in Europe and the United States—and came back, he says, "with my hair standing on end. I saw things . . . I heard things. . . . My God, if I could only print half of it!" He can't, of course. The hints you hear nervously dropped around laser installations are not meant for publication and, in any case, the hinter always clams up when you press him for corroborating information. What he has said, in its naked form, without the necessary clothing of tangible evidence, has an apocryphal sound and can hardly be passed on without embarrassment. Billion-watt beams. Beams that can slice buildings in half or cleave steel at a distance or vaporize aircraft or mow down men as a scythe mows grass. True or false? Reality or plan or merely dream? The facts cannot be had.

The laser business is like that. You can't easily tell what's apocryphal and what isn't, for the entire science seems apocryphal. It shouldn't exist; it is improbable and outrageous. Never before in history has a scientific invention gone from mental concepts to working hardware to world-wide practical application in so short a time. The raw idea of a laser was conceived only 11 years ago. The first working model was built only eight years ago. Today lasers are used in surgery, welding, drilling, surveying, weaponry. They are common items of technological hardware, available by mail order like Bunsen burners or microscopes. They are where they shouldn't really have arrived until the twenty-first century. "It's like being shot into the future," says Alan Haley, regional sales manager at Perkin-Elmer. "I'm selling a product that didn't exist when I got out of college—wasn't conceived, wasn't even dreamed of. I'm selling it like ordinary hardware. I carry samples of it around. What would a career counselor have said ten or twenty years ago if I'd told him I wanted to sell ray guns for a living?"

He would have laughed, of course. The entire short, brilliant history of the laser has been one of people laughing at one another. First A laughs scornfully at B, and a few weeks later B is laughing triumphantly at A as A struggles to pry his foot from his mouth. They laughed in 1959 at the man who said he could build a laser. He did it in 1960. They laughed in 1963 when some nut at the Bell Telephone Laboratories said it was possible to make a laser spit blue light instead of red. Maybe in the 1970's, they chuckled. It happened in 1964, and now lasers produce every color of the rainbow, plus a broad range of infrared and ultraviolet. In 1961, they had developed pulsed lasers that would spit brief bursts of high-energy light, but the continuous-beam lasers were barely achieving one watt of output and everybody laughed at the idea of more. In 1968, 3000 watt steady beams have been announced and more powerful beams have carefully not been announced. In 1965, they were grinning over a statement that had become a cliché: "The laser is a solution in search of a problem." Today, salesmen are unloading lasers on hardheaded industrialists who won't buy anything unless it coldly promises a profit. Now they laugh (but in an oddly hesitant way) at other far-out ideas: death rays, building slicers, jungle mowers. Who will laugh tomorrow? At whom? About what?

Nobody cares to guess. "This crazy industry has matured when it should still be in its bottle-sucking infancy," says Bill Bushor, an engineer who founded *Laser Focus* three years ago and is now the industry's acknowledged chief historian. One sign of an industry's maturity is the appearance of a self-supporting technical publication covering the field, and *Laser Focus* fits the description. Its history has resembled that of the laser: Like one of those TV reruns that are hacked apart to make room for commercials, all the connecting scenes seem to be missing and the chronology is bewilderingly compressed. Bushor started the publication in 1965 as a mousy little newsletter, but today it blooms as a four-color job on glossy paper, fat with ads. Its offices in Newton, Massachusetts, also resemble the laser business: cluttered, untidy, expanding too fast to pause and make order. "It's like being in one of those dreams where you can't stop running." says Bushor. "This is better than a two-hundred-million-dollar-a-year industry already, and some guess it could hit one billion by 1970."

So new is the science of lasers, in fact, that the inventors are still around to tell from recent memory their tales of lonely intellectual adventure—and not only still around but still young, still inventing.

The laser's history begins quietly on a dark bench in Washington, D.C. and weaves a strange, obscure path through a candy store in the Bronx and other dim, unlikely places. Columbia University physicist Charles Townes is credited with starting the drama. In Washington for a scientific conference in the spring of 1951, Townes strolled early one morning into Franklin Park and sat on a bench. There, he fitted together the pieces of an idea that had been churning about in his mind. The idea was the basis of an invention that Townes later dubbed "maser" (acronym for microwave amplification by stimulated emission of radiation)—a device which, by feeding on its own internal energies, generates a powerful beam of microwaves. The invention has since proved useful in radar, spacecraft guidance and other microwave applications.

During the late 1950s, another thought began to take form in several minds simultaneously, including Townes'. Microwaves belong to a broad spectrum of electromagnetic radiation in which our world is bathed. At one end of the spectrum are radio waves, which can be anywhere from several miles in length down to about 10 centimeters—at which point microwaves begin. At the other end of the spectrum are exotic forms such as X-rays, whose waves are measured in tenths of millionths of a milli-

meter. In between are radiations of every other conceivable wave length. To us, as humans, the most important part of this enormous spectrum is a narrow band of radiations whose wave lengths measure from roughly four to roughly seven ten-thousandths of a millimeter. For reasons about which biologists are still arguing, the earth's major animal life forms millenniums ago developed a remarkable organ that is sensitive to this particular little group of wave lengths. This organ is the eye. The wave lengths it senses are what we call light. (The human eye and some animal eyes are so cleverly made that they can even sense minute differences in wave lengths. The longer waves give us the peculiar visual sensation that we call red; the shorter ones, violet.)

It seemed to Charles Townes and several other physicists in the late 1950s that, since light waves are the same as microwaves except for being shorter, it should be possible—theoretically—to build a device that would do with light waves what the maser did with microwaves. By 1957, there was a name for this yet-unbuilt device. It was called a laser—the "L" standing for "light"—though Townes for a long time insisted on calling it an "optical maser." But nobody had any idea how such a device might work.

Townes was beginning to get some ideas, however. So were others, and a curious dramatis personae now began to assemble. One early arrival was a man who had married Townes' sister: physicist Arthur Schawlow, then at the Bell Telephone Laboratories and now at Stanford University. Another was a research associate and doctoral student at Columbia University's Radiation Laboratory, Gordon Gould.

Gould, out of Yale in 1943, had gone to work for the Manhattan District in the Corps of Engineers, developer of the first atomic bomb. He had met a girl and one night wandered into a Marxist discussion group with her. This cost him his job and his security clearance, subsequently made it difficult for him to get into scientific laboratories. He spent the next decade struggling to get jobs and continue his physics education. He was still struggling in 1957, when some startling thoughts about lasers occurred to him.

Independently and simultaneously, similar thoughts had occurred to the brother-in-law team of Townes and Schawlow. In essence, the thoughts were that it might be possible to take some fluorescent substance and "pump" its atoms up to an excited state by hitting them with a flash of light or a jolt of electric current. Normally, these excited atoms would calm down in random fashion, emitting photons one by one and making the substance glow dimly for a few minutes. But suppose you rigged up a trap of mirrors in such a way that some of the photons began to bounce back and forth. On each bounce, the photons would hit atoms that hadn't yet calmed down. These atoms would be jolted into releasing their photons sooner than normal and the new photons would join the gathering surge and hit still more excited atoms. In this way, perhaps, you could make all the excited atoms release their photons in a billionth of a second instead of several minutes. You might produce a blast of incredibly brilliant light. If you provided a way for some of the light to escape the mirror trap—maybe by making one mirror only partially reflective—you might get a beautiful strong beam.

In trying to explain these thoughts, Townes would sometimes ask puzzled listeners to think of a long swimming pool. At one end, rising from the water, you erect a thin wobbly pole, and atop the pole you build a platform. You hoist rocks onto the platform. This is analogous to the pumping up of atoms to their excited state. If the rocks release their stored energy (that is, fall into the pool) in random fashion, the result will be only a pool of choppy water. But suppose you rig the system to feed on its own energies in an orderly way. You let just one rock fall in. A nice tidy wave travels to the other end of the pool and bounces back. It jiggles the pole and this makes another rock fall in. This second rock hits the water at just the right time to amplify the existing

wave. In other words, its energies fall into step much like photons in a laser. The bigger wave travels down the pool and back, the pole wobbles, a third rock falls in and the wave grows still bigger. And so on.

Dr. Townes referred to this concept as "wave amplification by stimulated emission of rocks"—that is, a waser.

Gordon Gould, contemplatively commuting from Columbia University to a Bronx apartment, believed he had this field of thought to himself. But one night just before Halloween in 1957, Townes phoned him. The two had met occasionally on Columbia's campus. Townes wanted some data about certain high-intensity lamps with which Gould was working at the Radiation Lab. Townes' questions made Gould suddenly ask a question of his own: "Is Townes thinking what I'm thinking?"

Gould plunged into an undeclared race against Townes. He worked night and day on his laser calculations. One cold November night, Gould and his wife left their apartment and walked a few blocks to a candy store whose proprietor doubled as a notary public. Clutched in Gould's hand was a dirty gray laboratory notebook bearing the title "Some Rough Calculations on the Feasibility of a Laser." The notary witnessed it and dated it: Friday, the 13th of November, 1957.

Townes and Schawlow were making their own rough calculations about the same time. By mid-1958, they felt their figuring was specific enough to be patentable, and they and the Bell Labs applied for the law's protection. Gould, working alone with little equipment and a small budget, hampered by security restrictions, was farther behind. Seeking help, he left his university job and took his notes to a small scientific outfit named TRG now an affluent laser-making division of Control Data Corporation. Intrigued, TRG took him in and put him in a lab where no security clearance was required. He and TRG applied for their patent in early 1959.

A series of court battles then began. Some experts later said Townes and Schawlow's papers most nearly described the laser as it eventually came to be; some said Gould's. Gould's notarized notebook bore the earliest date, and it was mostly on this basis that Gould claimed to have conceived the invention first. But the court turned him down, largely because he hadn't proved "diligence" in going from general concept to specific calculations and thence toward hardware. The Bell Labs team was awarded a patent with the fetchingly rhythmic number 2,929,922.

Meanwhile, another interesting character had drifted on stage. This was Ted Maiman of Hughes Research Laboratories, who claimed that he, too, deserves credit for inventing the laser.

In 1959 and early 1960, though the patent battle was already joined, nobody had actually made a laser. Dozens of large corporations were trying: the Bell Labs, Westinghouse, General Electric, Raytheon. Many were trying it with potassium vapor and related gases, which seemed theoretically to promise the best results. They had great expensive, science-fictionish rigs on their lab tables. Every now and then, something would explode or an overloaded circuit would disintegrate and the scientists would curse and build a new, even less probable-looking contraption. And what of Ted Maiman?

It was laughable. Compared with the giant corporations that were thundering up and down the laser trail, Maiman was a mouse rustling in the weeds. He had a small, cramped, cluttered lab room at Hughes' Malibu Research Laboratories in California. Hughes supported him because he was considered a bright young fellow; but the hope was that he could eventually turn to something more promising. Maiman was pursuing a magnificently ridiculous notion. He was trying to make a laser out of a ruby.

A ruby? There were several impressive reasons you couldn't make a laser out of a ruby. Ruby had been shown to lack the required quantum efficiency—in lay terms, the go. Moreover, it obviously wouldn't be able to take the heat without cracking. Yet Maiman chose to ignore these facts. He had

a ruby crystal about the size and shape of a pencil stub. Its ends were ground parallel and were silvered, one more completely than the other. This was the heart of his proposed laser. The rest was like something from a five-and-dime store. Curled around the ruby was an ordinary helical flash tube, the kind photographers use. This was intended to provide the "pumping" light that would excite the ruby's atoms. Wrapped around the flash helix, in turn, was a dented aluminum reflector. That was all.

"Forget it, we've tried it, it won't work," some visiting Bell Labs physicists had assured Maiman. He had to admit the gadget didn't work yet, but he refused to admit it wouldn't. He was becoming the comic relief of the laser quest. He'd submitted a paper on his proposed laser to a technical journal, but the editor had rejected it. A photographer had come around to take a picture of the non-working laser but had found it so unimpressive ("like something a plumber might have screwed together," he said) that he asked Maiman to build a bigger, more scientific-looking mock-up. Another company had copied the picture to make its own ruby laser, and of course this laser didn't work, and this only increased Maiman's embarrassment.

"I'm sure it's just on the threshold of working," he said one day in June 1960 to a group of East Coast scientists. They'd come West for a convention and were indulging in the great new California sport of Dropping In On Maiman. They nodded politely as Maiman earnestly explained his reasoning. They left, nudging each other in the ribs. Maiman chomped his cigar gloomily.

Shortly afterward, his lab assistant, a man named Irnee D'Haenens, came in with a package. It contained three new ruby crystals of improved optical character, fabricated and polished with special loving care by the Linde Company, expert crystal maker. Maiman and D'Haenens mounted one of the new crystals in the gadget. They looked at each other. D'Haenens quietly closed the lab door. Maiman threw a switch and the helix flashed.

And a tiny spot of brilliant red light

appeared momentarily on the laboratory wall.

Bob Meyer, a Hughes publicity man, recalls being summoned to the Malibu lab building the next day. "The place was buzzing with excitement. People were standing around in the corridors babbling at each other. I couldn't understand what it was all about. I'd heard the word 'laser,' but I didn't really know what it was supposed to be."

The lab director, Dr. Lester Van Atta, tried to explain. "Great news!" he shouted at Meyer. "Maiman has achieved laser action!"

That was eight short years ago. Today, almost every newspaper-reading man in every industrial nation knows what a laser is. Literally thousands of lasers exist and literally hundreds of scientific groups throughout the world are working on improvements. "I knew I had something important," says Maiman, "but I never dreamed of anything like this."

The ruby laser today is the most powerful, though not in all respects the most useful, in the business. "We've developed continuous-wave ruby lasers, but most still operate in short pulses," says physicist Dr. Richard Daly of TRG, the outfit that took in the struggling Gordon Gould in his hour of need. "Pulsed operation isn't always what you want in every application. But it's right for many uses. Here—let me show you."

Daly has the typical laser man's fondness for showing off his gadgetry. In a large, windowless lab room, Daly tinkers with a thick metallic tube. He turns some dials. You adjust your goggles. He flips a switch and there is a sharp crack like a rifle shot. At the other end of the lab, a metal target seems to explode with a blinding white flash and a huge sunburst of sparks.

The metal is steel. In its center is a smoking hole about half an inch deep.

"This is a giant-pulse ruby laser," explains Daly. "You didn't actually see the laser light, because it was such a short pulse. It was a slug of light not much longer than your outstretched arms. But there was a lot of oomph in it. About a gigawatt."

A gigawatt is a billion watts. You can nearly go blind just thinking about that much light in that small a space. For comparison, consider the sun. On a clear summer day at high noon, the sun pours energy onto your head at a density of about one tenth of a watt per square centimeter. This much light can blind you if you look directly into it for long. But even an unfocused laser beam can deliver energy at literally millions of times that density. "Focused carefully," says Dr. R. D. Haun of Westinghouse, "a laser beam can deliver ten billion watts to a square centimeter."

It isn't only the power of laser beams that fascinates scientists. It's also the quality of utter neatness. A laser beam is "coherent"—meaning, in effect, orderly, like a good TV broadcast beam. Ordinary light is untidy. Its waves are of diverse sizes and never quite lined up right, and each photon behaves in a slightly different way as it passes through a focusing lens. Even the best lens can't focus this untidy light to a point, only to a fuzzy-edged blob. But laser light has been focused to a spot as small as 1/10,000 inch.

Jobs both of brute force and of microscopic tenderness can be done with light like this. TRG, for instance, sells a microscope-mounted laser that can deliver a pinprick of energy delicate enough to burn a single chromosome inside a living cell—a capability now being used in studies of genetics. An equally delicate Westinghouse laser recently drilled three neat round holes in a row across the breadth of a human hair.

Slightly more powerful beams are used in surgery. The American Optical Company, for example, makes a special laser instrument for operations inside the eye. By passing the beam through the eyes transparent cornea and lens, doctors can burn away a blood clot or weld a detached retina inside the eye without touching the outer parts. At the Children's Hospital of Cincinnati, doctors are experimenting with lasers in destroying cancers, drilling teeth and removing warts, tattoos and birthmarks—and, at the same time, they are trying to find exactly what laser light does to human skin and other tissues. Director of the laser laboratory Dr.

Leon Goldman, who has deliberately pricked himself with laser beams some 450 times, remarks that physicians as a group are often a decade late in taking advantage of scientific developments. But they began working with the laser almost as soon as it was invented.

Industrial engineers have also been quick to use the laser. Take the enormously varied, omnipresent industrial problem of hole drilling. "A laser beam can vaporize any substance on earth." says Ted Maiman. "That's why all kinds of engineers, drilling holes in all kinds of materials, have fallen in love with the laser." The Wurlitzer Company, for example, maker of pianos and other large musical instruments, for years has worn out bits by the hundreds drilling some 80,000 holes a day in hard work maple to a tolerance of 1/2000 inch. "*Please* say it can be done with a laser!" a Wurlitzer engineer begged TRG; and TRG, not having the heart to turn him away, is now designing a light-beam tool for the purpose. It shouldn't prove difficult, for a laser beam can even punch a hole through a diamond. Many types of fine wire are made by drawing soft metal through minuscule holes drilled in diamonds; and in the past, it used to take two to three days to drill such a hole. A laser tool made by Western Electric does it in a few minutes.

The US Department of Commerce and MIT are now dreaming about much larger holes. Late in 1966, two MIT sophomores, ignoring the amused chuckling of their professors, borrowed a powerful carbon-dioxide laser from Raytheon and trained its beam on a chunk of granite for 30 seconds. When they picked the granite up, it crumbled like dry mud. Engineering professor Robert Williams now believes this startling effect may be a key to fast, cheap tunneling. It may be possible to build a laser-headed boring machine that can eat its way through rock like a worm through cheese. The Commerce Department, interested in rapid-transit tunnels between cities, has encouraged MIT to probe further.

Because laser light is orderly, it can also be used for communication. Its waves can be modulated exactly like radio waves or micro-waves—and this is another application that brings a gleam to the eyes of scientists. The National Aeronautics and Space Administration has hired several companies to wonder about sending messages through interplanetary space on a laser beam. "A radio or radar beam fans out widely," says a Perkin-Elmer scientist who is working on this idea, "and after traveling millions of miles to get here from Mars, for example, its energy would be so far dissipated that we'd barely be able to pick it up. But a laser beam can be made so tight, so narrow, that it can get here from Mars and still be going strong." The US Army is also interested in the laser as an instrument of battlefield communication. Radio messages fan out and can easily be picked up by an enemy. But a message sent on a laser beam would go nowhere but to a single receiver—and even if the enemy saw the beam and tried to read it, he'd give himself away the instant he poked his receiving device into it.

The tidiness of laser light has also made possible a photographic technique called holography. A hologram is, in effect, a true three-dimensional image of an object. Instead of showing only some of the object's surfaces, as an ordinary photograph does, a hologram shows all surfaces—faithfully re-creates the entire object in light. By turning the hologram or walking around it, you can see the back of the object. This spooky effect depends on complex phenomena of diffraction and can be obtained only by illuminating the object with the coherent light of a laser. Industrial companies have begun using holograms to study, among other things, stresses in metals. Engineers might make a hologram of an aircraft-wing part, for example, while the part is at rest, then make another stop-action hologram while the part is vibrating as in flight. By comparing the two images, they see precisely how and in what dimensions the part was strained out of shape.

Bob Whitman, an artist with an ability to sell far-out ideas to large organizations, has discovered yet another use for the laser's eerie light. Working with Bell Labs engineers, Whitman this year built what he called "light drawings" at the Pace Gallery in New York.

These might be compared with line drawings on paper, except that the lines are thin, coloured, low-powered laser beams and the viewer stands within the three-dimensional drawing and "experiences" it instead of merely contemplating it from without. Whitman calls the drawings "articulations of space." Bell Labs people, who cooperated for reasons of publicity, are not quite sure what to call the drawings. Asked to comment, a company spokesman explained: "Well, this laser art is—it kind of—um. Well, we enjoyed working with Bob Whitman."

But of all the actual, possible and imaginable uses of the laser, the one that generates the most excitement is that of weaponry. The U.S. Army and Navy are known to be working strenuously on destructive light beans, but have kept the effort secret. The Air Force was less successful at first in keeping its lip zipped. General Curtis LeMay's speeches used to contain cryptic comments about "beam-directed energy weapons"; but in the past few years, the Air Force has declined to comment further on the subject. Similar secrecy shrouds laser research in Europe (and, of course, in Russia), though national pride occasionally forces security curtains to be lifted briefly. Late in 1966, Britain's Services Electronics Research Laboratory, a government science center, showed off some of its laser developments, and one item on display was a portable, battery-powered laser rifle. The cool-voiced British scientists referred to it smilingly as "a toy, of course," and showed how it could be used to pop balloons. Yet it is hard to believe that the frugal British government would spend its taxpayers' money to build balloon poppers for the lunch-hour entertainment of scientists. Such a rifle could be used in combat to blind enemy troops—and, at higher power, to do the same kinds of damage bullets do, or worse.

The new family of molecular lasers—particularly carbon-dioxide lasers—interests military men, because they offer high power as well as continuous beam operation. Such lasers work, in effect, by setting up a sort of vibration within molecules instead of dealing with excited atoms. Theoretically, they are capable of enormously higher power than anything yet developed. Raytheon, Westinghouse, Perkin-Elmer and dozens of other companies have military contracts to study molecular lasers—contracts surrounded with elaborate secrecy. One of the Perkin-Elmer's contracts has required the company to build an odd-shaped tall room; and although any visitor may peek into the room (after proving that he's a U.S. citizen), most of the company's employees and executives are mystified about the room's uses.

"If you think the laser business had produced surprises over the past few years, wait until the next few," says Gordon Gould. Like the laser, Gould since 1960 has risen rapidly from nowhere to prominence. Though he lost his fight for the basic laser patent, he now holds several other important patents in the field and has applied for others. These and a Control Data (including TRG) stock option have made him suddenly quite wealthy. Among his new possessions is a boat in which he periodically sails in the West Indies, gazes across vast ocean reaches and tries to see the future.

What he can see looks good to him. He left TRG in 1966 to become a professor at the Polytechnic Institute of Brooklyn. His basic job there is not to teach but to research and invent. He is interested in a new copper-vapor laser that produces green light of potentially colosal brightness. He is interested in picosecond pulses (a picosecond is a millionth of a millionth of a second). He is interested in more things than can conveniently be cataloged.

"It seems strange to say this when lasers are in such wide use," says Professor Gould, "but the laser is still a very young invention. It has only just begun. The possibilities ahead are—well, it's a word scientists don't like to use, but what else can I say? Fantastic."

CHAPTER 7

NUCLEI

7.1

The Discovery of Radioactivity
A. Sutcliffe and A. P. D. Sutcliffe

A few months after Röntgen's fortunate observation had led him to make his important discovery another scientist, after considering the production of X-rays, made an experiment which had an unexpected result. This unexpected result led to the discovery of radioactivity.

When X-rays are being produced, the glass wall of the Crookes's tube glows with a greenish shimmer where the cathode rays fall on it. But the glow stops immediately the cathode rays are cut off. The glowing part of the tube is said to be fluorescent.

Fluorescence is not uncommon. It is produced when sunlight falls on a few substances, which then glow with a bluish color. They lose their glow, however, immediately they are put in the dark. A few other substances glow in a similar way when exposed to sunlight but, unlike fluorescent substances, they continue to glow for a short time after being put in a dark place. These are said to be phosphorescent substances. Phosphorescence and fluorescence, however, are very similar in many ways.

E. Becquerel a well-known French scientist, and his son Henri, during the later part of the nineteenth century specialized in the study of substances containing the then rare metal called uranium. The father had written at length about the *fluorescence* of a few uranium salts. But Henri occasionally termed the glow *phosphorescence*. In this chapter, to save confusion, the term fluorescence will be used for both phenomena.

In January 1896 Henri Becquerel was one of many hundreds who visited an exhibition in Paris of the first X-ray photographs. He was particularly interested because another scientist had stated that X-rays were produced by the flourescent glass of the Crookes's tube.

Becquerel then had the idea that if fluorescent glass could produce X-rays so also might other fluorescent substances. He had in mind, of course, the salts of uranium which were by the way of being a family concern of the Becquerels. So he decided to test his idea by making the following simple experiment using the salt called uranium potassium

SOURCE. From *Stories from Science* (Cambridge Univ. Press, 1962) Vol. 2, Chapter 40.

sulphate which he himself had first prepared many years before 1896 for his father's experiments on fluorescence.

The experiment was based on the fact that a photographic plate wrapped in thick black paper is not affected by sunlight but is by X-rays. To the black paper surrounding such a plate he fixed a crystal of his uranium salt. Near to it he fixed a silver coin on top of which he put a similar crystal. Then he placed the plate so that the sunlight would fall on the crystals and make them fluoresce. He expected that the fluorescent crystals would produce X-rays and that those given off by the first crystal would produce a clear impression of the crystal on the developed plate. He also expected that the X-rays given off by the second fluorescing crystal would be stopped by the silver coin so that a dark coin-shaped patch would appear on the developed plate.

When Becquerel developed the plate he found that it looked exactly as he had expected; there was an impression of the first crystal and a well-defined dark patch beneath the spot where the coin had been. It therefore seemed to him that the fluorescing uranium salt gave off X-rays.

On February 26, 1896 he repeated the experiment. He put the wrapped plate out in the open air. The day was dull; so he left the crystals exposed to the light until the following day. This day was also very dull—indeed the amount of sunshine on both days put together had only been enough to make the crystals fluoresce very slightly. So he put the plate, with the crystals and coin still attached to its wrapper, into a dark cupboard with the intention of waiting for a brighter day when he would again expose the crystals. The next two days, as it chanced, were equally dull, so he decided to develop the plate. He expected only a faint impression as a result of the short exposure of the crystals to the sunlight, but to his great surprise, the impression of the crystal and the shadow of the coin on the plate were as clear and as well defined as those given before, even though the crystals had previously been exposed to bright sunlight for a long time!

This experiment seemed to prove that even faintly fluorescent crystals of the uranium salt gave out X-rays. Then he had the brilliant idea that crystals of this salt which had not been made fluorescent (by exposure to light) might give out X-rays. This idea called for further experimentation.

He therefore arranged a photographic plate with crystals and coin attached, as before, but this time he did not expose it to light but put it away in a dark cupboard for a few days. On developing the plate he again obtained the clear impression of the crystal and the dark shadow of the coin. This seemed to show that these crystals gave out X-rays although they had not been made fluorescent. Further experiments not only seemed to confirm this conclusion, but also showed that other compounds of uranium, and even the metal itself seemed to give out X-rays, irrespective of whether or not they were fluorescent.

Then came the other surprising discovery. The rays given off by uranium and its compounds were not X-rays at all, despite the resemblance in their effect on photographic plates! For experiments showed, without the slightest doubt, that they were a kind of rays hitherto unknown. These rays were called "Bécquerel rays" in honor of their discoverer.

Becquerel's chance discovery, as Sir Oliver Lodge said, opened up a new chapter of science. In 1897 Madam Curie began experiments to find out whether other substances gave out similar radiations. She examined almost every known substance and found that a few of them did, so she called them radioactive substances. Her most important discovery, however, was that a known weight of pitchblende, an ore containing uranium, gave off radiations that were more powerful than she expected according to the quantity of uranium in the weight of ore. Hence she suspected that the ore must contain a more radioactive substance than uranium. After a long and tedious process she obtained from a ton or more of ore a very tiny bit of a new, unknown element. She called it radium.

Before Becquerel's discovery, scientists

had firmly believed that the atom was the smallest particle of matter and that it could not be divided. Becquerel's news that an element was giving off *something* puzzled the scientists and made them wonder what these rays were made of. By 1902 research had given the answer: these rays contained very tiny particles of matter which must have been split off from the atoms of the elements. Hence it was established for the first time in history that there were smaller particles than atoms and that the atoms of radioactive elements underwent division of their own accord (or, as scientists put it, they underwent spontaneous disintegration).

In the loss of particles by the atoms—for example by radium atoms—it was shown that the energy liberated was tremendous, one estimate being that a gram of radium gave out as much energy as a ton of coal which was burning slowly. But—and this is most important—calculations showed that it took two to three thousand years for this energy to be liberated. Nevertheless, despite the long period involved, it was evident that matter was being changed into energy, a change which was contrary to the views which had been held for centuries.

The impact of Becquerel's discovery was referred to by Sir Henry Dale, who said:

At a meeting of the undergraduates' Natural Science Club in Cambridge in 1897 my contemporary, R. H. Strutt (later Lord Rayleigh, the eminent physicist and co-discoverer of the rare gases), gave us an account of Becquerel's discovery. I well remember the sceptical protest of one of us who was later to become world famous in theoretical physics and astronomy. "Why, Strutt," he said. "If this story of Becquerel's were true it would violate the law of conservation of energy." I like to remember the enterprising spirit of Strutt's rejoinder, "Well, all I can say is so much the worse for the law because I am quite sure that Becquerel is a trustworthy observer." And of course none of us there had then an inkling of the enormous expansion of knowledge of which such discoveries as these were to provide the points of origin, or of the whole armoury of physical resources which would thus be brought to the service of medicine.

Thus it was that Becquerel's discovery started research work which led to the discovery of radium, with its great benefits to which resulted not only in the production of the atomic bomb but in the gift to man of a tremendous source of energy for peaceful purposes.

It is a staggering thought that these important discoveries had their origin in the fact that the sun did not shine brightly on the last few days of February in the year 1896!

It is a more staggering thought, as mentioned by Professor Strutt, that Becquerel's research was carried out as a result of three false assumptions. The first of these was that X-rays were produced by fluorescent glass—and they are not. The second was that because fluorescent glass supposedly gave off X-rays so would other fluorescent substances—and they do not. The third one was that uranium salts gave off X-rays when they were *not* fluorescent—and they do not. Professor Strutt commented: "It seems a truly extraordinary coincidence that so wonderful a discovery should have resulted from the following-up of a series of false clues. And it may well be doubted whether the history of science affords any parallel to it."

7.2   *The Nature of the α Particle from Radioactive Substances*
                    Ernest Rutherford and T. Royds

The experimental evidence collected during the last few years has strongly supported the view that the α particle is a charged helium atom, but it has been found exceedingly difficult to give a decisive proof of the relation. In recent papers, Rutherford and Geiger have supplied still further evidence of the correctness of this point of view. The number of α particles from one gram of radium have been counted, and the charge carried by each determined. The values of several radioactive quantities, calculated on the assumption that the α particle is a helium atom carrying two unit charges, have been shown to be in good agreement with the experimental numbers. In particular, the good agreement between the calculated rate of production of helium by radium and the rate experimentally determined by Sir James Dewar, is strong evidence in favor of the identity of the α particle with the helium atom.

The methods of attack on this problem have been largely indirect, involving considerations of the charge carried by the helium atom and the value of e/m of the α particle. The proof of the identity of the α particle with the helium atom is incomplete until it can be shown that the α particles, accumulated quite independently of the matter from which they are expelled, consist of helium. For example, it might be argued that the appearance of helium in the radium emanation was a result of the expulsion of the α particle, in the same way that the appearance of radium A is a consequence of the expulsion of an α particle from the emanation. If one atom of helium appeared for each α particle expelled, calculation and experiment might still agree, and yet the α particle itself might be an atom of hydrogen or of some other substance.

We have recently made experiments to test whether helium appears in a vessel into which the α particles have been fired, the active matter itself being enclosed in a vessel sufficiently thin to allow the α particles to escape, but impervious to the passage of helium or other radioactive products.

The experimental arrangement is clearly seen in Figure 7.2–1. The equilibrium quantity of emanation from about 140 milligrams of radium was purified and compressed by means of a mercury-column into a fine glass tube A about 1.5 cm long. This fine tube, which was sealed on a larger capillary tube B, was sufficiently thin to allow the α particles from the emanation and its products to escape, but sufficiently strong to withstand atmospheric pressure. After some trials, Mr. Baumbach succeeded in blowing such fine tubes very uniform in thickness. The thickness of the wall of the tube employed in most of the experiments was less than 1/100 mm, and was equivalent in stopping power of the α particle to about 2 cm of air. Since the ranges of the α particles from the emanation and its products radium A and radium C are 4.3, 4.8, and 7 cm, respectively, it is seen that the great majority of the α particles expelled by the active matter escape through the walls of the tube.[1] The ranges of the α particles after passing through the glass were determined with the aid of a zinc-sulphide screen. Immediately after the introduction of the emanation the phosphorescence showed brilliantly when the screen was close to the tube, but practically disappeared at a distance of 3 cm. After an hour, bright phosphorescence was observable at a distance of

SOURCE. From the *Philosophical Magazine* **17**:281 (1909).

[1]The α particles fired at a very oblique angle to the tube would be stopped in the glass. The fraction stopped in this way would be small under the experimental conditions.

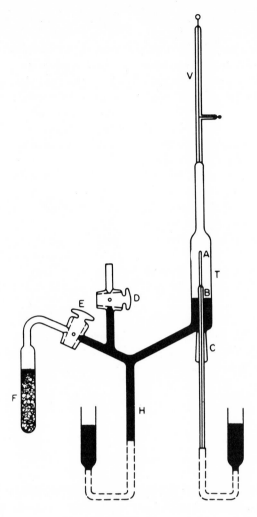

Figure 7.2—1.

5 cm. Such a result is to be expected. The phosphorescence initially observed was due mainly to the α particles of the emanation and its product radium A (period 3 min). In the course of time the amount of radium C, initially zero, gradually increased, and the α radiations from it of range 7 cm were able to cause phosphorescence at a greater distance.

The glass tube A was surrounded by a cylindrical glass tube T, 7.5 cm long and 1.5 cm diameter, by means of a ground-glass joint C. A small vacuum-tube V was attached to the upper end of T. The outer glass tube T was exhausted by a pump through the stop-cock D, and the exhaustion completed with the aid of the charcoal tube F cooled by liquid air. By means of a mercury column H attached to a reservoir, mercury was forced into the tube T until it reached the bottom of the tube A.

Part of the α particles which escaped through the walls of the fine tube were stopped by the outer glass tube and part by the mercury surface. If the α particle is a helium atom, helium should gradually diffuse from the glass and mercury into the exhausted space, and its presence could then be detected spectroscopically by raising the mercury and compressing the gases into the vacuum-tube.

In order to avoid any possible contamination of the apparatus with helium, freshly distilled mercury and entirely new glass apparatus were used. Before introducing the emanation into A, the absence of helium was confirmed experimentally. At intervals after the introduction of the emanation the mercury was raised, and the gases in the outer tube spectroscopically examined. After 24 hours no trace of the helium yellow line was seen; after 2 days the helium yellow [line] was faintly visible; after 4 days the helium yellow and green lines were bright; and after 6 days all the stronger lines of the helium spectrum were observed. The absence of the neon spectrum shows that the helium present was not due to a leakage of air into the apparatus.

There is, however, one possible source of error in this experiment. The helium may not be due to the α particles themselves, but may have *diffused* from the emanation through the thin walls of the glass tube. In order to test this point the emanation was completely pumped out of A, and after some hours a quantity of helium, about 10 times the previous volume of the emanation, was compressed into the same tube A.

The outer tube T and the vacuum-tube were removed and a fresh apparatus substituted. Observations to detect helium in the tube T were made at intervals, in the same way as before, but no trace of the helium spectrum was observed over a period of eight days.

The helium in the tube A was then pumped out and a fresh supply of emanation substituted. Results similar to the first experiment were observed. The helium yellow and green lines showed brightly after four days.

These experiments thus show conclusively that the helium could not have diffused through the glass walls, but must have been derived from the α particles which were fired through them. In other words, the experiments give a decisive proof that the α particle after losing its charge is an atom of helium.

7.3   *Some Personal Notes on the Discovery of the Neutron*
                                     Sir James Chadwick

A few months after his Bakerian Lecture of June 1920, in which he first mentioned what had been in his mind for some time, the possible existence of a neutral particle formed by the close combination of a proton and an electron, Rutherford invited me to join him in following up the experiments on the artificial disintegration of nitrogen which he had made in Manchester.

There were a number of reasons for this invitation, so welcome to me. Among them was that I had made some improvements in the technique of counting scintillations, better optical arrangements and a strict discipline; but also he wanted someone to talk to, to while away the tedium of working in darkness.

It was during the periods of waiting to begin counting that he expounded to me at length his views on the problems of nuclear structure, and in particular on the difficulty in seeing how complex nuclei could possibly build up if the only elementary particles available were the proton and the electron, and the need therefore to invoke the aid of the neutron. He freely admitted that much of this was pure speculation, and, being averse to speculation without some basis of experiment, he seldom mentioned these matters except in private discussion. Indeed, I believe that only on one occasion after the Bakerian Lecture did he again refer publicly to his views on the role of the neutron. He had not abandoned the idea, and he had completely converted me. From time to time in the course of the following years, sometimes

SOURCE. From *Ithaca*, 26 VIII–2 IX, 1962.

together, sometimes myself alone, we made experiments to find evidence of the neutron, both its formation and its emission from atomic nuclei. I shall mention some of the more respectable of these attempts; there were others which were so desperate, so far-fetched as to belong to the days of alchemy.

Immediately after the Bakerian Lecture Rutherford had asked J. L. Glasson to look for the production of neutrons when an electric discharge was passed through hydrogen, and a little later J. K. Roberts made a somewhat similar experiment. He could not really have expected that any evidence of the neutron would turn up in this way, but it had to be tried. Both the mass of hydrogen and the voltages used in these experiments were quite trivial.

It seemed to me not too unreasonable to look at hydrogen in the normal state, notwithstanding its apparent stability. If a close combination of proton and electron were possible at all, it might take place spontaneously; and the neutron so formed might break up again under the action of the cosmic radiation. With Rutherford's approval, I tried, in 1923, to detect the emission of gamma-radiation from the formation of neutrons in a large mass of hydrogenous material, using an ionization chamber and a point-counter as the means of detection.

A few years later, in 1928, Geiger and Müller devised what is now universally called the Geiger counter, which enormously increased the ability to detect gamma-radiation. Geiger very kindly sent me two of his new counters, as well as instructions for making them. Immediately, Rutherford and I used this new instrument to repeat the experiment with hydrogen. We went to all manner of tricks in the hope of finding some trace of the neutron. We also examined in the same way some of the rare gases, and any rare element we could lay our hands on, just in case any sign of the formation of the neutron or its emission might turn up. I mention these experiments only in a general way because some were quite wildly absurd.

After my first attempt in this way I had considered the possibility that the neutron could be formed, or exist, only in a strong electric field; and that perhaps one might find some evidence by firing fast protons into atoms, especially those of higher atomic number where some electrons were tightly bound. This was the vague idea behind the remark in a letter to Rutherford which is quoted in Eve's *Life* (p. 301)—"I think we shall have to make a real search for the neutron. I believe I have a scheme which may just work . . ." I thought that at least 200,000 volts would be necessary for the acceleration of the protons. No suitable transformer was available and, although Rutherford was mildly interested, there was no money to spend on such a wild scheme. (I might mention that the research grant of the Cavendish was about £2,000 a year, little even in those days for the amount of work which had to be supported.) I persisted with the idea for a year or two, and in the intervals of other work I tried to find a way of applying Tesla voltages to the acceleration of ions in a discharge tube. I had quite inadequate facilities, and no experience in such matters. I wasted my time, but not the Laboratory money.

During our work on the disintegration of the lighter elements by alpha-particles Rutherford and I had not been unmindful of the possibility of the emission of neutrons, especially from those elements which did not emit protons. We looked for faint scintillations due to a radiation undeflected by a magnetic field. The only specific reference to the search for the neutron in this way was made in a paper published in 1929, some years after the experiments.

The case of beryllium was interesting for two reasons. It did not emit protons under alpha-particle bombardment; and, though a false argument, the mineral beryl was known to contain an unusual amount of helium, suggesting that perhaps the Be nucleus split up under the action of the cosmic radiation into two alpha-particles and a neutron.

This matter intrigued me on and off for some years. I bombarded beryllium with alpha-particles, with beta-particles and with gamma-rays, generally using the scintillation method to detect any effect. In those days this was the only method of much use in the presence of the strong gamma-radiation of the radium active deposit, the chief source of radiation available to me. Quite early on, too early perhaps, I tried to devise suitable electrical methods of counting. I failed. Later, when the valve amplifier method had been developed by Greinacher, and put into use in the Cavendish by Wynn Williams, I was also able to make a polonium source, small but just enough for the purpose. With Constable and Pollard, I had another look at beryllium, and for a short but exciting time we thought we had found some evidence of the neutron. But somehow the evidence faded away. I was still groping in the dark.

The first indication of the neutron came in the work of H. C. Webster on the gamma-radiation excited in beryllium by alpha-particle collisions. I had had such work, the excitation of gamma-rays by bombarding light elements with alpha-particles, in mind for some years. An attempt had been made by L. H. Bastings, but this failed, because the polonium source was too weak and the instrument of detection, the electroscope, too insensitive. When the Geiger counter became available Webster took up this quest, but his first efforts were not very rewarding—I was still short of polonium.

This deficiency was overcome by the kind intercession of Dr. Feather, then in Baltimore, and the generosity of Dr. C. F. Burnam and Dr. F. West of the Kelly Hospital. They sent me, first by the hand of Dr. Feather and later by post, a number of old radon tubes which together contained what was, for me, a very large quantity of radium D and its product polonium. This gift was of immense value both immediately and later on.

In the meantime, Bothe and Becker had taken up this matter and they were the first to publish results. But Webster made a most interesting observation, that the radiation from beryllium which was emitted in the same direction as the incident alpha-particles was more penetrating than the radiation emitted in the backward direction. This fact, clearly established, excited me; it could only be readily explained if the radiation consisted of particles, and, from its penetrating power, of neutral particles. Believing that a neutral particle would produce tracks, though very sparsely ionized, I suggested that he should pass the radiation into an expansion chamber. To our dismay, for we were convinced that a neutral particle of some kind was involved, no such tracks were to be seen. We were very puzzled; we did not know how to reconcile the observations.

This near-miss occurred in June 1931. Shortly afterwards Webster left Cambridge for Bristol. I decided to take up the matter afresh, but my preparations were delayed, perhaps fortunately, by a change of my working quarters to another part of the laboratory. Then one morning I read the communication of the Curie-Joliots in the *Comptes Rendus*, in which they reported a still more surprising property of the radiation from beryllium, that of ejecting protons from matter containing hydrogen, a most startling property. Not many minutes afterwards Feather came to my room to tell me about this report, as astonished as I was. A little later that morning I told Rutherford. It was a custom of long standing that I should visit him about 11 A.M. to tell him any news of interest and to discuss the work in progress in the laboratory. As I told him about the Curie-Joliot observation and their views on it, I saw his growing amazement; and finally he burst out "I don't believe it." Such an impatient remark was utterly out of character, and in all my long association with him I recall no similar occasion. I mention it to emphasize the electrifying effect of the Curie-Joliot report. Of course, Rutherford agreed that one must believe the observations; the explanation was quite another matter.

It so happened that I was just ready to begin experiment, for I had prepared a beautiful source of polonium from the Baltimore material. I started with an open mind, though naturally my thoughts were on the neutron. I was reasonably sure that the Curie-Joliot observations could not be ascribed to a kind of Compton effect, for I had looked for this more than once. I was convinced that there was something quite new as well as strange. A few days of strenuous work were sufficient to show that these strange effects were due to a neutral particle and to enable me to measure its mass: the neutron postulated by Rutherford in 1920 had at last revealed itself.

I trust that I shall not be misunderstood if I add a postscript to this story. It is unnecessary to record my satisfaction, and delight, that the long search for the neutron had, in the end, been successful. The decisive clue had indeed been supplied by others. This after all is not unusual; advances in knowledge are generally the result of the work of many minds and hands. But I could not help but feel that I ought to have arrived sooner. I could offer myself many excuses—lack of facilities, and so on. But beyond all excuses I had to admit, if only to myself, that I had failed to think deeply enough about the properties of the neutron, especially about those properties which would most clearly furnish evidence of its existence. It was a chastening thought. I consoled myself with the reflection that it is much more difficult to say the first word on any subject, however obvious it may later appear, than the last word—a commonplace reflection, and perhaps only an excuse.

7.4   *Lawrence and the Cyclotron*
                         Nuel Pharr Davis

THE WESTERNER

. . . In the spring of 1929 Lawrence found himself doing well as college professors measure their lives. But time was going by, he was nearing thirty, and so far he had only picked at the periphery of the atom. Massive and tiny inside its veil of electrons, the nucleus remained unassailable. "Like a fly inside a cathedral," he called it at a lecture, never a man to varnish his similes. Physicists all over the world were talking about how to get at it. All agreed that the way was to stand above the roof of the cathedral and drop a hydrogen ion on the fly. Ions—atoms with an electron ripped off—were easy to drop. Being positively charged, they could be made to move toward a negative charge like a falling stone. The problem was to get enough voltage to make them fall fast enough. The nucleus protected itself behind a skin called the Coulomb barrier. To penetrate it an ion must move with an energy of many, many volts.[1] Lawrence did not know exactly how many. Like most physicists the world over, he thought vaguely in terms of a million. No apparatus in the world could generate such voltage, no insulation could contain it.

A man who lives at the Faculty Club has little to lose from an evening at the library. Lawrence formed the habit of going to it to meditate on the virgin nucleus and to keep up with the technical journals. One evening there his life was altered. "It is a curious thing," said Birge. "I have never been able to find out the day it happened. He did not keep notebooks then. The best guess is February 1929."

Overpersistent contributors haunt the technical journals much as they do the letters-to-the-editor columns of newspapers. The experienced reader recognizes their names and automatically skips. Such a contributor was Rolf Wideröe, a Norwegian living in Switzerland—volatile, restless, prolific of articles and ideas, often dubious. Coming

SOURCE. From *Lawrence and Oppenheimer* (New York: Simon and Schuster, 1968); reprinted by permission of the publisher.

[1]A nuclear physicist who hoped to be both precise and intelligible would say, "Ions must have a very high speed (velocity) in order to give them enough (kinetic) energy to penetrate the Coulomb barrier and for this, high accelerating voltages are required."

upon an article of Wideröe's in the *Archiv
für Electrotechnic*, Lawrence skipped the text
but looked at a diagram on the first page.

The diagram showed a pair of tubes set
end to end. Through the first tube an ion
could be made to fall from the positive to
the negative end with an energy of 25,000
volts. The idea was commonplace and the
energy contemptible. But at the time the ion
passed out of the first tube and entered the
second, the charge was to be reversed and
the ion given another 25,000 volts of energy.
Thus the diagram illustrated the concept
of resonance, which, in this context, means
giving repeated electric pushes at just the
right time to reaccelerate an accelerated
particle. The concept was already ten years
old, but Wideröe was the first to suggest this
specific apparatus.

Lawrence felt a stir of excitement. How
many tubes, he wondered, could one set end
to end? No limit on the number meant no
limit on the speed to which ions could accelerate. He sketched out a series in his mind.
There *was* a limit, he decided, at which the
ions would spread like a shotgun pattern
and hit the tube walls before they reached
the million-volt acceleration he wanted.

He laid down the magazine. Still his excitement intensified. He saw the ions curving
round as they fell—a vision as giddy and
dreamlike to him as that of the falling moon,
which also curved round, had been to
Newton. To make the ions go round, all that
was needed was to set them moving between
the poles of a magnet. More than that, the
time it took them to go round would be
always the same. An electric push repeated
at the right instant would send them spiraling out into bigger and bigger circles, traveling faster each time they went around. The
thought was big and new, a fundamental discovery. Here he could see no limit to the
speed that could be given them. He wrote
down a simple set of equations.

All this happened in a few minutes. Without reading Wideröe's article he walked
home to bed. Upon waking next morning, he
looked at his notes and found he had told
himself $MR\omega = eRH$. The mass of a particle

times the radius of the orbit in which it moves
times its angular velocity equals the electric
charge of the particle times the radius of the
orbit times the intensity of the magnetic field.
It was still true in the morning light, and so
beautiful he had to tell somebody.

"I may have been the first person Lawrence
mentioned it to," said Brady. "I will remember the event if not the date, because it
was then that I grew up as a physicist. He
came bursting into the lab at LeConte, where
I was working on another problem, his eyes
glowing with enthusiasm, and pulled me
over to the blackboard.

"He drew the equations of motion in a
magnetic field. 'Notice that R appears on
both sides,' he said. 'Cancels out. R cancels
R. Do you see what that means? The resonance condition is not dependent on the
radius of the circle [and therefore ions can
be accelerated to speeds without any apparent limit]. *Any* acceleration!'

"I studied the blackboard. 'Yes, I see,' I
answered. 'Looks good.'

"'R cancels R,' he said again. 'Do you see?'

"I got an ides this was going to go on all
day. 'About my problem of thin films,' I ventured. 'I have some questions.' He kept
repeating that R canceled R, but I persisted.
His eyes glazed. 'Oh, that,' he said. 'Well,
you know as much on that now as I do. Just
go ahead on your own.'

"I had never seen him this way before. He
had his problem, I had mine. Each to his
own. Nobody really cared. Gave me a terribly lonely feeling. So I grew up as a
physicist. Later I told him, and he said his
own turning point had come at Yale. Zeleny
was taking Robert Millikan, America's most
famous name in physics, on a tour of the
labs. Lawrence was in the dumps, thinking
he hadn't enough brains for physics and
ought to get out. Millikan asked so many fool
questions and had to try so hard to understand that Lawrence decided he himself
wasn't so bad after all, and from that time
he had confidence in his career.

"Afterward, when he told me this, it
helped, but I was feeling lost when he walked
away. He left in a rush, I suppose to tell

other people that R canceled R. While the door was closing behind him, I heard a voice from the corridor cadenced in a typical graduate-student whine: 'Gee! You know, the way he flings that door open, he's gonna kill somebody.'"

Through the spring and summer of 1929 Lawrence told a number of physicists how important it was that R canceled R. Their attitude he later summed up at a lecture as follows: "Don't forget that having an idea and making it work are two different things." This they would tell him just before, out of sheer boredom, they changed the subject.

. . . . .

During the Christmas holidays of 1929, Lawrence made a trip east and saw his brother John, then in his third year at Harvard Medical School. Otto Stern, an experimental physicist from Hamburg, was also at Harvard on a visit. To the two brothers he seemed very old, though he was only about fifty, and very august for his Nobel prize work on molecular beams. Stern did not understand Prohibition. He asked Lawrence to take him to a speakeasy. The three drank red wine from wet glasses that left intricate red rings on the table cloth.

It looked almost like a diagram. Lawrence lapsed into talk of accelerations in circles of increasing radius. Stern listened without ever giving the usual hint that he would welcome a change of subject. Enraptured, Lawrence drew a design for an experimental apparatus on the cloth. Stern bent to examine it so deliberately that he gave an impression of computer wheels clicking in his head. "Ernest," he said, "don't just talk any more. You must get back to California and get to work on that."

Stern ended Lawrence's long contemplation. Back in Berkeley, he surveyed available facilities. Edlefsen at just that time finished his thesis. Being a laboratory instructor, he stayed at Berkeley to wait for the formalities of his doctor's in the spring. Lawrence decided Edlefsen needed something to do. Once again he drew the design for the apparatus. The two started the laboratory glass

blower and the brass fitter to making parts for a model. Lawrence issued no formal announcement of his project. But by this time every large physics laboratory in the world was seeking a way to break open the nucleus of the atom with high voltage. Experimental physicists everywhere began talking of Lawrence's new approach.

. . . . .

[In September of 1930 Lawrence gave an address before the National Academy of Sciences about his accelerator. After the introductions] ... a ripple of excitement passed over the audience of scientists. On the platform behind Lawrence was his invention: a four-inch pillbox sprouting arms like an octopus, each joint so heavily daubed with red sealing wax that it was hard to tell the thing was made of glass. It was mounted between the poles of an electromagnet. Transformers, oscillator-tube equipment, and a vacuum pump strewed the floor.

Lawrence turned it on. In the pillbox, hydrogen ions were flung back and forth in an alternating electric field. The magnetic field curved their path and turned it into a spiral. An electrometer showed ions striking a collector with an energy far higher than the voltage of the electric field. This, said Lawrence, was evidence of resonance achieved.

To break into the nucleus of the atom, he went on, ions must be accelerated to about a million volts of energy. A straightforward way to accelerate them would be between two electric poles charged with a difference in potential of a million volts. But because of the high electric fields involved, the difficulties of constructing such an apparatus were obvious. His new method avoided these difficulties. To get a million-volt acceleration in the circling ions, he needed only a bigger magnet, a bigger circular chamber, and an oscillator tube of 10,000 volts potential. "Experiments indicate there are probably no serious difficulties in the way," he concluded.

"SPEED HYDROGEN IONS TO BREAK UP ATOMS," *The New York Times* headlined its story of the meeting. "A new apparatus to hurl particles at a speed of 37,000 miles a second in an effort to obtain a long-sought goal—the breaking up of the atom—was described here today by Professor Ernest O. Lawrence of the University of California."

Lawrence had put Berkeley on the map. "If I ever get a swelled head, promise you'll let me know," he asked Birge. He was still a year short of thirty, the critical time of fruition for American physicists.

THE ENCHANTED LABORATORY

In September of 1930 Lawrence entangled his career with that of a student, Stanley Livingston, who came to know him in a special way. "I shall never forget him as long as I live," says Livingston with conviction usually absent from such protests. Livingston needed a thesis project. About the first of the month he called on Lawrence while making a tour of LeConte to see what experimental physicists could offer him.

Having lost Edlefsen by graduation, Lawrence needed someone else to build a bigger model of his invention. He mentioned other projects to Livingston—he was still interested in photoelectric experiments—but pushed this one with intoxicating urgency. Livingston was not a man to be hurried.

"Afterward I went on shopping to several others of the faculty, including Brode and Leonard Loeb, an authority on electrical discharge in gases," he said. "During my talks with them I commented on Lawrence's suggestion and asked their advice." The response he got was cooler than competitiveness for graduate students would justify. The conversation turned into questions on the path of the speeding ions. What difference did the curved path make? Spiral, circle, or straight line, would not the shotgun pattern of ions spread out anyway? How could enough of them be focused in any one place to be worth reaccelerating and finally collecting? "I remember Loeb in particular was quite pessimistic," Livingston recalled.

The invention worked, Lawrence answered when Livingston went back to him. By temperament Livingston inclined toward the dark side. But he found that in Lawrence's presence doubt disappeared: "Lawrence was young, he was bursting with energy, his enthusiasm swept me off my feet."

About the time Lawrence's invention started getting unheard-of newspaper publicity, Livingston signed up as his graduate student. First, he looked at Lawrence's and Edlefsen's calculations. Then cautiously scraping superfluous red sealing wax off the apparatus, he examined it painstakingly. After making sure that all connections were tight, he turned on the power.

The resonance or circular reacceleration which the apparatus was designed for could at that time be determined only by correlating the voltage of the ions with the intensity of the magnetic field. Livingston could not make the apparatus work. He decided it never had worked. Ordinary unreaccelerated ions were arriving at the collector and electrons were leaving it by photo-ionization. A slip in mathematics could cause these phenomena to be mistaken for resonance. Going over the original calcultions, Livingston looked for the slip and found it. "Lawrence and Edlefsen had made an error in their calibration of the magnetic field," he said.

A few days later, in *Science* magazine, he read Lawrence's summary of his talk to the National Academy and was specially struck by the conclusion that probably no serious difficulty lay in the way of reaching a million volts. "The confidence of the paper was based on the error of thinking cyclotron resonance had been observed," he reflected. "The concluding statement was a sample of Lawrence's optimism."

.   .   .   .   .

LeConte's laboratory assistants were half-envious, half-amused to see him poking glumly at the famous merry-go-round that he could not make go round. Lawrence called it a device for producing high energy without requiring a high voltage. Livingston called it a magnetic resonance accelerator. For the proposed new model, Lawrence got him a giant voltage-oscillator vacuum tube trade-named Radiotron. The "-tron" suggested something fancy and mysterious to the assistants. To annoy Livingston, they formed the habit of asking him what results he was getting on the cyclotron.

For eight weeks Livingston tinkered with the apparatus he had inherited from Edlefsen. "By December of 1930," he said, "I observed the legitimate resonance effect in the glass-vial tube originally used." He believed he was the first man to see it. It was too infinitesimal to prove anything except that Loeb's doubts were well founded. But it freed Livingston to get on with his thesis project, the construction of a more substantial model.

He discarded almost everything but the electromagnet and built a new pillbox of brass. It was four inches wide like the glass one, but precise and different inside. Within it and insulated from it nested the halves of a smaller pillbox sliced down the middle from top to bottom. He called the halves "Dees" because they looked like two capital D's facing straight side to. The alternating electric field was set up in the quarter inch of space between them.

Lawrence emphasized to Livingston that the hollow inside each Dee must be kept free of the electric field. Thus an ion received its push in the space between the Dees, glided through one of them in a spiraling half-circle because of the magnetic field, then when it came out received another push into the other Dee.

To keep the electric field out of the Dees, Livingston mounted a perforated brass strip across the lips of the open side of each Dee. He made it so delicately that it was four fifths holes. Even so, he realized that a fifth of the revolving ions were being stopped each time they got halfway round. He took off the strips and replaced them with fine wires that would constitute less of a barrier.

In March of 1931 with this arrangement he accelerated ions to an energy of eighty-thousand volts with an alternating charge on the Dees of only two thousand volts. The ions were going round about twenty times, picking up speed at each half-revolution.[2]

The final acceleration and, more particularly, the number of ions accelerated were still too trifling even to hint of any practical use. But Lawrence, his eyes blazing with enthusiasm over Livingston's shoulder, did not see them that way. He ordered an eleven-inch magnet and set Livingston to building a new cyclotron—after fighting the name, the two had begun to accept it—larger in proportion, the third in the series.

"Lawrence stimulated and supported me," said Livingston afterward, a little wryly. Brady, who was crucifying himself over his own thesis project in accordance with Lawrence's prescription, got used to

---

[2]In more detail the cyclotron, which was about as complicated as an automobile, worked as follows:

Hydrogen gas was trickled into the vacuum chamber. There as a first step it had to be converted into positive ions—that is, the protons that constituted its atomic nuclei had to be stripped of their attendant electrons. To break the grip of an electron on a hydrgoen nucleus, one only needs to strike the clinging electron with a free electron moving at any speed over 13.5 volts, a feeble energy which is called the ionization potential of hydrogen. A small electric field pulled out free electrons at this energy from a hot filament located between the cyclotron Dees to turn the gas into ions. Lines of magnetic force channeled the electrons away in a stream that did not interfere with the operation of electric and magnetic forces upon the ions.

Commercial 10,000-volt high-frequency radio oscillator tubes gave each of the two Dees alternately a positive, then a negative electric charge. When a Dee was negatively charged, it attracted the freshly created positive ions. The ions plunged into it. Momentum would have carried them in a straight line until they hit the curved metal back of the Dee, except for the effect of the magnetic field. From this arose Lawrence's big discovery. The magnetic field exerted upon the moving ions a force continually at right angles to their direction of motion. Thus the ions were pulled sideways in a curve. Some of them curved all the way round and were already headed back in the opposite direction at the instant that the electric charge was alternated and the Dee they were traveling in became positively charged. The ions moved with increasing speed into the opposite Dee, where the process was repeated. As they gained speed, centrifugal force made them move in wider circles. But traveling faster, they could still cover the longer path in the same time and would theoretically be reaccelerated a third time, and then many more times thereafter. This was the reason for Lawrence's excitement over the discovery that the time of the ions' revolution was independent of the radius of their orbit, or that R canceled R.

watching a small evening drama. Livingston had a wife whom he had hardly seen since their marriage.

"Well, it's after six," Livingston would say. "Lois said she was planning a big dinner. She told me I had to get home this time sure."

"Just a few minutes," Lawrence would reply. "You told me yourself you were having trouble with the vacuum seal. It shouldn't take long."

Restimulated, Livingston would work far into the night. Vaguely he thought of himself as a permanency among transients whose movements at the periphery of his attention gave him some notion of the passage of time. Most would leave promptly at five, others like Brady a little later, as it seemed to him. Latest would be Lawrence, come back to check on the stimulation. "Had a night on the town," Livingston would think, looking at Lawrence's fresh collar with a sort of dull resentment.

Not even Lawrence could expect to drive Livingston to build the eleven-inch before the end of the academic year. Consequently in early April of 1931 Lawrence broke in on Livingston's concentration with a reminder of something Livingston had almost forgotten. "If you get your doctorate this spring, I can get you an instructorship next year," Lawrence said. "But you must get your doctorate."

The deadline for turning in a thesis and taking the required examination was then only three weeks away. "I don't have time to read the background material," Livingston protested.

"Don't worry," Lawrence replied. "We're making history. We can go back to the basic literature any time. Just write your thesis."

Livingston dreaded Birge, a pitiless examiner who was to sit on his committee and who so far had never let human considerations sway him. But the ordeal turned out all right: "When Birge asked me about Rutherford's historic developments in England, I could feel my ignorance showing. I hadn't any time to study the literature. Birge was a precisionist. I'm sure I flunked him

[flunked his examination]. Lawrence defended me. There's no doubt in my mind it was Lawrence who got me my pass. To me, this is a marvelous story of friendship between people doing exciting things together."

Though the cyclotron lacked visibly moving parts, everyone called it a machine, and though it was small, Lawrence was making it grow fast. First documentations were Lawrence's premature announcement in *Science* and Livingston's thesis, *The Production of High-Velocity Hydrogen Ions without the Use of High Voltage*. These inaugurated the country's most distinctive contribution to the nuclear age—big-machine physics. It began with a characteristic inaccuracy over the time element. "Around the lab they started calling Livingston the Ninety-day Wonder," recollects a graduate student. "But this was wrong. He didn't have anywhere near that much time to bone up on physics for his doctorate."

.   .   .   .   .

To be of much use as a research tool, the cycolotron would have to produce a beam of several million ions a second. Nothing like this was in sight, either for the four-inch or the unfinished eleven-inch which Livingston was working on. Furthermore, the energy of the ions was far too low in any reasonable projection one could make of the existing design. Each time the ions spiraled outward they gained energy, but each time a good proportion of them would also bump against the top or bottom of the hollow Dees and disappear. If the collector was placed far out to catch the fast ions, there would not be enough of them. If the collector was moved in to catch the slow ones, they would not have enough energy.

All this was what Loeb had pessimistically predicted, and Livingston still had no answer to make to him. Lawrence had stressed the need for keeping the hollow space inside the Dees free of the electric field. But Livingston had increased the yield of ions by replacing the perforated guard with a grid of wires.

In the unaccustomed stillness at Berkeley, Livingston cogitated. "I took the opportunity to strike out on my own. Hoping to get higher intensity, I removed the grid of wires. The intensity then became a hundred times what it was before. When Lawrence returned to Berkeley, he helped me to understand this was a consequence of focusing the ions by means of the electric field inside the Dees."

Ions spiraling up or down were pushed toward a central plane by the newly admitted electric field. The principle, electrostatic focusing, was the second of three fundamental discoveries necessary to make the cyclotron useful. Elsewhere, other scientists were making an independent discovery and application of electrostatic focusing. "About the same time," said Livingston, "it was being developed for electron lenses in electronics." But he had no conscious knowledge of this when he removed the grid of wires.

In July of 1931, Lawrence and Livingston worked on the eleven-inch. The two men did not yet know what it could do, but Lawrence already dreamed of something bigger. Each successive cyclotron had to start with an electromagnet. That of the eleven-inch weighed two tons. Leonard Fuller, chairman of electrical engineering, told Lawrence of one that weighed eighty-five tons rusting away in a storage dump in Palo Alto. Built for radio transmission across the Pacific, it had been made obsolete by the vacuum tube. Fuller, vice-president of the near-defunct Federal Telegraph Company, which owned it, offered it to Lawrence as a gift.

In the Flying Cloud[3] Lawrence and Livingston drove down to Palo Alto to look it over. "It was in bad shape," said Livingston. He thought of all the work required to re-sculpture and wind the core. Lawrence thought of how he could haul it to Berkeley and where he could put it. Already, the rest of the physics faculty had started to complain that his projects were crowding LeConte's second-floor workrooms. Not even he could have cheerful illusions about what they

would say if they saw him trying to hoist an eighty-five ton magnet in.

While Lawrence pondered this problem at Berkeley, a tall, gray-haired, unpressed eccentric inventor named Frederick Cottrell made a tour of the laboratory. From bulging packets he took a half-dozen pairs of dime-store glasses one after another and through them peered nearsightedly at the two cyclotrons. Cottrell was rich himself and distinguished for a knack of persuading other rich people to give money to science, of which he was the country's best-known patron. "I had the pleasure of conducting him through the physics laboratory and showing him our work on the development of means for accelerating charged particles to high velocities," said Lawrence. "He immediately appreciated the possibilities of the cyclotron principle and with infectious enthusiasm volunteered to help in any way he could."

Lawrence and Livingston took him to Palo Alto to see the big magnet. According to a legend hard to authenticate, Lawrence asked Cottrell to help him raise $500 for haulage. Cottrell replied that $5,000 would be easier to raise, since subsidizers of research judged the respectability of a request by its size.

Cottrell promised to see what he could do and went to Long Beach for a holiday before returning to his headquarters in New York. Lawrence and Livingston put the eleven-inch into operation. For the first time electrostatic focusing was tried out in a machine of usable size. Both men were astonished and excited, Lawrence exactly twice as much. A million volts, a half million volts: the report depended on who made it. Lawrence wrote two letters dated July 20, 1931.

One to Cottrell, Lawrence wrote alone:

I telegraphed you last night asking if it is convenient for me to come to see you Wednesday or Thursday . . . on my way east.

As you have probably surmised, I want to get your advice on the present best procedure

[3]Lawrence's automobile. [Ed.]

toward raising some more money for the high-speed proton experiments. The present developments in the work make it practically certain that we will be able to produce ten- or twenty-million-volt protons when the large magnet and a powerful oscillator are made available to us. With the present setup we have exceeded one million volts and the proton currents are very much more intense than we had expected. . . .

It seems to me that at least $10,000 and possibly $15,000 will be needed to cover adequately the expense of all the equipment needed for the work, and the money must be found somewhere . . . .

It appears that our difficulties are no longer of a physical but of financial nature.

The other, to the editor of the *Physical Review*, Lawrence and Livingston wrote together:

A method for the production of high-speed protons was described before the National Academy of Science last September. . . . Later before the American Physical Society . . . results were presented . . . and the conclusions . . . that there are no significant difficulties in the way of producing 1,000,000 volt protons. . . .

This important conclusion has now been confirmed. . . . Protons and hydrogen molecule ions having energies in excess of one half million volts have been produced. . . . Currents turned out to be surprisingly large. . . .

There can be little doubt that one-million-volt ions will be produced . . . when the present experimental tube is enlarged.

"Lawrence had the sort of genius which was 'way ahead of time," Livingston explained. Lawrence went east again. At Harvard he made his brother's medical associates uneasy by the way he talked cyclotrons. "Is he going out of his mind?" they asked. He was not. From Cottrell he obtained $5,000 and an introduction to Roger Adams, adviser to the Chemical Foundation, which had been created to subsidize research with money from war-confiscated patents. "Has vision," Adams decided, and he granted him $2,500. After a stop with Molly[4] at New Haven,

Lawrence returned to Berkeley in September for a passionate talk with President Sproul. He needed a home for the big magnet and a few thousands more toward the cost of turning it into a cyclotron. The university was about to tear down a shack just east of LeConte Hall. Lawrence showed in a letter to his family how lyrically he viewed the shack and how well he dealt with Sproul:

. . . The president has turned over a whole building (the old Civil Engineering laboratory) for our experiments and has put at our disposal $3,300 in addition to the $7,500 I got in N.Y. We are getting some more things as gifts, so the $10,800 will be adequate. I'm busy now in the process of spending it most effectively. We are going to have a great year of it. Only time can tell the outcome of our efforts, but I have hopes of doing some really important work in the not-too-distant future. . . .

Though I'm busy I naturally have time to write Molly every other day. Boy, I certainly am in love with her. . . . Molly had her engagement announced at the Lawn Club a week ago Wednesday. Now, I'm hoping she will decide on *next* June instead of a year later!

If it weren't for the unusually interesting state of our work I don't know what I would do without Molly! . . .

By the way, get the habit of using the air mail.

To Lawrence invention had been joy; to Livingston the need for it was a curse. The eleven-inch magnet was giving Livingston trouble. True, electrostatic focusing now held the ions flat as a watch spring in the first part of their spiraling course. But about a third of the way out, too many of them began to wobble and disappear. The loss was worse with the eleven-inch than with the four-inch, so the outlook for the twenty-seven-inch was cloudy.

Livingston interpreted the loss to mean there were irregularities in the magnetic field. No amount of truing the faces of the magnet poles ended the problem. Swearing, he gave up trying to make them perfectly flat and instead took an irrational step. He

---

[4]Mary Blumer, Lawrence's fiancée. [Ed.]

cut paper-thin sheet iron into pieces shaped like the silverfish moth that eats clothing. They were wide at one end and narrow at the other. To him they seemed exclamation points without the period. He laid them atop the pillbox vacuum chamber with the wide end toward the center.

Acceleration increased four times. The ions held their watch-spring path all the way out to the collector and arrived in a stream only a millimeter thick.

This was the invention that completed the cyclotron as the world knows it. The effect was immediately clear; the theory took much longer to work out than had that of electrostatic focusing. "Magnetic focusing came purely pragmatically," said Livingston. "Lawrence and I had not thought of it. I accomplished it by putting in shims of iron near the pole face with the narrow end outward. This had a large effect on small orbits and a smaller effect on large orbits. It was thus strictly empirical proof that cyclotrons had to have a magnetic field which decreased from the center outward. Eventually we realized that this was magnetic focusing."

Lawrence's sparkling eyes, pale blue and a little brighter than human—these are what Livingston remembers when at last he produced a million volts. He brought the eleven-inch, extensively rebuilt, to this peak in February of 1932. By now he was using protons, single hydrogen ions with more penetrating power than the paired molecular ions he had accelerated to half a million volts in July. "I wrote the figure on the blackboard," he said. "Lawrence came in late one evening. He saw the board, looked at the microammeter to check the resonance current, and literally danced around the room."

Lawrence spread the news next morning at breakfast in the Faculty Club. Livingston said: "We were busy all that day demonstrating million-volt protons to eager viewers."

Lawrence and Livingston offered the cyclotron to the world as the best and most effective machine for nuclear research. The cyclotron was now completely dependable,

they reported in a historic article in the *Physical Review*: "It is well to emphasize two particular features that have contributed more than anything else to the effectiveness of the method: *the focusing action of the electric and magnetic fields*, and the *simple means of empirically correcting the magnetic field* by the introduction of suitable iron shims."

"TO USE BIG MAGNET TO BREAK UP ATOM," ran an AP lead. "MAY TRY TO CREATE GOLD," said *The New York Times*. The *San Francisco Examiner* sent a reporter out to LeConte, and he returned awed.

Two University of California scientists are setting about trying to break up the atom and release its terrific energy. Working only with a two-ton magnet, the scientists, Professor Ernest O. Lawrence and M. Stanley Livingston, say they have been able to penetrate the outer husk of the atom. [This may be a confused reference to ionization experiments.] With the greater magnet, they hope to shatter the atom completely with an ultimate 25,000,000-volt impact. What wonders will result, only time can tell.

"Clearly . . . there are no difficulties in the way of producing one-million-volt ions," Lawrence had told the Physical Society in May of 1931 before the two focusing discoveries. Doubters like Loeb had based their objection on focusing. Now that Lawrence had a complete double-barreled answer to the objection, he phrased it himself forcefully: "Consideration of this matter might lead one to believe that it is a requirement impossible to achieve. It is therefore to be emphasized that this requirement has been obviated . . . by focusing action of the electric field . . . and by . . . focusing action due to curvature of the magnetic field."

This he wrote in an application for a patent on the cyclotron, which was received at the Patent Office on January 26, 1932. "I claim. . . ." he said, in the arrogant style which patent procedure forces on the solitary inventor. He did not mention Livingston.

Of course he did not need to; a graduate student is expected to contribute cheerfully not only his hands but such brains as he has. Lawrence gave the patent to Cottrell, who treated it as a mere symbol of achievement, a free license to all cyclotron builders. In the *Physical Review* Lawrence and Livingston had seemed to have a special relation. Outside it, Lawrence increasingly tended to credit the cyclotron's realization to many undifferentiated hands, to the whole circle of graduate and postdoctoral students enlarging around him. This was good for their morale, but not for Livingston's. He felt a youthful hurt that afterward turned into a lasting bitterness.

"A sorehead" Livingston is sometimes called by other physicists from California. The question of his share in the creation of the cyclotron divides physicists generally. Brady doubts Livingston was first to see resonance, electrostatic focusing, and magnetic focusing. "But even if this were true," he says, "the work was closely supervised by Lawrence. The impression I have is that Lawrence made all the major improvements. I think he worked out the whole focusing problem after the proton beam proved much larger and more intense than he had calculated. I don't think a day went by when Lawrence was in town that he didn't come to the lab to operate the controls.

"I'll have to admit my opinions are just those I picked up from the talk then going on in the lab. Anyway, Livingston seems to me to be afraid he won't get enough credit. He told me he had complained about this to Lawrence. Lawrence told him: 'If you're dissatisfied, you can drop out of the project, and I'll get someone else who can take your place.'"

"A matter of clashing personalities," say physicists who profess impartiality. Most accept Livingston's account. At Berkeley it came to be considered bad taste or worse to discuss the issue. As a result, Lawrence's definite original contributions to the cyclotron after the original idea have become unnecessarily clouded. There is a fuzzy overlap between the limited focusing action at the perimeter of a level magnetic field and the all-pervasive focusing action of a continually declining magnetic field. Of the former, which would not have helped much with large

cyclotrons, Lawrence may have planned to make use before Livingston discovered the latter.

As for Lawrence's chilling offer to Livingston to replace him, it is by no means clear how quickly this could have been done. "I stayed on, developing larger and larger cyclotrons," said Livingston. "I was the mechanic—no one else then could do it—and Lawrence promoted the money."

On May 3, 1932, the Berkeley *Gazette* reported:

Dr. Ernest O. Lawrence, world-renowned authority on the atom, will wed Miss Mary Blumer of New Haven, Connecticut, on May 14, in the eastern city, according to word received by friends in Berkeley. Dr. Lawrence left a fortnight ago to attend the meeting of the American Physical Society and read a paper there.

Brady at this time was working on the first research experiment ever attempted with the cyclotron. "It was because I had some free time," he explained. "Lawrence had held me a long while on my thin-films thesis project. Friends told me it was too long." ("He means me," interjected Mrs. Brady. "I told him. I wanted to get married.") "So I went to Lawrence and wailed, 'You could keep me at this for ten years,' and he let me finish and got me onto the faculty of St. Louis University for the next fall. Meanwhile, I was in LeConte with a fellowship and some free time. Livingston was over in the shack working hard on the twenty-seven-inch. But a graduate student named Milton White had got to wondering whether the eleven-inch really *was* accelerating high-speed particles. The only evidence so far had to be deduced from calculations of the magnetic and electric fields. He wanted something more concrete. So he got the idea of seeing whether the cyclotron could accelerate protons fast enough to shoot them through mica sheets of varying thickness. We mounted the sheets around the circumference of a wheel and set it inside the vacuum chamber just in front of the collector.

"Getting this done took us about to the end of July. I was fiddling with an improvement to the proton-injector gun, which had never worked well, just getting ready to shoot protons through mica. I never did. A telegram from Lawrence stopped me cold. It was about two physicists in England. 'Cock-croft and Walton have disintegrated the lithium atom,' it said. 'Get lithium from chemistry department and start preparations to repeat with cyclotron. Will be back shortly.'

"Lawrence was honeymooning on a boat in Connecticut. Must have put into shore some-where. I showed the telegram to the future Mrs. Brady and said, 'That's what physicists on their honeymoon think about.'"

"It was bad news to me." she recollects. "We'd planned to slip away the first week in August as soon as he did the mica thing. Now this meant we'd have to wait until they'd disintegrated the atom. I wanted to get married."

For Lawrence too it was bad news that he and the United States with such a unique and incomparable tool as the cyclotron could not have been first to penetrate the nucleus. Surprisingly, the English had done it with an old-fashioned conventional voltage gene-rator. "At that time it was generally thought that nothing much would happen in the way of nuclear reactions below about one mil-lion electron volts," Lawrence explained in a lecture. "In England Cockcroft and Walton at the Cavendish Laboratory . . . realizing the difficulties of reaching one million . . . with their apparatus decided to try with protons of energies of only several hundred thousand. They immediately discovered lithium and beryllium could be disintegrated with ener-gies as low as a quarter million. . . . We could have done so with our eleven-inch cyclotron almost a year before."

.   .   .   .   .

Lawrence and Oppenheimer were not the only Berkeley physicists active in research in 1933. Birge, that precise man, had long been critical of water. He complained in print that its principal ingredient, hydrogen, weighted 1/6000 more than all the textbooks said was its atomic weight. Gilbert Lewis, of the chemis-try department, who had helped bring about

Lawrence's promotion, began looking for what made a million hydrogen atoms weight more than a million times the theoretical atomic weight of a single one. Columbia University broke Lewis' and Birge's hearts by discovering the isotope deuterium, the odd one-in-five-thousand overweight hydrogen atom, just when Lewis was about to do so. Still Lewis went doggedly on with his investi-gation. At the start of 1933 he began pre-paring the world's first usable samples of heavy water (H_2^2O).

Lawrence glued himself to Lewis' shoulder and goaded him to produce faster. He wanted to accelerate deuterium in the cyclo-tron. All ions known up till then carried the burden of a repellent charge proportional to their mass. The lightest was, therefore, the most effective against the nucleus. Deuterium promised to become an exception. It carried no more repellent charge than an ordinary hydrogen ion or proton, but it had twice the mass and as a projectile might prove twice as effective. In 1932 an uncharged particle called the neutron had been discovered by an Englishman named Chadwick. Scientists speculated that the deuterium ion was a pro-ton and a neutron bound together.

Lawrence kept asking Lewis how much heavy water he had until about the first of March Lewis was able to show him a whole cubic centimeter. It was enough to accelerate, but at this point Lewis proved to be no physicist. Worried about whether he had manufactured a poison, he fed the whole sample to a mouse. It brought no good or harm to the mouse, but to Lawrence it almost brought apoplexy. "This was the most ex-pensive cocktail that I think mouse or man ever had!" he complained.

Lewis began accumulating another sample of heavy water. Livingston and the other re-search associates continued work on the twenty-seven-inch cyclotron. "We tuned it up slowly to higher energy by moving the col-lector outward," said Livingston. Often they would try placing the collector too far out and lose the beam.

.   .   .   .   .

The process was painful. The pillbox vacuum chamber was held together by Lawrence's red sealing wax. When taken apart, it had to be rewaxed carefully. To check for leaks, the wax had to be sprayed with alcohol, and then it would crumble. Everyone including Lawrence spent a good deal of time checking for pinpoint holes and resealing them with a gas burner.

At the center of the pillbox, an electric filament created the ions. The volume or intensity of the beam depended on the amount of current to the filament. It was held in place by a gob of wax, and when it burned out, no one knew exactly where to put the new one. The whole dismantling operation might have to be repeated several times.

"Lawrence loved to sit at the controls," says Aebersold. "It worried the rest of us. He always wanted to see how far he could push the meter up."

Livingood's recollections have the slangy elegance to be expected from a man with a classical education.

"Lawrence would say, 'Damn it, we've got this machine adjusted,'" Livingood recalls. "'Don't anybody go over the pencil mark I've put on the meter.'

"Then he'd reach for the controls himself. Bingo! Another filament burned out.

"'Well, I'll go play tennis,' he'd say. 'Fix it while I'm gone.'"

Through such crises the staff suffered until repetition and familiarity made them engineering problems, if only Lawrence had had the money to hire engineers. Late in March, Lewis brought pure science back into the Laboratory in the form of another sample of heavy water. While Lewis sat puffing a cigar and watching thoughtfully, they fed in the deuterium and tuned the beam against a lithium crystal. As before, it disintegrated into helium, but at a rate far faster than anyone had yet seen.

Now truly alchemists, they also converted aluminum, nitrogen, and half a dozen other light elements into helium. The helium nuclei, which are usually called alpha particles, flew off into the cloud chamber in glittering tracks well over three centimeters long. This showed the disintegration was proceeding with unexpected energy. Moreover, the targets also produced a stream of Chadwick's new particle, the neutron, on a surprising scale.

"Deuterium was a time of intense excitement when everything was new," said Livingston. "We bombarded targets with it and observed proton and neutron particles in the highest intensities ever seen by man." If one counted only the deuterium ions that actually hit and disintegrated a target nucleus, more energy was being produced than was consumed. They constituted what physicists call an exothermic system. Lawrence and his staff talked of a new possibility. If one could construct a large enough exothermic system, power might be generated to heat and light a city. Innocently, they did not think what else might be done to a city by such a system.

.   .   .   .   .

In that enchanted laboratory the Geiger counter was wired to the same switch as the cyclotron. By turning them both on and off at the same time, one could get on with operations faster. This is the reason the big new idea of 1934 had to make its way in from the outside.

On February 20, 1934, the staff was bombarding boron with deuterium ions. That afternoon Lawrence came running through the door waving a French journal with an article by Frédéric Joliot, who had saturninely watched his discomfiture at Brussels without adding to it. "It is now possible for the first time to create radioactivity in certain elements," Lawrence translated haltingly for the staff. Joliot had used a petty apparatus powered by a minute quantity of radium emanations. While writing his article, he had remembered Lawrence and closed with a postscript on what Lawrence could produce that was beyond Joliot's own power. "For example, Nitrogen 13, which according to our hypothesis is radioactive, could be obtained by bombarding carbon with deuterium ions."

Like puppets on a string, the staff changed the wiring of the Geiger counter and swung a carbon target into the beam. Five minutes later they turned off the cyclotron. "*Click . . . click . . . click* went the Geiger counter," said Livingston. "It was a sound that no one who was there would ever forget." Thus the first Nitrogen 13 ever known to exist in the world announced its presence.

The implications for medicine, biology, and the approaching nuclear age were enormous. The staff had often bombarded carbon, but always before had switched off the Geiger counter with the beam. Now they stared at each other with an identical expression on their faces.

"We looked pretty silly," said Thornton. "We could have made the discovery any time."

Livingood, the most highly educated, interpreted most tersely: "We felt like kicking each other's butts."

7.5   *Disintegration of Uranium by Neutrons: A New Type of Nuclear Reaction*
Lise Meitner and O. R. Frisch

On bombarding uranium with neutrons, Fermi and collaborators found that at least four radioactive substances were produced, to two of which atomic numbers larger than 92 were ascribed. Further investigations demonstrated the existence of at least nine radioactive periods, six of which were assigned to elements beyond uranium, and nuclear isomerism had to be assumed in order to account for their chemical behavior together with their genetic relations.

In making chemical assignments, it was always assumed that these radioactive bodies had atomic numbers near that of the element bombarded, since only particles with one or two charges were known to be emitted from nuclei. A body, for example, with similar properties to those of osmium was assumed to be eka-osmium ($Z = 94$) rather than osmium ($Z = 76$) or ruthenium ($Z = 44$).

Following up an observation of Curie and Savitch, Hahn and Strassmann found that a group of at least three radioactive bodies, formed from uranium under neutron bombardment, were chemically similar to barium and, therefore, presumably isotopic with radium. Further investigation, however, showed that it was impossible to separate these bodies from barium (although mesothorium, an isotope of radium, was readily separated in the same experiment), so that Hahn and Strassmann were forced to conclude that *isotopes of barium ($Z = 56$) are formed as a consequence of the bombardment of uranium ($Z = 92$) with neutrons.*

At first sight this result seems very hard to understand. The formation of elements much below uranium has been considered before, but was always rejected for physical reasons, so long as the chemical evidence was not entirely clear cut. The emission, within a short time, of a large number of charged particles may be regarded as excluded by the small penetrability of the "Coulomb barrier," indicated by Gamow's theory of alpha decay.

On the basis, however, of present ideas about the behavior of heavy nuclei, an entirely different and essentially classical picture of these new disintegration processes suggests itself. On account of their close packing and strong energy exchange, the particles in a heavy nucleus would be expected to move in a collective way which has some resemblance to the movement of a liquid drop. If the movement is made sufficiently violent by adding energy, such a drop may divide itself into two smaller drops.

In the discussion of the energies involved in the deformation of nuclei, the concept of surface tension of nuclear matter has been

SOURCE. From *Nature* **143**: 239 (1939).

used and its value has been estimated from simple considerations regarding nuclear forces. It must be remembered, however, that the surface tension of a charged droplet is diminished by its charge, and a rough estimate shows that the surface tension of nuclei, decreasing with increasing nuclear charge, may become zero for atomic numbers of the order of 100.

It seems therefore possible that the uranium nucleus has only small stability of form and may, after neutron capture, divide itself into two nuclei of roughly equal size (the precise ratio of sizes depending on finer structural features and perhaps partly on chance). These two nuclei will repel each other and should gain a total kinetic energy of $\sim$ 200 MeV, as calculated from nuclear radius and charge. This amount of energy may actually be expected to be available from the difference in packing fraction between uranium and the elements in the middle of the periodic system. The whole "fission" process can thus be described in an essentially classical way, without having to consider quantum-mechanical "tunnel effects," which would actually be extremely small, on account of the large masses involved.

After division, the high neutron/proton ratio of uranium will tend to readjust itself by beta decay to the lower value suitable for lighter elements. Probably each part will thus give rise to a chain of disintegrations. If one of the parts is an isotope of barium the other will be krypton $(Z = 92 - 56)$, which might decay through rubidium, strontium and yttrium to zirconium. Perhaps one or two of the supposed barium-lanthanum-cerium chains are then actually strontium-yttrium-zirconium chains.

It is possible, and seems to us rather probable, that the periods which have been ascribed to elements beyond uranium are also due to light elements. From the chemical evidence, the two short periods (10 sec and 40 sec) so far ascribed to ^{239}U might be technium isotopes $(Z = 43)$ decaying through ruthenium, rhodium, palladium and silver into cadmium.

In all these cases it might not be necessary to assume nuclear isomerism; but the different radioactive periods belonging to the same chemical element may then be attributed to different isotopes of this element, since varying proportions of neutrons may be given to the two parts of the uranium nucleus

By bombarding thorium with neutrons, activities are obtained which have been ascribed to radium and actinium isotopes. Some of these periods are approximately equal to periods of barium and lanthanum isotopes resulting from the bombardment of uranium. We should therefore like to suggest that these periods are due to a "fission" of thorium which is like that of uranium and results partly in the same products. Of course it would be especially interesting if one could obtain one of these products from a light element, for example, by means of neutron capture.

It might be mentioned that the body with half-life 24 min which was chemically identified with uranium is probably really ^{239}U, and goes over into an eka-rhenium which appears inactive but may decay slowly, probably with emission of alpha particles. (From inspection of the natural radioactive elements, ^{239}U cannot be expected to give more than one or two beta decays; the long chain of observed decays has always puzzled us.) The formation of this body is a typical resonance process; the compound state must have a life-time a million times longer than the time it would take the nucleus to divide itself. Perhaps this state corresponds to some highly symmetrical type of motion of nuclear matter which does not favor "fission" of the nucleus.

7.6   *Letter to President Roosevelt*
                                  Albert Einstein

                                              Albert Einstein
                                              Old Grove Road
                                              Nassau Point
                                              Peconic, Long Island
                                              August 2, 1939

F. D. Roosevelt
President of the United States
White House
Washington, D.C.

Sir:

Some recent work by E. Fermi and L. Szilard, which has been communicated to me in manuscript, leads me to expect that the element uranium may be turned into a new and important source of energy in the immediate future. Certain aspects of the situation seem to call for watchfulness and, if necessary, quick action on the part of the Administration. I believe, therefore, that it is my duty to bring to your attention the following facts and recommendations.

In the course of the last four months it has been made probable—through the work of Joliot in France as well as Fermi and Szilard in America—that it may become possible to set up nuclear chain reactions in a large mass of uranium, by which vast amounts of power and large quantities of new radium-like elements would be generated. Now it appears almost certain that this could be achieved in the immediate future.

This new phenomenon would also lead to the construction of bombs, and it is conceivable—though much less certain—that extremely powerful bombs of a new type may thus be constructed. A single bomb of this type, carried by boat or exploded in a port, might very well destroy the whole port together with some of the surrounding territory. However, such bombs might very well prove to be too heavy for transportation by air.

The United States has only very poor ores of uranium in moderate quantities. There is some good ore in Canada and the former Czechoslovakia, while the most important source of uranium is the Belgian Congo.

In view of this situation you may think it desirable to have some permanent contact maintained between the Administration and the group of physicists working on chain reactions in America. One possible way of achieving this might be for you to entrust with this task a person who has your confidence and who could perhaps serve in an unofficial capacity. His task might comprise the following:

a) To approach Government Departments, keep them informed of the further developments, and put forward recommendations for Government action, giving particular attention to the problem of securing a supply of uranium ore for the United States.

SOURCE. From *Einstein on Peace*, edited by Otto Nathan, and Heinz Norden (New York: Simon and Schuster, 1960). Reprinted by permission of the Estate of Albert Einstein.

b) To speed up the experimental work which is at present being carried on within the limits of the budgets of University laboratories, by providing funds, if such funds be required, through his contacts with private persons who are willing to make contributions for this cause, and perhaps also by obtaining the cooperation of industrial laboratories which have the necessary equipment.

I understand that Germany has actually stopped the sale of uranium from the Czechoslovakian mines which she has taken over. That she should have taken such early action might perhaps be understood on the ground that the son of the German Under-Secretary of State, von Weizsäcker, is attached to the Kaiser Wilhelm Institut in Berlin, where some of the American work on uranium is now being repeated.

Yours very truly,

A. Einstein

7.7   *The First Atomic Pile*

Corbin Allardice and Edward R. Trapnell

On December 2, 1942, man first initiated a self-sustaining nuclear chain reaction, and controlled it.

Beneath the West Stands of Stagg Field, Chicago, late in the afternoon of that day, a small group of scientists witnessed the advent of a new era in science. History was made in what had been a squash-rackets court.

Precisely at 3:25 P.M., Chicago time, scientist George Weil withdrew the cadmium-plated control rod and by his action man unleashed and controlled the energy of the atom.

As those who witnessed the experiment became aware of what had happened, smiles spread over their faces and a quiet ripple of applause could be heard. It was a tribute to Enrico Fermi, Noble Prize winner, to whom, more than to any other person, the success of the experiment was due.

Fermi, born in Rome, Italy, on September 29, 1901, had been working with uranium for many years. Awarded the Nobel Prize in 1938, he and his family went to Sweden to receive the prize. The Italian Fascist press severely criticized him for not wearing a Fascist uniform and failing to give the Fascist salute

when he received the award. The Fermis never returned to Italy.

From Sweden, having taken most of his personal possessions with him, Fermi proceeded to London and thence to America where he remained.[1]

The modern Italian explorer of the unknown was in Chicago that cold December day in 1942. An outsider looking into the squash court where Fermi was working would have been greeted by a strange sight. In the center of the 30-by-60-foot room, shrouded on all but one side by a gray balloon-cloth envelope, was a pile of black bricks and wooden timers, square at the bottom and a flattened sphere on top. Up to half of its height, its sides were straight. The top half was domed, like a beehive. During the construction of this crude-appearing but complex pile (the name which has since been applied to all such devices) the standing joke among the scientists working on it was: "If people could see what we're doing with a million and a half of their dollars, they'd think we are crazy. If they knew why we were doing it, they'd be sure we are."

In relation to the atomic bomb program,

SOURCE. From *The new Treasury of Science*, Edited by H. Shapley, S. Rapport, and H. Wright (New York: Harper and Row, 1965).

[1]Until his death in 1954. [Ed.]

of which the Chicago pile experiment was a key part, the successful result, reported on December 2, formed one more piece for the jigsaw puzzle which was atomic energy. Confirmation of the chain reactor studies was an inspiration to the leaders of the bomb project, and reassuring at the same time because the Army's Manhattan Engineer District had moved ahead on many fronts. Contract negotiations were under way to build production-scale nuclear chain reactors, land had been acquired at Oak Ridge, Tennessee, and millions of dollars had been obligated.

Three years before the December 2 experiment it had been discovered that when an atom of uranium was bombarded by neutrons, the uranium atom sometimes was split, or fissioned. Later it had been found that when an atom of uranium fissioned, additional neutrons were emitted and became available for further reaction with other uranium atoms. These facts implied the possibility of a chain reaction, similar in certain respects to the reaction which is the source of the sun's energy. The facts further indicated that if a sufficient quantity of uranium could be brought together under the proper conditions, a self-sustaining chain reaction would result. This quantity of uranium necessary for a chain reaction under given conditions is known as the critical mass, or more commonly, the "critical size" of the particular pile.

Further impetus to the work on a uranium reactor was given by the discovery of plutonium at the Radiation Laboratory, Berkeley, California, in March, 1940. This element, unknown in nature, was formed by uranium 238 capturing a neutron, and thence undergoing two successive changes in atomic structure with the emission of beta particles. Plutonium, it was believed, would undergo fission as did the rare isotope of uranium, U^{235}.

Meanwhile at Columbia Fermi and Walter Zinn and their associates were working to determine operationally possible designs of a uranium chain reactor. Among other things, they had to find a suitable moderating material to slow down the neutrons traveling at relatively fast velocities. In July, 1941, experiments with uranium were started to obtain measurements of the reproduction factor (called "k"), which was the key to the problem of a chain reaction. If this factor could be made sufficiently greater than 1, a chain reaction could be made to take place in a mass of material of practical dimensions. If it were less than 1, no chain reaction could occur.

Since impurities in the uranium and in the moderator would capture neutrons and make them unavailable for further reactions, and since neutrons would escape from the pile without encountering uranium 235 atoms, it was not known whether a value for "k" greater than unity could ever be obtained.

Fortunate it was that the obtaining of a reproduction factor greater than 1 was a complex and difficult problem. If Hitler's scientists had discovered the secret of controlling the neutrons and had obtained a working value of "k," they would have been well on the way toward producing an atomic bomb for the Nazis.

One of the first things that had to be determined was how best to place the uranium in the reactor. Fermi and Leo Szilard suggested placing the uranium in a matrix of the moderating material, thus forming a cubical lattice of uranium. This placement appeared to offer the best opportunity for a neutron to encounter a uranium atom. Of all the materials which possessed the proper moderating qualities, graphite was the only one which could be obtained in sufficient quantity of the desired degree of purity.

The study of graphite-uranium lattice reactors was started at Columbia in July, 1941, but after reorganization of the uranium project in December, 1941, Arthur H. Compton was placed in charge of this phase of the work, under the Office of Scientific Research and Development, and it was decided that the chain reactor program should be concentrated at the University of Chicago. Consequently early in 1942 the Columbia and Princeton groups were transferred to Chicago, where the Metallurgical Laboratory was established.

At Chicago, the work on sub-critical size piles was continued. By July, 1942, the measurements obtained from these experimental piles had gone far enough to permit a choice of design for a test pile of critical size. At that time, the dies for the pressing of the uranium oxides were designed by Zinn and ordered made. It was a fateful step, since the entire construction of the pile depended upon the shape and size of the uranium pieces.

It was necessary to use uranium oxides because metallic uranium of the desired degree of purity did not exist. Although several manufacturers were attempting to produce the uranium metal, it was not until November that any appreciable amount was available.

Although the dies for the pressing of the uranium oxides were designed in July, additional measurements were necessary to obtain information about controlling the reaction, to revise estimates as to the final critical size of the pile, and to develop other data. Thirty experimental sub-critical piles were constructed before the final pile was completed.

Meantime, in Washington, Vannevar Bush, Director of the Office of Scientific Research and Development, had recommended to President Roosevelt that a special Army Engineer organization be established to take full responsibility for the development of the atomic bomb. During the summer, the Manhattan Engineer District was created, and in September, 1942, Major General L. R. Groves assumed command.

Construction of the main pile at Chicago started in November. The project gained momentum, with machining of the graphite blocks, pressing of the uranium oxide pellets, and the design of instruments. Fermi's two "construction" crews, one under Zinn and the other under Herbert L. Anderson, worked almost around the clock. V. C. Wilson headed the instrument work.

Original estimates as to the critical size of the pile were pessimistic. As a further precaution, it was decided to enclose the pile in a balloon-cloth bag which could be evacuated to remove the neutron-capturing air.

The bag was hung with one side left open; in the center of the floor a circular layer of graphite bricks was placed. This and each succeeding layer of the pile was braced by a wooden frame. Alternate layers contained the uranium. By this layer-on-layer construction a roughly spherical pile of uranium and graphite was formed.

Facilities for the machining of graphite bricks were installed in the West Stands. Week after week this shop turned out graphite bricks. This work was done under the direction of Zinn's group, by skilled mechanics led by millwright August Knuth. In October, Anderson and his associates joined Zinn's men.

Describing this phase of the work, Albert Wattenberg, one of Zinn's group, said: "We found out how coal miners feel. After eight hours of machining graphite, we looked as if we were made up for a minstrel. One shower would remove only the surface graphite dust. About a half hour after the first shower the dust in the pores of your skin would start oozing. Walking around the room where we cut the graphite was like walking on a dance floor. Graphite is a dry lubricant, you know, and the cement floor covered with graphite dust was slippery."

Before the structure was half complete, measurements indicated that the critical size at which the pile would become self-sustaining was somewhat less than had been anticipated in the design.

Day after day the pile grew toward its final shape. And as the size of the pile increased, so did the nervous tension of the men working on it. Logically and scientifically they knew this pile would become self-sustaining. It had to. All the measurements indicated that it would. But still the demonstration had to be made. As the eagerly awaited moment drew nearer, the scientists gave greater and greater attention to details, the accuracy of measurements, and exactness of their construction work.

At Chicago during the early afternoon of December 1, tests indicated that critical size was rapidly being approached. At 4 P.M. Zinn's group was relieved by the men working under Anderson. Shortly afterward the last layer of graphite and uranium bricks was placed on the pile. Zinn, who remained, and Anderson made several measurements of the activity within the pile. They were certain that when the control rods were withdrawn, the pile would become self-sustaining. Both had agreed, however, that should measurements indicate the reaction would become self-sustaining when the rods were withdrawn, they would not start the pile operating until Fermi and the rest of the group could be present. Consequently, the control rods were locked and further work was postponed until the following day.

That night the word was passed to the men who had worked on the pile that the trial run was due the next morning.

About 8:30 on the morning of Wednesday, December 2, the group began to assemble in the squash court.

At the north end of the squash court was a balcony about ten feet above the floor of the court. Fermi, Zinn, Anderson, and Compton were grouped around instruments at the east end of the balcony. The remainder of the observers crowded the little balcony. R. G. Noble, one of the young scientists who worked on the pile, put it this way: "The control cabinet was surrounded by the 'big wheels;' the 'little wheels' had to stand back."

On the floor of the squash court, just beneath the balcony, stood George Weil, whose duty it was to handle the final control rod. In the pile were three sets of control rods. One set was automatic and could be controlled from the balcony. Another was an emergency safety rod. Attached to one end of this rod was a rope running through the pile and weighted heavily on the opposite end. The rod was withdrawn from the pile and tied by another rope to the balcony. Hilberry was ready to cut this rope with an ax should something unexpected happen, or in case the automatic safety rods failed. The third rod, operated by Weil, was the one which actually held the reaction in check until withdrawn the proper distance.

Since this demonstration was new and different from anything ever done before, complete reliance was not placed on mechanically operated control rods. Therefore

a "liquid-control squad," composed of Harold Lichtenberger, W. Nyter, and A. C. Graves, stood on a platform above the pile. They were prepared to flood the pile with cadmium-salt solution in case of mechanical failure of the control rods.

Each group rehearsed its part of the experiment.

At 9:45 Fermi ordered the electrically operated control rods withdrawn. The man at the controls threw the switch to withdraw them. A small motor whined. All eyes watched the lights which indicated the rods' position.

But quickly the balcony group turned to watch the counters, whose clicking stepped up after the rods were out. The indicators of these counters resembled the face of a clock, with "hands" to indicate neutron count. Nearby was a recorder, whose quivering pen traced the neutron activity within the pile.

Shortly after ten o'clock, Fermi ordered the emergency rod, called "Zip," pulled out and tied.

"Zip out," said Fermi. Zinn withdrew "Zip" by hand and tied it to the balcony rail. Weil stood ready by the "vernier" control rod which was marked to show the number of feet and inches which remained within the pile.

At 10:37 Fermi, without taking his eyes off the instruments, said quietly:

"Pull it to 13 feet, George." The counters clicked faster. The graph pen moved up. All the instruments were studied, and computations were made.

"This is not it," said Fermi. "The trace will go to this point and level off." He indicated a spot on the graph. In a few minutes the pen came to the indicated point and did not go above that point. Seven minutes later Fermi ordered the rod out another foot.

Again the counters stepped up their clicking, the graph pen edged upwards. But the clicking was irregular. Soon it leveled off, as did the thin line of the pen. The pile was not self-sustaining—yet.

At 11 o'clock, the rod came out another six inches; the result was the same: an increase in rate, followed by the leveling off.

Fifteen minutes later, the rod was further

withdrawn and at 11:25 was moved again. Each time the counters speeded up, the pen climbed a few points. Fermi predicted correctly every movement of the indicators. He knew the time was near. He wanted to check everything again. The automatic control rod was reinserted without waiting for its automatic feature to operate. The graph line took a drop, the counters slowed abruptly.

At 11:35, the automatic safety rod was withdrawn and set. The control rod was adjusted and "Zip" was withdrawn. Up went the counters, clicking, clicking, faster and faster. It was the clickety-click of a fast train over the rails. The graph pen started to climb. Tensely, the little group watched and waited, entranced by the climbing needle.

Whrrrump! As if by a thunderclap, the spell was broken. Every man froze—then breathed a sigh of relief when he realized the automatic rod had slammed home. The safety point at which the rod operated automatically had been set too low.

"I'm hungry," said Fermi. "Let's go to lunch."

Perhaps, like a great coach, Fermi knew when his men needed a "break."

It was a strange "between halves" respite. They got no pep talk. They talked about everything else but the "game." The redoubtable Fermi, who never says much, had even less to say. But he appeared supremely confident. His "team" was back on the squash court at 2:00 P.M. Twenty minutes later, the automatic rod was reset and Weil stood ready at the control rod.

"All right, George," called Fermi, and Weil moved the rod to a predetermined point. The spectators resumed their watching and waiting, watching the counters spin, watching the graph, waiting for the settling down, and computing the rate of rise of reaction from the indicators.

At 2:50 the control rod came out another foot. The counters nearly jammed, the pen headed off the graph paper. But this was not it. Counting ratios and the graph scale had to be changed.

"Move it six inches," said Fermi at 3:20. Again the change—but again the leveling off.

Five minutes later, Fermi called: "Pull it out another foot."

Weil withdrew the rod.

"This is going to do it," Fermi said to Compton, standing at his side. "Now it will become self-sustaining. The trace will climb and continue to climb. It will not level off."

Fermi computed the rate of rise of the neutron counts over a minute period. He silently, grim-faced, ran through some calculations on his slide rule.

In about a minute he again computed the rate of rise. If the rate was constant and remained so, he would know the reaction was self-sustaining. His fingers operated the slide rule with lightning speed. Characteristically, he turned the rule over and jotted down some figures on its ivory back.

Three minutes later he again computed the rate of rise in neutron count. The group on the balcony had by now crowded in to get an eye on the instruments, those behind craning their necks to be sure they would know the very instant history was made. In the background could be heard William Overbeck calling out the neutron count over an annunciator system. Leona Marshall (the only girl present), Anderson, and William Sturm were recording the readings from the instruments. By this time the click of the counters was too fast for the human ear. The clickety-click was now a steady brrrrr. Fermi, unmoved, unruffled, continued his computations.

"I couldn't see the instruments," said Weil. "I had to watch Fermi every second, waiting for orders. His face was motionless. His eyes darted from one dial to another. His expression was so calm it was hard. But suddenly, his whole face broke into a broad smile."

Fermi closed his slide rule——

"The reaction is self-sustaining," he announced quietly, happily. "The curve is exponential."

The group tensely watched for twenty-eight minutes while the world's first nuclear chain reactor operated.

The upward movement of the pen was leaving a straight line. There was no change to indicate a leveling off. This was it.

"O.K., 'Zip' in," called Fermi to Zinn,

who controlled that rod. The time was 3:53 P.M. Abruptly, the counters slowed down, the pen slid down across the paper. It was all over.

Man had initiated a self-sustaining nuclear reaction—and then stopped it. He had released the energy of the atom's nucleus and controlled that energy.

Right after Fermi ordered the reaction stopped, the Hungarian-born theoretical physicist Eugene Wigner presented him with a bottle of Chianti wine. All through the experiment Wigner had kept this wine hidden behind his back.

Fermi uncorked the wine bottle and sent out for paper cups so all could drink. He poured a little wine in all the cups, and silently, solemnly, without toasts, the scientists raised the cups to their lips—the Canadian Zinn, the Hungarians Szilard and Wigner, the Italian Fermi, the Americans Compton, Anderson, Hilberry, and a score of others. They drank to success—and to the hope they were the first to succeed.

A small crew was left to straighten up, lock controls, and check all apparatus. As the group filed from the West Stands, one of the guards asked Zinn:

"What's going on, Doctor, something happen in there?"

The guard did not hear the message which Arthur Compton was giving James B. Conant at Harvard, by long distance telephone. Their code was not prearranged.

"The Italian navigator has landed in the New World," said Compton.

"How were the natives?" asked Conant.

"Very friendly."

7.8   *An Interview with Brigadier General Thomas F. Farrell*

The scene inside the shelter was dramatic beyond words. In and around the shelter were some twenty-odd people concerned with last-minute arrangements. Included were Dr. Oppenheimer, the director who had borne the great scientific burden of developing the weapon from the raw materials made in Tennessee and Washington, and a dozen of his key assistants, Dr. Kistiakowsky, Dr. Bainbridge, who supervized all the detailed arrangements for the test; the weather expert, and several others. Besides those, there were a handful of soldiers, two or three army officers and one naval officer. The shelter was filled with a great variety of instruments and radios.

For some hectic two hours preceding the blast, General Groves stayed with the director. Twenty minutes before the zero hour, General Groves left for his station at the base camp, first because it provided a better observation point and second, because of our rule that he and I must not be together in situations where there is an element of danger which existed at both points.

Just after General Groves left, announcements began to be broadcast of the interval remaining before the blast to the other groups participating in and observing the test. As the time interval grew smaller and changed from minutes to seconds, the tension increased by leaps and bounds. Everyone in that room knew the awful potentialities of the thing that they thought was about to happen. The scientists felt that their figuring must be right and that the bomb had to go off but there was in everyone's mind a strong measure of doubt.

We were reaching into the unknown and we did not know what might come of it. It can safely be said that most of those present were praying—and praying harder than they had ever prayed before. If the shot were successful, it was a justification of the several

SOURCE. From "The War Department Release on the New Mexico Test," July 16, 1945.

years of intensive effort of tens of thousands of people—statesmen, scientists, engineers, manufacturers, soldiers, and many others in every walk of life.

In that brief instant in the remote New Mexico desert, the tremendous effort of the brains and brawn of all these people came suddenly and startlingly to the fullest fruition. Dr. Oppenheimer, on whom had rested a very heavy burden, grew tenser as the last seconds ticked off. He scarcely breathed. He held on to a post to steady himself. For the last few seconds, he stared directly ahead and then when the announcer shouted "Now!" and there came this tremendous burst of light followed shortly thereafter by the deep growling roar of the explosion, his face relaxed into an expression of tremendous relief. Several of the observers standing back of the shelter to watch the lighting effects were knocked flat by the blast.

The tension in the room let up and all started congratulating each other. Everyone sensed "This is it!" No matter what might happen now all knew that the impossible scientific job had been done. Atomic fission would no longer be hidden in the cloisters of the theoretical physicists' dreams. It was almost full grown at birth. It was a great new force to be used for good or for evil. There was a feeling in that shelter that those concerned with its nativity should dedicate their lives to the mission that it would always be used for good and never for evil.

Dr. Kistiakowsky threw his arms around Dr. Oppenheimer and embraced him with shouts of glee. Others were equally enthusiastic. All the pent-up emotions were released in those few minutes and all seemed to sense immediately that the explosion had far exceeded the most optimistic expectations and wildest hopes of the scintists. All seemed to feel that they had been present at the birth of a new age—The Age of Atomic Energy—and felt their profound responsibility to help in guiding into right channels the tremendous forces which had been unlocked for the first time in history.

As to the present war, there was a feeling that no matter what else might happen, we now had the means to insure its speedy conclusion and save thousands of American lives. As to the future, there had been brought into being something big and something new that would prove to be immeasurably more important than the discovery of electricity or any of the other great discoveries which have so affected our existence.

The effects could well be called unprecedented, magnificent, beautiful, stupendous and terrifying. No man-made phenomenon of such tremendous power had ever occurred before. The lighting effects beggared description. The whole country was lighted by a searing light with the intensity many times that of the midday sun. It was golden, purple, violet, gray and blue. It lighted every peak, crevasse and ridge of the nearby mountain range with a clarity and beauty that cannot be described but must be seen to be imagined. It was that beauty the great poets dream about but describe most poorly and inadequately. Thirty seconds after, the explosion came first, the air blast pressing hard against the people and things, to be followed almost immediately by the strong, sustained, awesome roar which warned of doomsday and made us feel that we puny things were blasphemous to dare tamper with the forces heretofore reserved to the Almighty. Words are inadequate tools for the job of acquainting those not present with the physical, mental and psychological effects. It had to be witnessed to be realized.

CHAPTER 8
ELEMENTARY PARTICLES AND
ELEMENTARY PROCESSES

8.1   C. D. Anderson: The Positive Electron
8.2   F. Reines and C. L. Cowan, J.: The Neutrino
8.3   An Interview with Victor F. Weisskopf
8.4   M. Gardner: The Fall of Parity
8.5   D. H. Frisch and A. M. Thorndike: The First Observation of an Artificially-Produced Hyperon

---

8.1   *The Positive Electron*

Carl D. Anderson

On August 2, 1932, during the course of photographing cosmic-ray tracks produced in a vertical Wilson chamber (magnetic field of 15,000 gauss) designed in the summer of 1930 by Professor R. A. Millikan and the writer, the tracks shown in Figure 8.1–1 were obtained, which seemed to be interpretable only on the basis of the existence in this case

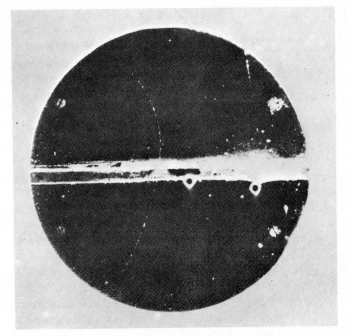

Figure 8.1–1.   A 63 MeV positron ($B\rho = 2.1 \times 10^5$ gauss-cm) passing through a 6 mm lead plate and emerging as a 23 MeV positron ($B\rho = 7.5 \times 10^{-4}$ gauss-cm. The length of this latter path is at least ten times greater than the possible length of a proton path of this curvature.

SOURCE. From *Physical Review* **43**: 491 (1933).

of a particle carrying a positive charge but having a mass of the same order of magnitude as that normally possessed by a free negative electron. Later study of the photograph by a whole group of men of the Norman Bridge Laboratory only tended to strengthen this view. The reason that this interpretation seemed so inevitable is that the track appearing on the upper half of the figure cannot possibly have a mass as large as that of a proton for as soon as the mass is fixed the energy is at once fixed by the curvature. The energy of a proton of that curvature comes out 300,000 volts, but a proton of that energy according to well established and universally accepted determinations has a total range of about 5 mm in air while that portion of the range actually visible in this case exceeds 5 cm without a noticeable change in curvature. The only escape from this conclusion would be to assume that at exactly the same instant (and the sharpness of the tracks determines that instant to within about a fiftieth of a second) two independent electrons happened to produce two tracks so placed as to give the impression of a single particle shooting through the lead plate. This assumption was dismissed on a probability basis, since a sharp track of this order of curvature under the experimental conditions prevailing occurred in the chamber only once in some 500 exposures, and since there was practically no chance at all that two such tracks should line up in this way. We also discarded as completely untenable the assumption of an electron of 20 MeV entering the lead on one side and coming out with an energy of 60 MeV on the other side. A fourth possibility is that a photon, entering the lead from above, knocked out of the nucleus of a lead atom two particles, one of which shot upward and the other downward. But in this case the upward moving one would be a positive of small mass so that either of the two possibilities leads to the existence of the positive electron.

In the course of the next few weeks other photographs were obtained which could be interpreted logically only on the positive-electron basis, and a brief report was then published with due reserve in interpretation in view of the importance and striking nature of the announcement.

MAGNITUDE OF CHARGE AND MASS

It is possible with the present experimental data only to assign rather wide limits to the magnitude of the charge and mass of the particle. The specific ionization was not in these cases measured, but it appears very probable, from a knowledge of the experimental conditions and by comparison with many other photographs of high- and low-speed electrons taken under the same conditions, that the charge cannot differ in magnitude from that of an electron by an amount as great as a factor of two. Furthermore, if the photograph is taken to represent a positive particle penetrating the 6 mm lead plate, then the energy lost, calculated for unit charge, is approximately 38 MeV, this value being practically independent of the proper mass of the particle as long as it is not too many times larger than that of a free negative electron. This value of 63 MeV per cm energy-loss for the positive particle it was considered legitimate to compare with the measured mean of approximately 35 MeV for negative electrons of 200–300 MeV of energy since the rate of energy-loss for particles of small mass is expected to change only very slowly over an energy range extending from several MeV to several hundred MeV. Allowance being made for experimental uncertainties, an upper limit to the rate of loss of energy for the positive particle can then be set at less than four times that for an electron, thus fixing, by the usual relation between rate of ionization and charge, an upper limit to the charge less than twice that of the negative electron. It is concluded, therefore, that the magnitude of the charge of the positive electron which we shall henceforth contract to positron is very probably equal to that of a free negative electron which from symmetry considerations would naturally then be called a negatron.

It is pointed out that the effective depth of the chamber in the line of sight which is the same as the direction of the magnetic lines of force was 1 cm and its effective diameter at right angles to that line 14 cm, thus insuring that the particle crossed the chamber practically normal to the lines of force. The change in direction due to scattering in the lead, in this case about 8° measured in the plane of the chamber, is a probable value for a particle of this energy though less than the most probable value.

The magnitude of the proper mass cannot as yet be given further than to fix an upper limit to it about twenty times that of the electron mass. If [8.1–1] represents a particle of unit charge passing through the lead plate then the curvatures, on the basis of the information at hand on ionization, give too low a value for the energy-loss unless the mass is taken less than twenty times that of the negative electron mass. Further determinations of Bρ for relatively low energy particles before and after they cross a known amount of matter, together with a study of ballistic effects such as close encounters with electrons, involving large energy transfers, will enable closer limits to be assigned to the mass.

To date, out of a group of 1300 photographs of cosmic-ray tracks 15 of these show positive particles penetrating the lead, none of which can be ascribed to particles with a mass as large as that of a proton, thus establishing the existence of positive particles of unit charge and of mass small compared to that of a proton. In many other cases due either to the short section of track available for measurement or to the high energy of the particle it is not possible to differentiate with certainty between protons and positrons. A comparison of the six or seven hundred positive-ray tracks which we have taken is, however, still consistent with the view that the positive particle which is knocked out of the nucleus by the incoming primary cosmic ray is in many cases a proton . . .

8.2

The Neutrino

Frederick Reines and Clyde L. Cowan, Jr.

Each new discovery of natural science broadens our knowledge and deepens out understanding of the physical universe; but at times these advances raise new and even more fundamental questions than those which they answer. Such was the case with the discovery and investigation of the radioactive process termed "beta decay." In this process an atomic nucleus spontaneously emits either a negative or positive electron, and in so doing it becomes a different element with the same mass number but with a nuclear charge different from that of the parent element by one electronic charge. As might be expected, intensive investigation of this interesting alchemy of Nature has shed much light on problems concerning the atomic nucleus. A new question arose at the beginning, however, when it was found that accompanying beta decay there was an unaccountable loss of energy from the decaying nucleus, and that one could do nothing to the apparatus in which the decay occurred to trap this lost energy. One possible explanation was that the conservation laws (upon which the entire structure of modern science is built) were not valid when applied to regions of subatomic dimensions. Another novel explanation, but one which would maintain the integrity of the conservation laws, was a proposal by Wolfgang Pauli in 1933 which hypothesized a new and fundamental particle to account for the loss of energy from the nucleus. This particle would

SOURCE. From *Nature* 178: 446 (1956).

be emitted by the nucleus simultaneously with the electron, would carry with it no electric charge, but would carry the missing energy and momentum—escaping from the laboratory equipment without detection.

The concept of this ghostly particle was used by Enrico Fermi (who named it the "neutrino") to build his quantitative theory of nuclear beta decay. As is well known, the theory, with but little modification, has enjoyed increasing success in application to nuclear problems and has itself constituted one of the most convincing arguments in favor of the acceptance of Pauli's proposal. Many additional experimental tests have been devised, however, which have served to strengthen the neutrino hypothesis; and also to provide information as to its properties. The very characteristic of the particle which makes the proposal plausible—its ability to carry off energy and momentum without detection—has limited these tests to the measurement of the observable details of the decay process itself: the energy spectra, momentum vectors and energy states associated with the emitted electron and with the recoiling daughter nucleus. So, for example, an upper limit has been set on the rest mass of the neutrino equal to $1/500$ of the rest mass of the electron by careful measurement of the beta-energy spectrum from tritium decay near its end point, and it is commonly assumed that the neutrino rest mass is identically zero.

.　　.　　.　　.　　.

The Pauli-Fermi theory not only requires the neutrino to carry energy and linear momentum from beta-decaying nuclei but also angular momentum, or "spin." The simplest of beta-decay proccesses, the decay of the free neutron, illustrates this:

$$n \rightarrow p + \beta^- + \bar{v} \qquad (1)$$

As the neutron, proton and beta particle all carry half-integral spin, it is necessary to assign a spin quantum number of $1/2$ to the neutrino to balance the angular momenta

of equation (1), where any two of the three product particles must be oriented with spin vectors antiparallel. As all four of the particles in equation (1) are, therefore, fermions and should obey the Dirac relativistic wave equations for spin $1/2$ particles, there are presumably antiparticles corresponding to each, of which as yet only the anti-electron (or positron) and the antiproton have been identified. The antiparticle corresponding to the neutrino in equation (1) may be obtained by rearrangement of the terms in the following manner:

$$p \rightarrow n + \beta^+ + v \qquad (2)$$

This process is observed in positron decay of proton-rich radioactive nuclides where the proton and daughter neutron are both constituent nucleons. Further rearrangement results in the reaction:

$$\beta^- + p \rightarrow n + v \qquad (3)$$

This is discriptive of the capture of an electron from one of the inner atomic shells by a nuclear proton and is equivalent to equation (2). The question of the identity of the neutrino, v, appearing in equations (2) and (3) with the neutrino, $\bar{v}$, appearing in equation (1) thus arises. With no finite mass or magnetic moment yet measured for either of the neutrinos, one is under no compulsion to assume that they are not in fact identical. The rule of algebraic conservation of fermions, which states that fermions are produced or disappear in particle-antiparticle pairs, requires the $\bar{v}$ of equation (1) to be named "antineutrino," since it is emitted with a negative electron. The identity or non-identity of the neutrino, v, and the antineutrino, $\bar{v}$, although of no observable significance in single beta decay, should be amenable to test by measurement of the decay constant for double beta decay of certain shielded isotopes. This process was studied theoretically by M. Goeppert-Mayer for the case in which neutrinos are not identical with antineutrinos and by Furry for the case in which the two neutrinos are identical, as proposed by Majorana. Double beta decay is typified by the possible decay of neodymium-150:

$$^{150}\text{Nd} \rightarrow {}^{150}\text{Sm} + 2\beta^- + 2\bar{\nu} \quad \text{(Dirac-Mayer)}$$
$$(4a)$$
$$^{150}\text{Nd} \rightarrow {}^{150}\text{Sm} + 2\beta^- \quad \text{(Majorana-Furry)}$$
$$(4b)$$

If the neutrino and antineutrino are identical, then the virtual emission of one neutrino and its immediate re-absorption by the nucleus are equivalent to the real emission of two neutrinos, and equation (4b) is applicable. This cancellation is not possible if the neutrino and antineutrino differ. The half-lives for processes such as equation (4) have been shown by Primakoff and by Konopinski to be quite different in the two cases, of the order 10^{19} years for equation (4a) and 10^{15} years for equation (4b), where 5.4 MeV. is available for the decay. Furthermore, a line spectrum for the total energy of the two beta particles is to be expected for the Majorana-Furry case (equation 4b).

That a decay period consistent with equation (4b) does not exist has been shown for a number of shielded isotopes, first by Kalkstein and Libby, then by Fireman and Schwartzer for tin-124; by Awschalom for calcium-48; and our associates and us for neodymium-150. In the neodymium-150 experiment, a lower limit of 4×10^{18} years (corresponding to one standard deviation in the background) was set on the mean life against Majorana-Furry decay. This limit is to be compared with a reasonable value on this hypothesis of 1.3×10^{15} years and one calculated for identical neutrinos (using most severe assumptions) to be 6×10^{17} years. The conclusion remains that the neutrino and antineutrino are distinct particles with an as yet undetected "difference." This conclusion is further supported by the negative results of an experiment recently reported by R. Davis employing the reaction:

$$^{37}\text{Cl} + \nu \rightarrow {}^{37}\text{Ar} + \beta^- \quad (5)$$

The chlorine target was supplied by 1,000 gallons of carbon tetrachloride placed near a large reactor, and the liquid was tested for the presence of argon-37. Fission fragments, being rich in neutrons, should emit only the antineutrino, $\bar{\nu}$.

While careful reasoning from experimental evidence gathered about all terms in the beta-decay process—except the neutrino—may support the inference that a neutrino exists, its reality can only be demonstrated conclusively by a direct observation of the neutrino itself. If the neutrino is a real particle carrying the missing energy and momentum from the site of a beta decay, then the discovery of these missing items at some other place would demonstrate its reality. Thus, if negative beta decays as in equation (1) could be associated at another location with the inverse reaction:

$$\bar{\nu} + p \rightarrow \beta^+ + n \quad (6)$$

which is observed to occur at the predicted rate, the case would be closed. An expression for this reaction cross-section has been obtained by application of the principle of detailed balancing to equation (1), knowing the decay constant and electron energy spectrum for the beta decay of free neutrons:

$$\sigma = \left(\frac{G^2}{2\tau}\right)\left(\frac{h}{mc}\right)^2\left(\frac{p}{mc}\right)^2 \frac{1}{(v/c)} (\text{cm}^2) \quad (7)$$

where σ is the cross-section in cm^2; $G^2 (= 44 \times 10^{-24})$ is the dimensionless lumped beta-coupling constant based on neutron decay; and p, m and v are the momentum, mass and speed of the emitted positron, respectively, c is the speed of light, and $2\pi\hbar$ is Planck's constant, all in c.g.s. units. For neutrinos of 3-MeV energy incident on free protons, this cross-section is 10^{-43} cm^2. Explicit solution of equation (6) for the cross-section as a function of the neutrino energy yields:

$$\sigma = 1.0 \times 10^{-44}$$
$$\times (E - a)\sqrt{(E - a)^2 - 1} \ (\text{cm}^2) \quad (8)$$

where $a + 1 (= 3.53)$ is the threshold for the reaction and E is the neutrino energy, both in units of $m_e c^2$. The threshold for a proton bound in a nucleus is higher by an amount equal to the energy difference between the target and daughter nuclei. It is interesting to note that the penetrability of matter is given by equation (8) to be infinite for neutrinos with low energies ($E < a + 1$) and is very large for neutrinos of only a few MeV, the mean free path for absorption being measured in the latter case in terms comparable to the radius of the universe.

Equation (6) may be employed in an experiment in which a large number of hydrogen atoms are provided as targets for an intense neutrino flux and are watched by a detector capable of recording the simultaneous production of a positron and a neutron. Such a direct experiment is made possible by the availability of high beta-decay rates of fission fragments in multi-megawatt reactors and advances in detection techniques through the use of liquid scintillators. An estimate of the neutrino flux available from large reactors shows that a few protons should undergo reaction (6) per hour in 50 liters of water placed near the reactor. The problem, then, is to observe these events with reasonable efficiency against the background of reactor neutrons and gamma-rays, natural radioactivity and cosmic rays. In an experiment conducted at the Hanford Plant of the Atomic Energy Commission by us in 1953, an attempt was made in this direction. The target protons were supplied by 300 liters of liquid scintillator (toluene plus trace amounts of terphenyl, and alpha-naphtha-phenyloxayole in which cadmium

propionate was dissolved). A delayed coincidence-rate of pairs of pulses, the first of each pair being assignable to the positron and the second to a neutron capture in cadmium, of 0.4 ± 0.2 counts per minute was observed, in agreement with the predicted rate, and with a large reduction in the backgrounds mentioned above. The signal-to-total-background ratio, however, was still very low (1/20), rendering further testing of the signal impractical and leaving the results tentative. On the basis of the Hanford experience it was felt that the detection problem was soluble in a definitive manner, and a second experiment was designed with the view of further reduction of backgrounds and providing means for checking each term of equation (6) independently.

Figure 8.2–1 is a schematic diagram of the detection scheme employed in this experiment. The sequence of events pictured is as follows: a neutrino from the decay of a fission fragment in a reactor causes a target proton to be changed into a neutron with the simultaneous emission of a positron. The positron is captured by an electron in the target water, emitting two 0.51-MeV annihilation gamma-rays, which are detected simultaneously by counters I and II. The neutron moderates and diffuses for several microseconds and is finally captured by the cadmium giving a few gamma rays (totalling 9 MeV), which are again detected by I and II. Thus we have a prompt coincidence

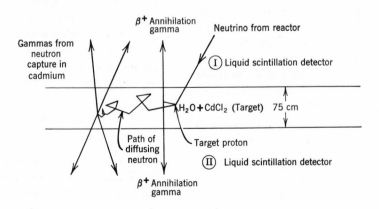

Figure 8.2–1.   Schematic diagram of neutrino detector.

followed in several microseconds by a second prompt coincidence, providing a very distinctive sequence of events.

The over-all size of the equipment was set by the number of events expected per hour per liter of water, and the detection efficiency one could hope to achieve. A primary factor in the design geometry and detection efficiency was the absorption of the positron annihilation radiation by the target water itself. Experimentation and calculations showed that an optimum water thickness was 7.5 cm. Since the over-all efficiency dictated a target volume of about 200 liters to yield several counts per hour, two target tanks were used, each measuring 1.9 m × 1.3 m × 0.07 m. The depth of the liquid scintillation detector (61 cm) was such as to absorb the cadmium-capture gamma-rays with good efficiency and transmit the resultant light to the ends of the detector with minimal loss. The scintillating liquid (triethylbenzene,

terphenyl and POPOP wavelength shifter) were viewed from the ends of each detector tank by 110 5-in. Dumont photomultiplier tubes, a number determined primarily by the amount of light emitted in a scintillation. The complete detector consisted of a "club sandwich" arrangement employing two target tanks between three detector tanks, comprising two essentially independent triads which used the center detector tank in common. The entire detector was encased in a lead-paraffin shield and located deep underground near one of the Savannah River Plant production reactors of the United States Atomic Energy Commission. Signals from the detectors were transmitted via coaxial cables to an electronics trailer located outside the reactor building. The pulses were analysed by pulse-height and time-coincidence circuits and, when acceptable, were recorded photographically as traces on triple-beam oscilloscopes. Figure 8.2–2 is

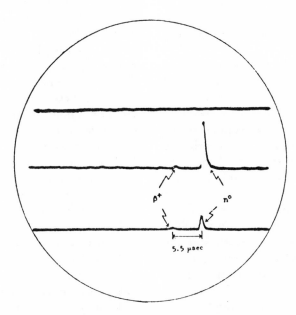

Figure 8.2–2. A characteristic record. Each of the three oscilloscope traces shown corresponds to a detector tank. The event recorded occurred in the bottom triad. First seen in coincidence are the "positron" annihilation gamma-ray pulses in each tank followed in 5.5 μsec by the larger "neutron" pulses. The amplification was chosen in this case to enable measurement of the neutron pulses. A second oscilloscope with higher amplification was operated in parallel to enable measurement of the positron pulses.

a record of an event in the bottom triad. The entire system was calibrated using a plutonium-beryllium neutron source and a dissolved copper-64 positron source in the target tanks; and standardized pulsers were used to check for stability of the electronics external to the detector itself. The response of the detector to cosmic ray μ-mesons was also employed as a check on its performance. After running for 1,371 hr, including both reactor-up and reactor-down time, it was observed that:

(1) A signal dependent upon reactor-power, 2.88 ±0.22 counts/hr in agreement with the predicted cross-section (6 × 10^{-44} cm^2), was measured with a signal-to-reactor associated accidental background in excess of 20/1. The signal-to-reactor independent background ratio was 3/1.

(2) Dilution of the light water solution in the target tank with heavy water to yield a proton density of one-half normal caused the reactor signal to drop to one-half its former rate. The efficiency of neutron detection measured with the plutonium-beryllium source was unchanged.

(3) The first pulse of the pair was shown to be positron annihilation radiation by subjecting it to a number of tests: its spectrum agreed with the spectrum of positron annihilation radiation from copper-64 dissolved in the water, and it was absorbed in the expected manner by thin lead sheets inserted between the target tank and one detector.

(4) The second pulse of the pair was identified as due to the capture in cadmium of a neutron born simultaneously with the positron by virtue of its capture-time distribution as compared both with calculations and observations with a neutron source. The second pulse spectrum was consistent with that of cadmium-capture gamma-rays, and removal of the cadmium resulted in disappearance of the reactor signal.

(5) Reactor-associated radiations such as neutrons and gamma-rays were ruled out as the source of the signal by two kinds of experiment. In the first, a strong americium-beryllium neutron source was placed outside the detector shield and was not only found very inefficient in producing acceptable delayed coincidences but was also found to produce a first-pulse spectrum which was unlike the required signal in that it was monotonically decreasing with increasing energy. In the second experiment an additional shield, which provided an attenuation factor of at least 10 for reactor neutrons and gamma-rays, was observed to cause no change in the reactor signal outside the statistical fluctuations quoted in (1).

Completion of the term-by-term checks of equation (6) thus demonstrated that the free neutrino is observable in the near vicinity of a high-power fission reactor.

.   .   .   .   .

Since the proposal of the neutrino hypothesis by Pauli and its success in Fermi's theory of nuclear beta decay, the particle has been called upon to play similar parts in the observed decay of a number of different mesons. The question arises as to the identity of these neutrino-like particles with the neutrino of nucleon decay. It is to be noted that in nuclear beta decay the initial and final nuclei both quite obviously interact strongly with nuclei. This is not the case in (π, μ) decay, where the emission of a "neutrino" converts the interaction of the heavy particles with nuclei from strong to weak. Furthermore, despite the apparent equality of the nuclear beta-decay matrix elements with those associated with (μ, β) decay, both the initial and final products of the latter interact weakly with nuclei.

The neutrino is the smallest bit of material reality ever conceived of by man; the largest is the universe. To attempt to understand something of one in terms of the other is to attempt to span the dimension in which lie all manifestations of natural law. Yet even now, despite our shadowy knowledge of these limits, problems arise to try the imagination in such an attempt. If nuclear reactions played a part in a cataclysmic birth of the universe as we assume, what fraction of the primordial energy was quickly drained into the irreversible neutrino field? Are these neutrinos— untouched by anything from almost the beginning of time—trapped by the common gravitational field of the universe, and if so, what is their present density, their energy spectrum and angular distribution? Do neutrinos and antineutrinos exist in equal numbers? If the neutrino has zero rest mass, is it to be considered with "matter" particles in discussing its gravitational potential, or with electromagnetic radiation? The problem of detecting these cosmic end-products of all nuclear energy generation processes and the measurement of their characteristics presents a great challenge to the physics of to-day.

8.3   *An Interview with Victor F. Weisskopf*

Quantum physics is very different from classical physics. How do you see the difference?

I like to say it in the following way: Before we got to quantum theory our understanding of nature did not correspond at all to one of the most obvious characters of nature, namely, the definite and specific properties of things. Steam is always steam, wherever you find it. Rock is always rock. Air is always air. This property of matter whereby it has characteristic properties seems to me one of the most obvious facts of nature. Yet classical physics has no way of accounting for it. In classical physics, the properties are all continuous.

What do you mean by "continuous"?

There are no two classical systems that are really identical. Take the planetary systems of stars, of which we all know that there are billions. According to our present knowledge, you can be sure that no two of them are exactly identical. In some the sun will be a little larger, in some the planets would be a little larger, the orbits would be a little different. . . . Why? . . . Classical physics allows us an immense range of possibilities. The behavior of things depends on the initial conditions, which can have a continuum of values.

Now quantum theory changes this fundamentally, because things are quantized. No longer is "any" orbit possible, only certain ones, and all the orbits of a particular kind are the same. Thus in quantum theory it makes sense to say that two iron atoms are "exactly" alike because of the quantized orbits. So, an iron atom here and an iron atom in Soviet Russia are exactly alike. Quantum theory brought into physics this idea of identity.

I'm struck that you stressed the word "exactly," because for many people it's a certain inexactness that characterizes quantum physics. They remember the uncertainties.

I'm an old fighter against this interpretation of the uncertainty relation. Quantum theory brought in just that exactness.

The classical Greek approach was apparently based on sound intuition. The Greeks had a picture of discreteness in nature.

Yes, but the Greeks postulated the existence of atoms; they did not explain it. One cannot understand on the basis of classical physics how it is possible to have a mechanical system of one kind and a mechanical system of another kind and no mechanical system in between.

So that atoms, which were axiomatic to the Greeks, remained unexplained assumptions through the nineteenth century. You had experiments on atomic weights and had the kinetic theory of gases but that did not mean that atoms were understood.

SOURCE. From *The Way of the Scientist* (New York: Simon and Schuster, 1966) p. 87; reprinted by permission of the publisher.

They remained axiomatic up until 1913. If you look a little under the surface, you hit the same problem always. You mentioned gas theory. Now before quantum theory, people rightly looked at gases as colliding atoms. Yet how come such collisions don't change the nature of atoms? Atoms must have a structure, a mechanism, inside, and the collision must leave some change in it. Yet we know that it isn't so. The stability of the atom is something that is not understandable in classical theory.

Was this question posed at all before 1913?

Oh yes, in the famous Boltzmann paradox: Classical mechanics leads you to expect that, for a system of atoms in thermal equilibrium at a given temperature, the thermal energy should be shared equally among all the possible modes of motion. *All* modes. In a piece of heated material the electrons should run around faster, the protons should vibrate more rapidly within the nuclei, the parts of which the protons are made should vibrate more strongly within their bounds, and so forth. Thus the specific heat in any ordinary piece of matter should be extremely large. In actual fact the specific heat has just the size that can be accounted for by the external motions of the atoms alone. It was not understandable how the heat energy doesn't get into the atom and excite the internal degrees of freedom. This Boltzmann paradox came in 1890, well before quantum theory. There was no explanation.

Your implication is that there had already been advance beyond the Greek idea of the atom as the uncuttable one. Did the nineteenth-century physicists imagine there was structure within the atom?

Yes. The atom emits light, and after the discovery of the electromagnetic nature of light, it was clear there must have been some motion inside the atom that emits the light, so there must have been internal structure. And there was also a philosophical idea behind it; namely, the concept of an imaginary atom without internal structure doesn't make much sense. One must ask, "What's inside?" Now, one could have said it is solid, but

even if it is solid, we know the solid has a structure.

So the philosophical question of what happens if you cut the atom remains.

To my mind, quantum theory for the first time indicates how one has to deal with problems of this kind. Quantum theory tells us that an atom is a non-divisible entity, *if* the energies applied to it are below a certain threshold. If the processes inflicted upon the atom are below a certain threshold, the atom is really indivisible, in the real sense of the word. It means that if atoms collide with energies less than the threshold, they bounce off completely unaffected, in exactly the same state that they were before. This is the new idea. That's the quantum idea.

However, when you are way above the threshold, the atoms go to pieces, and they behave like ordinary classical systems containing parts and particles. For example, at very high temperatures an atom is completely decomposed into its parts, the nucleus and the electrons. Consider a sodium atom and a neon atom. The former has eleven electrons, the latter ten. Below the threshold they are in their characteristic quantum states; very different. One is a metal, the other a gas. Above the threshold—at high temperatures—they are both a gas of nuclei and electrons. It is what one calls a plasma. In this state there is not much difference between a sodium plasma and a neon [plasma].

And just as there are threshold energy levels for disrupting an atom, so there are levels above which the nuclei would be split.

True enough, there is also a threshold above which the nucleus goes to pieces. This threshold is much higher than the atomic threshold. The atomic thresholds are of the order of a few electron volts; the nuclear threshold is at much higher energies—a few million electron volts.

I like to use the term "quantum ladder" for this. These are two steps of the ladder.

The quantum ladder has made it possible to discover the structure of the natural world step by step. When we investigate phenomena at energies characteristic of atoms we need not worry about the internal structure of their nuclei. And when we study the behavior of gases at normal temperatures and pressures we need not worry about details of the internal structure of the atoms that make up the gas. In that way the quantum ladder solves the Boltzmann paradox. The finer structure of matter does not participate in the exchange of energy until the average energy has reached that rung on the ladder.

Our whole experience in daily life is down low on your ladder, within the atomic level of the quantum ladder.

Yes. That or even lower. I started with the atom, but there are also steps farther down the quantum ladder, which are important for our life. Molecules, macromolecules, crystals. All life consists of macromolecules. The lower you go on the quantum ladder the more pronounced becomes the specificity of the structures: nucleus—atom—molecule—macromolecule—life.

How do these specific structures come about? This is central point, it seems.

The quantum is an important precondition of the structuralization of nature. The particles fall into definite pattern—the quantum orbits of the nucleus, the quantum orbits of the atom, the quantum orbits of the molecule, the quantum orbits of the macromolecule. Our hereditary properties are nothing else than the quantum states of the parts of a nucleic-acid chain, the so-called DNA. In some way the recurrence every spring of a flower of a certain shape is an indirect expression of the existence of certain quantum orbits in the DNA molecule—a consequence of the identity and uniqueness of quantum orbits.

Is it an accident or is there some deeper reason for the spacing of rungs on the quantum ladder?

Oh yes, there is a very good reason. It lies in the size-energy relation, that the smaller the system is, the higher the quantum energies are. For example, it is not an accident that the quantum energy of the outer electron

shell in an atom is only a few volts, whereas the quantum energy of a nuclear system is a few million volts. It's because of their size.

But, there must be a reason that atoms and nuclei have that size.

Well, the reason the electron shell exists is the electric attraction between the nucleus and the electrons, and the reason that the nucleus exists as a unit is the nuclear force between nucleons, that is, protons and neutrons.

Up to 1930 we dealt with two forces in nature. They are very well known to us on a macroscopic level—namely gravity and electricity. Bohr had found in 1913 that the chemical forces—the forces within the atom—are electric. Only in 1930 when we first experimented on the inner structure of the nucleus did a new force come in, the nuclear force. It is the force that holds the protons and neutrons together when they form the atomic nucleus.

So the answer on the particular level of your question is, I think, the existence of these two force fields. Why these force fields? That is an unsolved question, and a question which I have a very definite feeling will be solved in our high-energy research.

There is a next step on the quantum ladder higher than the one which breaks the nucleus.

Yes, there is. Recently experiments with big accelerators have shown us that the proton and the neutron have a structure too. These particles can be changed into different states, they can absorb energy; in short, a world within the proton has been discovered. That is the next higher rung of the ladder.

What is the energy threshold for that step?

You can get at it only if you go to energies way beyond the temperatures and energies in the center of the stars. We don't know where the universe displays such energies . . . well, we have it in cosmic rays—very rare events. We have it at the target of our accelerator at CERN, and maybe the center of the galaxy is of such type. We just don't know.

With the 30-billion-volt machines you have just broken into a new highest level of energy. The immediate effect has been the discovery of thirty or forty particles. The new machines seem to cause more confusion than anything else.

I don't accept your premise. The statement that high-energy physics has found thirty or forty particles has brought this field into disrepute. But that reputation is wrong—for several reasons. One is that everybody counts the antiparticles as extra particles, which is as if you would double the number of animal species by calling the mirror image of an animal another animal. And there's more than that. I think it is wrong to call an excited state of a system a new system. It's as if we would say that the excited hydrogen atom is another atom. It's becoming clearer and clearer that many of these particles are nothing else than the excited states of other particles.

Which particles are fundamental and which are excited states?

For example, the sigma particle, the lambda particle, and the xi particle are all excited states of the proton. I would go so far as to say we have only two elementary particles, the baryon and the lepton, and these particles have different states, different configurations, just like the hydrogen atoms. The proton and neutron are two states of the baryon, just like the spin up and spin down of the electron in the ground state of hydrogen. The lambda, the xi, and the sigma are excited states. The systematics of all these new states is what I like to call the "third spectroscopy." We have atomic spectroscopy, the quantum levels of the atom, nuclear spectroscopy, the quantum levels of the nucleus, and now we have the third spectroscopy, which is the quantum levels of the nucleon.

The leptons also occur in different forms: as electrons, as neutrinos, and as heavy electrons (sometimes called mu mesons).

Where do the pi mesons and the K mesons fit in your picture? Aren't they also particles?

I would rather not call them particles. They are field quanta. Just as the light quantum is a quantum of the electromagnetic field, so the pi meson and the K meson are quanta of the nuclear field.

What is a field?

Fields began as a way of expressing the force between particles. The attraction between two unlike electric charges can also be expressed in terms of the action of the field of one on the other. But the field is not just a mathematical fiction, it is as real as its particle sources, and we can speak of its energy, and so forth.

Now, every field has a quantum. When a field propagates in space, when it is emitted in form of radiation, it propagates in form of quanta. The very best known field quantum is the photon, the quantum of the electromagnetic field. The nuclear force seems to require two quanta—one is the pi meson and the other is the K meson. They both play an important role. The pi meson is responsible for the outer reaches of the nuclear field, and the K meson for the force at very close distances.

You're not troubled that there are two quanta for this field?

I'm troubled, but not as much as one might think. The field is just somewhat more complicated. The electric field falls off inversely with distance, it's just straight $1/r$; you see, the nuclear field is a complicated field, so no wonder two quanta. What is complicated is that the quanta carry isotopic spin and another quantum number which is called hypercharge or "strangeness."

It's not just as if you had the two quanta of that field, but also you have to adduce new quantum numbers.

Yes. These quantum numbers play an important role in the "third spectroscopy" we mentioned before. The excited states of the baryon can be classified according to these new quantum numbers.

But the basic idea is the same as in the spectrum of excited states of any atom, say hydrogen. The ground state is of course the proton or the neutron—the neutron has a little more energy. Then there are several excited states—the names originally given them make no sense any more so we call them by their various quantum numbers.

There's a nice historical parallel, isn't there? Just as the early history of optical spectroscopy was marked by the naming of spectral lines according to their appearance, so the names of nuclear particles have grown with their experimental discovery.

Exactly . . . now there are also excited states which have a different strangeness quantum number. And they have names—the lambda, for example. But they have names only because of the fact that their different strangeness makes them metastable and they last long enough to be observed as an apparently different particle. It's like an atomic state which is metastable.

You get these extreme energies from CERN and Brookhaven accelerators at intensities that apparently do not exist in the universe otherwise. When you are doing physics of a sort that nature doesn't do, what are you doing?

I am deeply convinced that nature has such a variety that any process we find on earth will be of importance somewhere. And that's why I think that the experiments we are now doing on these highest rungs of the quantum ladder will have significance in one of those unsolved problems, such as the problem of the expansion of the universe, the creation of matter, or the fundamental structure of matter. It may be that the problems of the creation of the universe are connected.

You are now perhaps experimenting with conditions which, 20 billion years ago . . . .

That is a matter of interpretation. As you know there are two views—which I would like to call two "religions." One is the Big Bang theory where the universe started billions of years ago with tremendous pressure and energies in a small volume, and the other is the Continuous Creation theory. I'm sure that the true answers to the questions would be neither one nor the other. But it is correct to think that if the Big Bang has something to do with reality, some of the early phases might have had to do with the latest rung on

the quantum ladder. My religion probably is Continuous Creation religion—but I'm not sure you had better put that in.

You say "religion" to indicate a real difference between those parts of physics which have the authority of physical experiment and those parts that are extrapolating to conditions that in no sense we can duplicate.

Exactly. One must really draw a serious line there. Although I also believe that these speculations are the most exciting one can imagine. But they are really different from physics itself. Sometime, when we shall know more about these things, they might become true physics.

Is the line of progress inevitably toward the still higher and higher levels of energy; or does your ladder have a top rung?

A last quantum rung? I cannot tell. It is always the highest hope, maybe a hope only, of physicists that at the next step of the quantum ladder you will find the all-embracing principle. Heisenberg thought so. In every physicist there is an element of belief that you will sometime come to the recognition of some fundamental facts which close it, from which you can explain everything. I'm not so sure. It may be that nature is inexhaustible. But it might not be. How do we know?

You're saying there would be a last step if you could predict what would happen if you used a hundred times as much energy?

Yes. But in order to reach that state of affairs we must also have a Heisenberg or a Bohr of the future who gives us the theory that explains all phenomena in terms of what we know. Until we have that theory, we will never have a guarantee that there is not a new world coming up. We would have to build higher-energy accelerators to find out.

Earlier you said that atoms are philosophically unsatisfying because you can always ask what is within them. If physics is not inexhaustible, then at some point you will truly have elementary, non-divisible particles.

I wouldn't call it the elementary particle,

it might be something else. A field, or even some new thing which is as far from the field as the field is from the particle, consequently something new, but that embraces the whole. What it will be we don't know—we are just at the beginning.

8.4   *The Fall of Parity*

Martin Gardner

As far as anyone knows at present, all events that take place in the universe are governed by four fundamental types of forces (physicists prefer to say "interactions" instead of "forces," but there is no harm in using here the more common term):

1. Nuclear force.
2. Electromagnetic force.
3. Weak interaction force.
4. Gravitational force.

The forces are listed in decreasing order of strength. The strongest, nuclear force, is the force that holds together the protons and neutrons in the nucleus of an atom. It is often called the "binding energy" of the nucleus. Electromagnetism is the force that binds electrons to the nucleus, atoms into molecules, molecules into liquids and solids. Gravity, as we all know, is the force with which one mass attracts another mass; it is the force chiefly responsible for binding together the substances that make up the earth. Gravitational force is so weak that unless a mass is enormously large it is extremely difficult to measure. On the level of the elementary particles its influence is negligible.

The remaining force, the force involved in "weak interactions," is the force about which the least is known. That such a force must exist is indicated by the fact that in certain decay interactions involving particles (such as beta-decay, in which electrons or positrons are shot out from radioactive nuclei), the speed of the reaction is much slower than it would be if either nuclear or electromagnetic forces were responsible. By "slow" is meant a reaction of, say, one ten-billionth of a second. To a nuclear physicist this is an exceedingly lazy effect—about a ten-trillionth the speed of reactions in which nuclear force is involved. To explain this lethargy it has been necessary to assume a force weaker than electromagnetism but stronger than the extremely weak force of gravity.

The "theta tau puzzle," over which physicists scratched their heads in 1956, arose in connection with a weak interaction involving a "strange particle" called the K-meson.

.   .   .   .   .

There appeared to be two distinct types of K-mesons.[1] One, called the theta meson, decayed into two pi mesons. The other, called the tau meson, decayed into three pi mesons. Nevertheless, the two types of K-mesons seemed to be indistinguishable from each other. They had precisely the same mass, same charge, same lifetime. Physicists would have liked to say that there was only one K-meson; sometimes it decayed into two, sometimes into three pi mesons. Why didn't they? Because it would have meant that parity was not conserved. The theta meson had even parity. A pi meson has odd parity. Two pi mesons have a total parity that is even, so parity is conserved in the decay of the theta meson. But *three* pi mesons have a total parity that is odd.

Physicists faced a perplexing dilemma

SOURCE. From *The Ambidextrous Universe* (Basic Books: New York, 1964); copyright 1964 by Martin Gardner.
[1]The theta meson is now called the K_1° and the tau meson is now called the K_2°:

$$K_1^\circ \rightarrow 2\pi; \; K_2^\circ \rightarrow 3\pi. \; \text{[Ed.]}$$

with the following horns:

1. They could assume that the two K-mesons, even though indistinguishable in properties, were really two different particles: the theta meson with even parity, the tau meson with odd parity.

2. They could assume that in one of the decay reactions parity was not conserved.

To most physicists in 1956 the second horn was almost unthinkable.... It would have meant admitting that the left-right symmetry of nature was being violated; that nature was showing a bias for one type of handedness. The conservation of parity had been well established in all "strong" interactions (that is, in the nuclear and electromagnetic interactions). It had been a fruitful concept in quantum mechanics for thirty years.

In April, 1956, during a conference on nuclear physics at the University of Rochester, in New York, there was a spirited discussion of the theta-tau puzzle. Richard Phillips Feynman, a physicist at the California Institute of Technology, raised the question: Is the law of parity sometimes violated? In corresponding with Feynman, he has given me some of the details behind this historic question. They are worth putting on record.

The question had been suggested to Feynman the night before by Martin Block, an experimental physicist with whom Feynman was sharing a hotel room. The answer to the theta-tau puzzle, said Block, might be very simple. Perhaps the lovely law of parity does not always hold. Feynman responded by pointing out that if this were true, there would be a way to distinguish left from right. It would be surprising, Feynman said, but he could think of no way such a notion conflicted with known experimental results. He promised Block he would raise the question at next day's meeting to see if anyone could find anything wrong with the idea. This he did, prefacing his remarks with, "I am asking this question for Martin Block." He regarded the notion as such an interesting one that, if it turned out to be true, he wanted Block to get credit for it.

Chen Ning Yang and his friend Tsung Dao Lee, two young and brilliant Chinese-born physicists, were present at the meeting. One of them gave a lengthy reply to Feynman's question.

"What did he say?" Block asked Feynman later.

"I don't know," replied Feynman. "I couldn't understand it."

"People teased me later," writes Feynman, "and said my prefacing remark about Martin Block was made because I was afraid to be associated with such a wild idea. I thought the idea unlikely, but possible, and a very exciting possibility. Some months later an experimenter, Norman Ramsey, asked me if I believed it worth while for him to do an experiment to test whether parity is violated in beta decay. I said definitely yes, for although I felt sure that parity would *not* be violated, there was a possibility it would be, and it was important to find out. 'Would you bet a hundred dollars against a dollar that parity is not violated?' he asked. 'No. But fifty dollars I will.' 'That's good enough for me. I'll take your bet and do the experiment.' Unfortunately, Ramsey didn't find time to do it then, but my fifty dollar check may have compensated him slightly for a lost opportunity."

During the summer of 1956 Lee and Yang thought some more about the matter. Early in May, when they were sitting in the While Rose Cafe near the corner of Broadway and 125th Street, in the vicinity of Columbia University, it suddenly struck them that it might be profitable to make a careful study of all known experiments involving weak interactions. For several weeks they did this. To their astonishment they found that although the evidence for conservation of parity was strong in all strong interactions, there was no evidence at all for it in the weak. Moreover, they thought of several definitive tests, involving weak interactions, which would settle the question one way or the other. The outcome of this work was their now-classic paper "Question of Parity Conservation in Weak Interactions."

"To decide unequivocally whether parity

is conserved in weak interactions," they declared, "one must perform an experiment to determine whether weak interactions differentiate the right from the left. Some such possible experiments will be discussed."

Publication of this paper in *The Physical Review* (October 1, 1956) aroused only mild interest among nuclear physicists. It seemed so unlikely that parity would be violated that most physicists took the attitude: Let someone else make the tests. Freeman J. Dyson, a physicist now at the Institute for Advanced Study in Princeton, writing on "Innovation in Physics" (*Scientific American*, September 1958) had these honest words to say about what he called the "blindness" of most of his colleagues.

"A copy of it [the Lee and Yang paper] was sent to me and I read it. I read it twice. I said, 'This is very interesting,' or words to that effect. But I had not the imagination to say, 'By golly, if this is true it opens up a whole new branch of physics.' And I think other physicists, with very few exceptions, at that time were as unimaginative as I."

Several physicists were prodded into action by the suggestions of Lee and Yang. The first to take up the gauntlet was Madame Chien-Shiung Wu, a professor of physics at Columbia University and widely regarded as the world's leading woman physicist. She was already famous for her work on weak interactions and for the care and elegance with which her experiments were always designed. Like her friends Yang and Lee, she, too, had been born in China and had come to the United States to continue her career.

The experiment planned by Madame Wu involved the beta-decay of cobalt-60, a highly radioactive isotope of cobalt which continually emits electrons. In the Bohr model of the atom, a nucleus of cobalt-60 may be thought of as a tiny sphere which spins like a top on an axis labeled north and south at the ends to indicate the magnetic poles. The beta-particles (electrons) emitted in the weak interaction of beta-decay are shot out from both the north and the south ends of nuclei. Normally, the nuclei point in all

directions, so the electrons are shot out in all directions. But when cobalt-60 is cooled to near absolute zero (-273 degrees on the centigrade scale), to reduce all the joggling of its molecules caused by heat, it is possible to apply a powerful electromagnetic field which will induce more than half of the nuclei to line up with their north ends pointing in the same direction. The nuclei go right on shooting out electrons. Instead of being scattered in all directions, however, the electrons are now concentrated in two directions: the direction toward which the north ends of the magnetic axes are pointing, and the direction toward which the south ends are pointing. If the law of parity is not violated, there will be just as many electrons going one way as the other.

To cool the cobalt to near absolute zero, Madame Wu needed the facilities of the National Bureau of Standards, in Washington, D.C. It was there that she and her colleagues began their historic experiment. If the number of electrons divided evenly into two sets, those that shot north and those that shot south, parity would be preserved. The theta-tau puzzle would remain puzzling. If the beta-decay process showed a handedness, a larger number of electrons emitted in one direction than the other, parity would be dead. A revolutionary new era in quantum theory would be under way.

At Zurich, one of the world's greatest theoretical physicists, Wolfgang Pauli, eagerly awaited results of the test. In a now famous letter to one of his former pupils, Victor Frederick Weisskopf (then at the Massachusetts Institute of Technology), Pauli wrote: "I do *not* believe that the Lord is a weak left-hander, and I am ready to bet a very high sum that the experiments will give symmetric results."

Whether Pauli (who died in 1958) actually made (like Feynman) such a bet is not known. If he did, he also lost. The electrons in Madam Wu's experiment were *not* emitted equally in both directions. Most of them were flung out from the south end; that is, the end toward which a majority of the cobalt-60 nuclei pointed their south poles.

At the risk of being repetitious, and possibly boring readers who see at once the full implication of this result, let us pause to make sure we understand exactly why Madam Wu's experiment is so revolutionary. It is true that the *picture* (Figure 8.4–1) of the cobalt-60 nucleus, spinning in a certain direction around an axis labeled N and S, is an asymmetric structure not superposable on its mirror image. But this is just a picture. As we have learned, the labeling of N and S is purely conventional. There is nothing to prevent one from switching N and S on all the magnetic fields in the universe. The north ends of cobalt-60 nuclei would become south, the south ends north, and a similar exchange of poles would occur in the electromagnetic field used for lining up the nuclei. Everything prior to Madame Wu's experiment suggested that such a switch of poles would not make a measurable change in the experimental situation. If there were some intrinsic, observable difference between poles—one red and one green, or one strong and one weak—then the labeling of N and S would be more than a convention. The cobalt-60 nuclei would possess true spatial asymmetry. But physicists knew of no way to distinguish between the poles except by

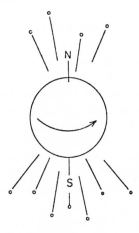

Figure 8.4–1. An electron is more likely to be flung out from the south end of a cobalt-60 nucleus than from its north end.

testing their reaction to other magnetic axes. In fact, as we have learned, the poles do not really exist. They are just names for the opposite sides of a spin.

Madame Wu's experiment provided for the first time in the history of science a method of labeling the ends of a magnetic axis in a way that is not at all conventional. *The south end is the end of a cobalt-60 nucleus that is most likely to fling out an electron!*

The nucleus can no longer be thought of as analogous to a spinning sphere or cylinder. It must now be thought of as analogous to a spinning cone. Of course, this is no more than a metaphor. No one has the slightest notion at the moment of why or how one end of the axis is different, in any intrinsic way, from the other. But there is a difference! "We are no longer trying to handle screws in the dark with heavy gloves," was the way Sheldon Penman of the University of Chicago put it (*Scientific American*, July 1961), "we are being handed the screws neatly aligned on a tray, with a little searchlight on each that indicates the direction of its head."

It should be obvious now that here at long last is a solution to the Ozma problem—an experimental method of extracting from nature an unambiguous definition of left and right. We say to the scientists of Planet Xs "Cool the atoms of cobalt-60 to near absolute zero. Line up their nuclear axes with a powerful magnetic field. Count the number of electrons flung out by the two ends of the axes. The end that flings out the most electrons is the end that we call 'south.' It is now possible to label the ends of the magnetic axis of the field used for lining up the nuclei, and this in turn can be used for labeling the ends of a magnetic needle. Put such a needle above a wire in which the current moves away from you. The north pole of this needle will point in the direction we call 'left.'"

We have communicated precisely and unambiguously to Planet X our meaning of the word "left." Neither we nor they will be observing in common any single, particular asymmetric structure. We will be observing in common a universal law of nature. In the weak interactions, nature herself, by her own intrinsic handedness, has provided an operational definition of left and right![2] It is easy to understand why Pauli and other physicists did not expect Madame Wu's experiment to overthrow parity. It would have meant that nature is not ambidextrous!

. . . . . .

The cobalt-60 experiment provides a method by which the puzzled astronauts could tell whether they were reversed. Of course they would have to find some cobalt on the unknown planet, convert it to its radioactive isotope by bombarding it with neutrons, and so on. But assuming that they had the equipment and could find the necessary materials, they would be able to test their handedness.[3]

. . . . .

Although evidence against the conservation of parity was strongly indicated by Madame Wu's work in late 1956, the experiment was not finally completed until early in January 1957. Results were formally announced by Columbia University's distinguished physicist Isador Rabi on January 15, 1957. The announcement also included the results of a confirming experiment conducted by Columbia physicists at the Nevis Cyclotron Laboratories at Irvington-on-Hudson in Westchester County, New York. This confirming test, made with mu mesons, showed an even stronger handedness. The mu mesons shot out twice as many electrons in one direction as in the other. Independent of both experiments, a third test was made at the University of Chicago using the decay of pi and mu mesons. It, too, showed violation of parity. All over the world physicists began testing parity in other weak interactions. By 1958 it was apparent that parity is violated in *all* such interactions. The theta-tau puzzle was solved. There is only *one* K-meson. Parity is *not* conserved.

"A rather complete theoretical structure

has been shattered at the base," declared Rabi (quoted by the *New York Times*, January 16, 1957), "and we are not sure how the pieces will be put together." An unnamed physicist was reported by the *Times* as saying that nuclear physics had been battering for years at a closed door only to discover suddenly that it wasn't a door at all—just a picture of a door painted on a wall. Now, he continued, we are free to look around for the true door. O. R. Frisch, the physicist who was a co-discoverer of nuclear fission, reports in his book *Atomic Physics Today* (Basic, 1961) that on January 16, 1957, he received the following air letter from a friend:

Dear Robert:
HOT NEWS. Parity is not conversed. Here in Princeton they talk about nothing else; they say it is the most important result since the Michelson experiment. . . .

The Michelson experiment was the famous Michelson-Morley test in 1887 which established the constant velocity of light regardless of the motion of source and observer—a historic experiment which paved the way for Einstein's theory of relativity. Madame Wu's experiment may well prove to be equally historic.

The two tests were very much alike in their shattering element of surprise. Everybody expected Albert Michelson and Edward Morley to detect a motion of the earth relative to a fixed "ether." It was the negative result of this test that was so upsetting. Everybody expected Madame Wu to find a left-right symmetry in the process of beta-decay. Nature sprang another surprise! It was surprising enough that certain particles had a handedness; it was more surprising that handedness seemed to be observable only in weak interactions. Physicists felt a shock even greater than Mach had felt when he first encountered the needle-and-wire asymmetry.

---

[2]But what if "Planet X" is composed of anti-matter?! (See footnote 3.) [Ed.]

[3]C. Michael has pointed out that there is a much simpler way to communicate the idea of right-handedness—one merely transmits a right-circularly polarized radio wave![Ed.]

"Now after the first shock is over," Pauli wrote to Weisskopf on January 27, after the staggering news had reached him, "I begin to collect myself. Yes, it was very dramatic. On Monday, the twenty-first, at 8 P.M. I was supposed to give a lecture on the neutrino theory. At 5 P.M. I received three experimental papers [reports on the first three tests of parity] . . . . I am shocked not so much by the fact that the Lord prefers the left hand as by the fact that he still appears to be left-handed symmetric when he expresses himself strongly. In short, the actual problem now seems to be the question: Why are strong interactions right-and-left symmetric?"

The Indian physicist Abdus Salam (from whose article on "Elementary Particles" in *Endeavor*, April 1958, the extracts from Pauli's letters are taken) tried to explain to a liberal-arts-trained friend why the physicists were so excited about the fall of parity. "I asked him," wrote Salam in this article, "if any classical writer had ever considered giants with only the left eye. He confessed that one-eyed giants have been described, and he supplied me with a full list of them; but they always sport their solitary eye in the middle of the forehead. In my view, what we have found is that space is a weak left-eyed giant."

Physicist Jeremy Bernstein, in an article on "A Question of Parity" which appeared in *The New Yorker*, May 12, 1962, reveals an ironic sidelight on the story of parity's downfall. In 1928 three physicists at New York University had actually discovered a parity violation in the decay of a radioactive isotope of radium! The experiment had been repeated with refined techniques in 1930. "Not only in every run," the experimenter reported, "but even in all readings in every run, with few exceptions," the effect was observable. But this was at a time when, as Bernstein puts it, there was no theoretical context in which to place these results. They were quickly forgotten. "They were," writes Bernstein, "a kind of statement made in a void. It took almost thirty years of intensive research in all branches of experimental and theoretical physics, and, above all, it took

the work of Lee and Yang, to enable physicists to appreciate exactly what those early experiments implied."

In 1957 Lee and Yang received the Nobel prize in physics for their work. Lee was then 30, Yang 34. The choice was inevitable. The year 1957 had been the most stirring in modern particle physics, and Lee and Yang had done most of the stirring. Today the two men have adjacent offices at the Institute for Advanced Study in Princeton, where they continue to collaborate. Both live in Princeton with their attractive wives and children, proud of their Chinese heritage, deeply committed to science, and with a wide range of interests outside of physics and mathematics. If you are curious to know more about these two remarkable men, look up Bernstein's excellent *New Yorker* article.

It is worth pausing to note that, like so many other revolutions in physics, this one came about as the result of largely abstract, theoretical, mathematical work. Not one of the three experiments that first toppled parity would have been performed at the time it was performed if Lee and Yang had not told the experimenters what to do. Lee had had no experience whatever in a laboratory. Yang had worked briefly in a lab at the University of Chicago, where he was once a kind of assistant to the great Italian physicist Enrico Fermi. He had not been happy in experimental work. His associates had even made up a short rhyme about him which Bernstein repeats:

> Where there's a bang,
> There's Yang.

Laboratory bangs can range all the way from an exploding test tube to the explosion of a hydrogen bomb. But the really Big Bangs are the bangs that occur inside the heads of theoretical physicists when they try to put together the pieces handed to them by the experimental physicists.

John Campbell, Jr., the editor of *Analog Science Fiction*, once speculated in an editorial that perhaps there was some difference in the intellectual heritage of the Western and Oriental worlds which had predisposed two

Chinese physicists to question the symmetry of natural law. It is an interesting thought. I myself pointed out, in my Mathematical Games column in *Scientific American*, March 1958, that the great religious symbol of the Orient (it appears on the Korean national flag) is the circle divided asymmetrically as shown in Fig 8.4–2. The dark and light areas are known respectively as the Yin and Yang. The Yin and Yang are symbols of all the fundamental dualities of life: good and evil, beauty and ugliness, truth and falsehood, male and female, night and day, sun and moon, heaven and earth, pleasure and pain, odd and even, left and right, positive and negative . . . the list is endless. This dualism was first symbolized in China by the odd and even digits that alternate around the perimeter of the *Lo shu*, the ancient Chinese magic square of order 3. Sometime in the tenth century the *Lo shu* was replaced by the divided circle, which soon became the dominant Yin-Yang symbol. When it was printed or drawn, black and white was used, but when painted, the Yang was made red instead of white. The two small spots were (and still are) usually added to symbolize the fact that on each side of any duality there is always a bit of the other side. Every good act contains an element of evil, every evil act an element of good; every ugliness includes some beauty, every beauty includes some ugliness, and so on. The spots remind the scientist that every "true" theory contains an element of falsehood. "Nothing is perfect," says the Philosopher in James Stephens' *The Crock of Gold*. "There are lumps in it."

The history of science can be described as a continual, perhaps never-ending, discovery of new lumps. It was once thought that planets moved in perfect circles. Even Galileo, although he placed the sun and not the earth at the center of the solar system, could not accept Kepler's view that the planetary orbits were ellipses. Eventually it became clear that Kepler had been right: the orbits are *almost* circles but not quite. Newton's theory of gravity explained why the orbits were perfect ellipses. Then slight deviations in the Newtonian orbits turned up and were in turn explained by the correction factors of relativity theory that Einstein introduced into the Newtonian equations "The real trouble with this world of ours," comments Gilbert Chesterton in *Orthodoxy*, "is not that it is an unreasonable world, nor even that it is a reasonable one. The commonest kind of trouble is that it is nearly reasonable, but not quite. . . . It looks just a little more mathematical and regular than it is; its exactitude is obvious, but its inexactitude is hidden; its wildness lies in wait."

To illustrate, Chesteron imagines an extraterrestrial examining a human body for the first time. He notes that the right side exactly duplicates the left: two arms, two legs, two ears, two eyes, two nostrils, even two lobes of the brain. Probing deeper he finds a heart on the left side. He deduces that there is another heart on the right. Here of course, he encounters a spot of Yin within the Yang. "It is this silent swerving from accuracy by an inch," Chesterton continues, "that is the uncanny element in everything. It seems a sort of secret treason in the universe . . . Everywhere in things there is this element of the quiet and incalculable."

Feynman, with no less reverence than Chesterton, says the same thing this way at the close of a lecture on symmetry in physical laws (Lecture 52 in *The Feynman Lectures on Physics*, Addison-Wesley, 1963):

Why is nature so nearly symmetrical? No one has any idea why. The only thing we might suggest is something like this: There is a gate in Japan, a gate in Neiko, which is sometimes called by the Japanese the most beautiful gate in all

Figure 8.4–2. The asymmetric Yin-Yang symbol of the Orient.

Japan; it was built in a time when there was great influence from Chinese art. This gate is very elaborate, with lots of gables and beautiful carving and lots of columns and dragon heads and princes carved into the pillars, and so on. But when one looks closely he sees that in the elaborate and complex design along one of the pillars, one of the small design elements is carved upside down; otherwise the thing is completely symmetrical. If one asks why this is, the story is that it was carved upside down so that the gods will not be jealous of the perfection of man. So they purposely put the error in there, so that the gods would not be jealous and get angry with human beings.

We might like to turn the idea around and think that the true explanation of the near symmetry of nature is this: that God made the laws only nearly symmetrical so that we should not be jealous of His perfection!

Note that the Yin-Yang symbol is asymmetrical. It is not superposable on its mirror image. The Yin and Yang are congruent shapes, each asymmetrical, each with the same handedness. By contrast the Christian symbol, the cross, is left-right symmetrical. So is the Jewish six-pointed Star of David, unless it is shown as an interlocking pair of triangles that cross alternately over and under each other. It is a pleasant thought that perhaps the familiar asymmetry of the oriental symbol, so much a part of Chinese culture, may have played a subtle, unconscious role in making it a bit easier for Lee and Yang to go against the grain of scientific orthodoxy; to propose a test which their more symmetric minded Western colleagues had thought scarcely worth the effort.

8.5   *The First Observation of an*
Artificially-Produced Hyperon

D. H. Frisch and A. M. Thorndike

Most of the recently discovered elementary particles were found in cosmic rays before being produced by man-made accelerators. The primary cosmic rays are protons, and to a lesser extent heavier nuclei. These protons fall on the earth's atmosphere, hit the protons and neutrons in the molecules of nitrogen and oxygen in the air, and make various elementary particles, mainly π mesons. To explore the properties of these new particles thoroughly, however, it was necessary to have a more intense and controllable source of high-energy particles than cosmic rays.

The first cyclotron gave proton energies of less than 1 MeV in 1932, but larger machines capable of higher energies were soon built.

.   .   .   .   .

Despite the interruption of World War II,

a 184-inch-diameter cyclotron was completed at the University of California Radiation Laboratory in 1947. In 1948, E. Gardner and C. Lattes showed that π mesons were produced artificially by this cyclotron. It then became possible to create these new π mesons in much greater numbers than in cosmic rays and study them under controlled conditions in the laboratory. In the next few years several other cyclotrons with energies above the π meson threshold were completed, and many physicists began the study of the processes by which π mesons are created and also of the reactions that they can make.

During the same years, a circular proton accelerator of the "synchrotron" type was under construction at Brookhaven National Laboratory on Long Island in New York. It was designed for a much higher energy—3000 MeV. By the summer of 1952, a beam of 2200 MeV protons had been obtained. Now

SOURCE. From *The Ambidextrous Universe* (Basic Books: New York, 1964); copyright 1964 by Martin Gardner. Publishing, Inc., reprinted by permission of Van Nostrand-Reinhold Co.

for the first time a beam of particles was available with a great deal more energy than was required for π meson production in the simple processes of the previous equations. Such energies had previously been found only in the occasional high-energy particles in cosmic rays, and therefore the accelerator was named the "Cosmotron."

A great variety of new experiments became possible. Would π mesons be produced in collisions between π mesons and nucleons? Could two or more π mesons be produced at once in a collision between two nucleons? And, most exciting, would the machine also be a source of the short-lived hyperons which had been seen only in cosmic rays? All sorts of questions which had been in the minds of physicists studying the behavior of elementary particles could now be answered by experimentation.

One of the most important questions concerned a paradox that had been puzzling the physicists investigating hyperons and K mesons. Stated briefly, the trouble was this: the rate of decay of these particles is much too slow to be consistent with their rate of production. As an example consider the Λ° hyperon, which appeared to be made in the reaction

$$\pi^- + p \to \Lambda^\circ$$

Actually, no reaction in which two particles combine to form one particle can take place unless there is still another particle nearby to take up the extra momentum, so this reaction can only take place if the proton is in a nucleus, or if some other particle is formed.

A Λ° decays via the apparently reverse reaction

$$\Lambda^\circ \to \pi^- + p$$

Such a decay reaction normally can take place

without another particle present, provided, of course, that the sum of the rest-masses of the products is less than the mass of the decaying particle.

·   ·   ·   ·   ·

The decay reaction takes about 10^{-10} second. This is really a very long time by nuclear standards. If this decay reaction were typical of nuclear decays, the Λ° should disintegrate in about the time it takes to form it, some 10^{-22} second.[1]

A. Pais, a theoretical physicist working at the Institute for Advanced Study in Princeton, suggested that the puzzle of the long decay time of the Λ° could be explained if another particle is produced along with the Λ°. He proposed that the production reaction be given by

$$\pi^- + \rho \to \Lambda^\circ + K^\circ$$

Then, since the K° is no longer nearby when the Λ° tries to decay, the true reverse reaction

$$\Lambda^\circ + K^\circ \to \pi^- + K^\circ$$

can't take place. The decay happens only because there is another kind of force, called a "weak interaction," which allows the decay $\Lambda^\circ \to \pi^- + p$ to proceed without using a K°, but much more slowly. Thus Pais' hypothesis of "associated production" would allow Λ° production in a large fraction of the collisions between a π^- meson and a proton, with a K° made along with it at the same time by means of a "strong interaction." By contrast, the decay of the Λ° would be slow because only a weak interaction would cause it to occur in the absence of the K°.

Meanwhile, back at the Cosmotron, a group of experimental physicists—Fowler, Shutt, Thorndike, and Whittemore—had

---

[1]The π^- meson moving at nearly the speed of light, 3×10^8 meters per second, crosses a proton, which has a diameter of 2×10^{-15} meter, in $\frac{2}{3} \times 10^{-23}$ second. The chance of a π^- making a Λ° in one transit is about 1/15th, so the "formation time" for Λ° by π^-s striking protons is $15 \times \frac{2}{3} \times 10^{-23} = 10^{-22}$ second. Such a number is so much less than any time that enters our own sensations that it has no association with anything in our personal experience. To an elementary particle, however, 10^{-22} second is quite a considerable length of time.

built a new diffusion cloud chamber. This cloud chamber had the advantage that it could operate with almost pure hydrogen in its track-sensitive volume, at a pressure of 300 pounds per square inch. The pictures taken at the Cosmotron with this cloud chamber, they hoped, would give direct and unambiguous answers to some of these questions just proposed, since there would be a great deal of information about each reaction that was observed. Of course, many other interesting experiments were undertaken at the Cosmotron besides those of the Cloud Chamber Group. The work of this group, however, provides a good example of some exciting discoveries made in recent years and also of the long hours of work and cooperative effort by many people that characterize such large-scale experiments. Let us reconstruct part of their work during part of the year 1953.

It was early afternoon, March 13, the day of the first real Cosmotron run for the cloud chamber. In the experimental area the 15-ton bulk of the chamber and its magnet had been lined up to a fraction of an inch in its position in front of a 1-inch diameter hole passing through the Cosmotron's concrete shielding wall, eight feet thick. High energy neutrons were supposed to emerge from this opening, but there wasn't much information about this "neutron beam." No one even knew for sure that there were any neutrons in it.

Particle beams from accelerators move in straight lines unless deflected by a magnetic or electric field. Gravity exerts a negligible force because elementary particles emerging from accelerators move so fast that they don't have time to fall very far under the force of gravity as they cross even a large experimental room. A particle going with half the speed of light falls only 5×10^{-14} meter, very much less than an atomic diameter, in going 15 meters horizontally. Thus if left alone these elementary particles really act just as beams of light, going straight out from the place of their production to where they are observed.

The cloud chamber had been filled with hydrogen early in the morning. By ten o'clock it had reached equilibrium, ready for operation. Cameras had been checked out, flash-

lamps worked, the generator for the magnet was up to power, and since Long Island is cold in March, the generator wasn't overheating.

Getting the Cosmotron into operation was a far more complicated job for its operating crew. Its main generator had been started first, then the magnet current switched on. Vacuum pumps were inspected; vacuum tank pressure was satisfactory. The Van de Graaff accelerator, used to accelerate the protons for injection into the bigger machine, was running smoothly; high voltage was on the inflector plates, ready to push the injected particles into their orbits in the circular synchrotron for further acceleration. The injected beam trace on the console oscilloscope looked strange, and so there was a brief conference in the Control Room. Cliff Swartz, who had designed the pick-up plates, said it was probably all right that way, and so R. F. power was turned on.

No beam.

For several hours, the men at the console kept adjusting the R. F. frequency control knobs, and finally a few circulating protons were found. By one o'clock this mysterious process of "tuning up the machine" had increased the number of protons to the usual value.

Having worked all morning to increase the Cosmotron's intensity, it was then necessary to reduce it! The cloud chamber was flooded with too many stray particles, particularly electrons. Shutt and Whittemore kept the cloud chamber in operation and watched for interesting tracks. Fowler and Bill Tuttle, the group's electrical engineer, hovered over the pulsed generator used to power the cloud chamber magnet. Thorndike, in the Cosmotron control room, received reports from them via a private three-way telephone and passed on the necessary information to the Cosmotron operating crew.

The cloud chamber had been in operation for a few minutes, taking pictures once every eight seconds. The beam pulse repetition rate had been reduced, since the maximum repetition rate of one pulse every five seconds did not seem to allow enough time to clear out background ions from the previous pulse. A short piece of film, known as a "test strip" had just been developed to see whether the actual photographs were satisfactory or not. Shutt and Whittemore were inspecting it. The three-way telephone was silent while everyone waited for the verdict. Some minutes passed. In the control room Thorndike tried for the third time to relight his empty pipe, having nothing more useful to do. Every eight seconds the gong rang, sounding out that the Cosmotron magnet was still pulsing. Then:

"The intensity may be a little high, but let's try it this way."

"Turn on the R. F."

"R. F. is on."

"Take pictures."

With everything in operation, the day's run was finally really under way. Apart from minor interruptions for checking and servicing equipment, the run continued until midnight. By that time eleven 100-foot rolls of 35 mm film had been filled with cloud-chamber photographs—about 4000 in all. Several additional days' runs followed soon afterwards. Altogether some 20,000 neutron beam pictures were taken.

Within a few days some film had been developed, and the long task of searching carefully through the thousands of pictures to find events of interest had begun. About one picture in a hundred showed particles emerging from a nuclear interaction caused by a neutron striking a proton in the hydrogen, the rest merely "background" tracks coming from the walls of the chamber. Analysis of the interactions in hydrogen soon showed that production of two π mesons in a single neutron-proton collision was a common occurrence. One of our questions had been answered.

At the end of the afternoon of April 3 the four physicists were checking over the new events found in the pictures that had been looked at during the day. There seemed to be a considerable number of "good" events, so it was clear that the next day would be a busy one. Noticing that it was just five o'clock, the punctual Whittemore left to pick up his wife, who worked in another building, and the

group broke up. Thorndike lingered on, feeling the universal curiosity about what will happen next, and idly turned ahead to look at the next few pictures, as one often looks ahead a few pages before putting down a fast-moving book. He did not turn far, though. On the first new picture a track junction caught his eye. He looked at it for a minute—certainly that characteristic V-shaped pair of tracks could be nothing else but a Λ° hyperon! This case, shown in Figure 8.5–1, was the first definite example of an artificially produced hyperon.

This event and another found soon afterwards showed that hyperons could indeed be produced by the neutral beam. However, the Λ°'s came from collisions occurring in the wall of the chamber, and as a result not much could be said about the production reaction. A much more valuable example would be one in which the Λ° was produced in a collision visible in the sensitive region of the cloud chamber. Unfortunately, the experimenters found no case of this type in these pictures taken with the neutron beam.

Later in the spring a similar series of runs was begun with a negative π meson beam rather than the neutron beam. The π mesons had an energy of 1.4 BeV, which was somewhat lower than the average energy of the neutrons used previously, but much higher than that of any π meson that had previously been studied. The first few events from this run showed that incident π mesons could indeed make secondary π mesons. The most common reactions were those given by

$$\pi^- + p \rightarrow \pi^- + p + \pi^\circ$$
$$\pi^- + p \rightarrow \pi^- + n + \pi^+$$

On June 7 an event with a V-shaped pair of tracks was found, and the experimenters felt certain that this represented a Λ° hyperon again. They noted that the track of the incoming π^- meson disappeared suddenly in the middle of the chamber. This could be accounted for by a reaction with entirely neutral products

$$\pi^- + p \rightarrow \Lambda^\circ + K^\circ$$

After traveling 0.65 cm the Λ° decayed to

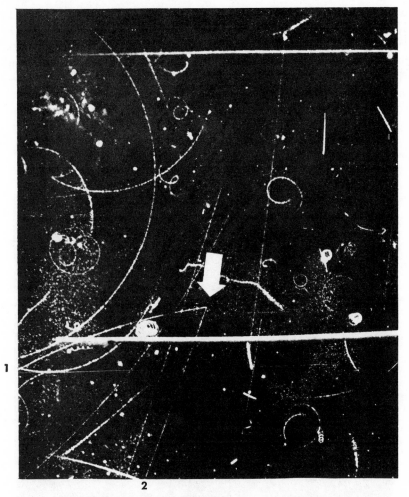

Figure 8.5–1.   The first Λ° hyperon produced artificially. This diffusion cloud-chamber photograph shows that the Λ° decays below the point of the arrow into a π^- meson (track 1) and a proton (track 2) in the process $\Lambda^\circ \to \pi^- + p$. The Λ° was presumably produced in a collision between a neutron from the Cosmotron and a proton or neutron in an iron nucleus in the wall of the chamber above the area shown, but the tracks observed in the chamber do not give any information about this collision.

two charged particles, identified as a proton and a π^- meson. Their tracks start at 0.65 cm from the end of the original π^- track.

The presence of the other neutral secondary particle—perhaps a K° or perhaps one or more others—was indicated in order to conserve momentum, but since its decay was not observed, there was little information about it. If, indeed, a *single* neutral particle was emitted in addition to the Λ°, the mass of the unknown neutral particle could be com-

puted from energy and momentum balance. Such a computation gave a mass approximately equal to the known mass of the K°.

As time went on the experimenters found a number of similar events, all of which seemed to indicate that every time a Λ° was produced a K° was also produced. This was just what Pais had suggested. The most convincing evidence, of course, would be an example in which a Λ° and K° could both be seen to decay in the chamber. By the end of that summer of 1953, one such example had been found, and the process of "associated production," as the reaction came to be called, could hardly be questioned. More recent examples of the associated production of a $\Lambda°$ and a K° are Plate V, a reaction in a bubble chamber, and Figure 8.5–2, the same, as seen in a spark chamber.

The associated production of particles in the Cosmotron experiments showed that the theoretical ideas introduced by Pais and by Gell-Mann and Nishijima were indeed correct, or at least on the right track. The idea that certain combinations of particles were necessary for the easy creation of—and decay into—other elementary particles began to look like a new conservation law. The quantity that is to be conserved in order to describe the rules of associated production is now called "strangeness."

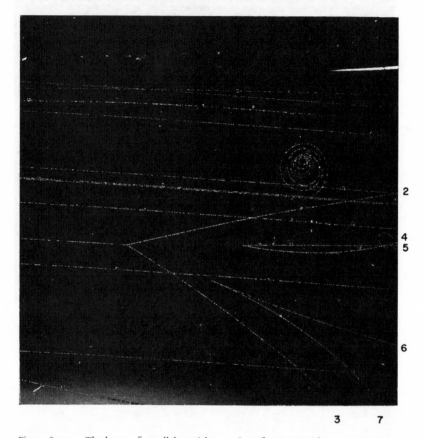

Figure 8.5–2. The beam of parallel particles consists of protons with energy 2.85 BeV passing from left to right. An incoming proton (1) collides with a proton in the hydrogen of the bubble chamber. A proton (2) and π^+ meson (3) are emitted, and in addition two neutral particles. The first neutral particle is a $\Lambda°$ hyperon, which disintegrates to give a π meson (4) and a proton (5). The second neutral particle is a $K°$ meson, which disintegrates to give a π^- meson (6) and a π^+ meson (7).

CHAPTER 9
ASTROPHYSICS AND COSMOLOGY

9.1   *The Earth Among the Stars*

                           Ernst J. Öpik

The earth as our abode of life has impressed itself upon our mind so much that in everyday life we treat it as our Universe. When we refer to "the world," we mean the earth, forgetting that our planet is only a tiny speck in the world of planets, suns, and galaxies.

In itself, the smallness of the earth is no reason for denying its importance. One should not be impressed too much by mere quantity; great dimensions and heavy mass have no merit by themselves; they cannot compare in value with immaterial things, such as thoughts, emotions, and other expressions of the soul. To us the earth *is* the most important of all celestial bodies, because it has become the cradle and seat of our spiritual values. Yet, at the same time, we should be able to assess its place in the material world in correct perspective.

In antiquity, judging naively by appearance, man saw in heaven and earth two equal parts of the material cosmos. Many advanced brains were able to perceive the falseness of this judgment, but they could not, or dared not, insist upon their true views against public opinion; they preferred to teach officially what people wanted to be taught. Passages in the Bible which, if taken literally, are in conflict with modern science, probably originated under pressure, or as the expression of naive public opinion. Saint Augustine, in the fifth century, evidently referred to this when he wrote: "The Lord did not say 'I send you the Spirit [Holy Ghost], to instruct you about the courses of Sun and Moon'; He wanted you to become Christians, not astronomers." Similarly, in the thirteenth century, Thomas Aquinas pointed out that, in questions of the natural sciences, one should not subject the Bible to a crucial test; even if it could be clearly shown that statements in the Bible were scientifically incorrect, one should not proclaim this, "lest the Holy Scripture should be ridiculed by the Unfaithful and, thus, their way to Faith blocked."

In our modern age we have overcome this impediment; we can freely proclaim our scientific findings without endangering spiritual truths.

The most important scientific revelation since Copernican times is the recognition of the fact that the earth belongs to the heavens, being part of them. Instead of using the ancient antithesis between heaven and earth, by which undue material importance was attributed to our planet, we would now say—

SOURCE. From *The Oscillating Universe* (New York: New American Library, 1960) Chapter 1; copyright by Ernst J. Öpik, reprinted by arrangement with The New American Library, Inc.

the heavens, including that tiny corner of them which we inhabit and call earth. To us the earth has become a celestial body, one among numberless others.

The earth is a rocky globe, freely suspended in space and moving around the sun. It is kept in its position relative to the sun by the gravitational force of the sun's attraction; gravitation counteracts the centrifugal force of orbital motion. The two forces are in equilibrium, and the earth swings around the sun on its annual course—a slightly elliptical orbit. The earth can stay on this orbit for thousands of millions of years to come, as it has done for similar intervals of time in the past. This means that the orbit of the earth is practically stable, or that our planet will not go away far enough from, or come near enough to, the sun to endanger terrestrial life. This stable position of the earth, as well as of its sister planets, depends upon a peculiar feature of the solar system: The central body, the sun, is almost 1,000 times more massive than all its planets taken together; it therefore rules supreme, preventing the planets from considerably disturbing one another's motions. In a system in which the planets are large as compared with their sun, or in which there are several suns, the mutual gravitation of the members of the system plays havoc with the orbits, which change within all imaginable limits. Were our earth a member of such a system—there are numerous such cases among the stars— life could hardly develop on its surface, as it would be extinguished by either heat or cold every time it started.

Thus, happily, the solar system is a monarchy where the earth—a tiny subject— enjoys security, being made to move on its prescribed, almost fixed, path.

Although small among other celestial bodies, by human standards the earth is a body of enormous size. Yet its dimensions are within the reach of unaided human locomotion: it would take only 7,000 hours to walk "around the world," and many a postman has covered a several times greater distance during his lifetime. The mass of the earth is 6,000,000,000,000,000,000,000 tons,

a figure which transcends our powers of imagination. Yet the sun is 333,000 times heavier, and its circumference is 109 times greater than that of the earth. No wonder it keeps the earth in obedience from a distance of 93 million miles, although these millions upon millions of tons of our globe are racing along with a speed of 18 miles per second.

If, by some magic, we could detach ourselves from the earth and stop in space (remaining at rest relative to the sun) on an evening, we would see the surface of the earth moving away downwards with terrific speed. Within half a minute we would be well outside the atmosphere, the terrestrial globe appearing half illuminated, half in darkness, and still covering the greater portion of a hemisphere. The apparent size of the earth would shrink quickly, like a railway train racing away. Soon the earth would assume the appearance of a half-moon, of a beautiful bluish glare, covered with brilliant white spots—clouds illuminated by the sun—and darker areas where clouds are absent and the continents and oceans are visible through the atmospheric haze. Although similar to the view presented to us by the moon at first quarter, the spectacle of the earth as seen from space would be much more colorful and glamorous. As the distance increased, fewer details on the surface of the earth would be discernible, and the moon (if not already visible) would emerge from behind the shrinking edge of the earth. After 4 hours we would enjoy the sight of the twin planet earth-moon, the moon about the same size as it appears to us now, the earth $3\frac{1}{2}$ times bigger and 60 times brighter than the moon. While they continued to be carried away in their orbital motion, their apparent sizes and mutual distance would seem to decrease. Within a fortnight the eye would no longer discern the shape or dimensions of these bodies; they would look like two sparkling points—a double star of extreme brilliance: the earth in brightness exceeding Venus, the moon equal to Sirius and at an apparent distance of about half a degree from the earth (equal to the apparent diameter of the moon for a terrestrial observer). In the background, the black sky of interplanetary space would be studded with numberless stars no longer mindful of the presence of the sun; the sun would be shining in another corner of the sky with undiminished brilliance, yet its rays, not being reflected by an atmosphere, would not interfere with the visibility of the stars.

From our extraterrestrial vantage point the smallness of the earth, as compared with the sun, becomes obvious. The earth looks like a little star beside the giant sun. The comparison, of course, is somewhat misleading. The fixed stars (not planets), although they appear small to us, are not so small at all; they are themselves suns, many of them bigger than ours. Being millions of times farther away than the sun, they look deceptively faint. The 572 million miles which represent the annual path of the earth around the sun encompass a corner of space which is as small compared to stellar distances as the dimensions of a little town are in comparison with those of the solar system.

To take a bird's-eye view of the sun and its planets, let us speed up 10,000 times and travel at 186,000 miles a second, the velocity of light—the greatest velocity attainable by a material object according to relativity. At such a speed we could go $7\frac{1}{2}$ times around the earth in one second, and reach the moon in $1\frac{1}{4}$ seconds. The distance from earth to sun would be covered in 8 minutes. Let us depart at right angles to the plane of the earth's orbit, or the ecliptic. After one hour's flight the earth will appear as a first-magnitude star, only 8° from the sun—so close that the distance sun-to-earth can be covered by the palm of the outstretched hand; the other planets will be seen at their appropriate distance: Venus and Mercury close to the sun and hardly visible, Jupiter at a distance of 45° from the sun and almost as bright as we see it from earth. After 2 to 3 hours of flight, the earth will become invisible to the naked eye; after 5 days it will no longer be seen in a telescope. Yet, to reach another star—another sun—we should have to travel for 5 years at this speed. The distance attained would then be 5 light years—5 years of travel with the velocity of light. If the solar system were observed from such a distance, nothing but the sun itself, as a star of first magnitude, would be observable with the most powerful human means. To see the earth from a nearby star, or, which is the same, to observe planets of this star from earth, telescopes one mile across and ten miles long would be required: clearly, this appears to be impossible by present human standards. Thus, from the interstellar viewpoint, the earth is nonobservable, or, for all practical purposes, nonexistent. Looking at the starry sky, we have to overlook the planets; it is suns alone that count now, our sun— a brilliant speck embodying the entire solar system—simply one among millions. Yet, to our mental eye, planets, other earths, are ever present in the universe, although physically we cannot see them.

Through its master, the sun, our earth is made a member of the stellar universe. With respect to the swarm of stars in our neighborhood (within a distance of a few hundred light years) our sun is moving with a velocity of 12 miles per second towards a point between the constellations Hercules and Lyra, carrying along all its planets, including the earth. Thus, our globe spirals in interstellar space on a screw-line produced by the combination of the earth's orbital motion around the sun and the translational motion of the solar system towards Hercules-Lyra. With the translational motion it would take some 60,000 years for the sun to cover the distance to the nearest star; on the interstellar scale it is a relatively slow and unimportant motion. Other stars are moving in different directions, without much regularity in their motions. These small relative motions of the millions of stars in our neighborhood remind us of a swarm of dancing gnats.

Yet, by way of the sun, our earth has other loyalties to obey. The swarm of stars which presents itself to us at night, and to which our sun belongs, is not as structureless as it appeared to the astronomers of half a century ago, to whom only the irregular gnat-dance relative motions of its members were known. All these stars belong to a major system, the Milky Way. This galactic system has a flattened shape, like a lens; it contains about 100,000 million stars, about 50 times the human population of the earth, each star either a planetary system or a system of two or more suns (double or multiple stars). It is over 100,000 light years across and thousands of light years thick. Along its equator there are so many stars seen in the same direction that their combined light appears to us as the Milky Way belt. This system is rotating around a center with great speed; the velocity of rotation which carries the swarm of surrounding stars (including our sun) around the center is 150 miles per second. Our sun is at a distance of 28,000 light years from the center of the Milky Way, and it would take us 250 million years to complete one revolution. Unlike the solar system, the Milky Way has no central governing body; it is a republic where the motions of single members are ruled by the combined attraction of the entire population.

Yet the Milky Way, or the Galaxy, is not the only system of its kind in space: there are millions and millions of them in all directions at distances of millions and hundreds of millions of light years. The earth, through our sun, belongs to one of them, our Milky Way.

The contemplation of the above-described systems, or steps, into which our universe is organized, and to which the earth belongs, may give an idea of the material insignificance of our globe and of the vastness of the universe. At the same time, we may feel elated at being able to grasp the magnificence of the world by our minds—we, tiny creatures on a tiny globe, nay, we, citizens of the heavens through our earth, a celestial body among the stars.

THE ORIGIN OF THE EARTH

The origin of the celestial bodies is as fascinating a problem as it is a difficult one. We know only the present state of a small portion of the universe; yet from this limited amount of knowledge we are trying to reconstruct the past history of the whole, in much the same manner as from the outward appearance, the general expression and wrinkles of the face of a stranger, we might try to decipher his life history and origin. The task seems to be almost hopeless.

With respect to the origin of the earth we are much more on solid ground (literally and figuratively speaking) than in the case of other, distant celestial bodies. The study of the rocky surface layers tells us something about the past history of our planet; this is the task of geology, which, among other things, is concerned with the deciphering of fossil remains of past life found in the rock strata. Other sciences—geophysics, seismology, astronomy—help in disclosing the present structure of our globe. The chemist and physicist, by studying radioactivity, are able to tell the age of the rocks, thus putting terrestrial chronology on a solid basis. This, perhaps, is the most spectacular modern achievement in the study of the history of the earth. Radioactive elements slowly change into other, stable elements; thus, uranium and thorium are, in the course of time, converted into certain kinds (isotopes) of lead. The amount of uranium decreases, that of lead increases, gradually. From the relative amounts of these two elements, uranium as the original substance and lead as its end product, one can calculate the age of the rocks which contain them.

Of course, more recent stages in the history of the earth are better known than the earlier ones. The fossil remains found in rocks are subjected to erosion and volcanic activity; the older they are, the more likely they are to have changed or become completely erased.

Our knowledge of the history of the earth's surface for the last 500 million years is relatively complete, whereas for the earlier period of 2,000 to 3,000 million years the data are rather scanty. This is due not only to the difference in age and the loss of old "records," but also to the circumstance that in more recent times fossils were much more numerous and durable. Animals with hard crusts or shells appeared on earth only about 500 million years ago, and these were much better preserved as fossils than their soft predecessors. Therefore, although there was life on earth long before then, few remains of it are found from ages earlier than 500 million years ago.

The oldest rocks so far explored belong to the continent of Africa: their age is about 2,900 million years. A comparative study of the radioactivity of rocks of different ages and origin leads to the conclusion that the earth's solid crust itself is not much older than 3,500 million years. At this remote epoch the beginning of our globe must be looked for.

It is believed that the earth condensed out of a nebula—a cloud composed of gas and dust—almost simultaneously with the other members of the solar system, the sun and the planets. The dust contained in the solar nebula obscured completely the sun's radiation in the plane of the ecliptic. Except in the nearest surroundings of the sun, it became terribly cold there, perhaps 260 degrees below zero centigrade ($-260\,^{\circ}$C). At this low temperature various gases, such as water vapor, ammonia, nitrogen, carbon dioxide, carbon monoxide, and methane, formed "snow flakes" or "icicles" around the dust particles. This cosmic snow began to accumulate into larger pieces, and from the mutual collision of these pieces or "planetesimals" the earth was built. As the earth was growing, its gravitational force increased, and the planetesimals were soon falling on it with velocities sufficient to generate a considerable amount of heat. The growing earth became hot, the icy stuff melted, evaporated, and was lost back to space, whereas the stony or metallic substance remained. This explains why there is so much rock and so little gas on earth, although the nebulae of interstellar space contain 100 times more gas than rocky substance or dust.

Opinions have been expressed that the earth, in the process of formation, although it warmed up considerably, never got hot enough to become completely molten. These opinions were based on the calculation of radiation losses of heat to space from the surface of the earth. However, in these calculations the shielding effect of dust has never been taken into account. The dust filling the space around the growing earth would act as an extremely efficient blanket, preventing heat losses to space, and allowing the surface of the earth (as well as its interior) to become completely molten. There can be little doubt that the earth was liquid at this first, although short, stage of its life. This stage may have lasted "only" about one million years.

The stony material probably separated from metallic nickel-iron while it was in the form of dust in the solar nebula. In the molten earth these substances did not mix permanently: Stony lava, like oil on water, rose to the surface and later became the crust of the earth (when sufficiently cooled); molten iron sank to the bottom forming the earth's core, which is still molten.

The space around the earth was soon (within about one million years) swept clean of dust by the earth itself; the other planets did their share in cleaning up their own surroundings. For the first time the rays of the sun burst through the darkness and shone on our young, still glowing globe. The protective blanket of dust disappeared, and the surface of the earth began cooling quickly. A thin skin of lighter minerals, which now is known as the granitic-basaltic layer, solidified at the surface, but the chief solidification of the earth's crust began at the bottom, 1,800 miles beneath the surface, at the boundary of the molten iron core. At this depth the pressure, over one million atmospheres, is so great that rocks would solidify at temperatures over 4,000° C, while iron could remain molten.

In meteorites, which are broken-off fragments of ancient planets destroyed by collisions, stone is separated from iron, as is believed to be the case in the earth's interior. Radioactive analysis of meteorites, as compared with a similar analysis of the earth's crust, indicates an age of 4,500 million years for the time of separation of iron from stone. This figure is of a good degree of precision; only, we cannot tell for certain to what stage of the evolution of the solar system it refers.

Most probably it is the time which just preceded the formation of the earth and other planets. Hence, the over-all age of the earth is 4,500 million years, or somewhat less. This is in good accord with the estimate of not less than 3,500 million years for the age of the solid crust. The solidification of the crust, protected on the outside by the granitic-basaltic skin, may indeed have taken a long interval of time, say 1,000 million years. The cooling must have been slow in the beginning, because of heat generated in the interior by large quantities of radioactive materials (especially radioactive potassium). These materials were only gradually used up, or transported to their present location near the surface of the earth, whence their heat is more readily lost to space and does not much affect the interior.

Not so long ago it was believed that the planets were formed from splashes of solar material, caused by the tidal action of some star that happened to pass near the sun. This theory, still upheld in many textbooks and encyclopedias, has lost ground. The splashes, consisting of extremely hot gas, could not possibly have condensed into the planets; they would have "exploded" into space, becoming a thin gas spread over a large volume without any chance of condensation. On the contrary, the old nebular theory of the origin of the solar system is now gaining support from all sides, after having been in abeyance for half a century.

9.2   *Next—The Planets*

Arthur C. Clarke

It has been said that history never repeats itself but that historical situations recur. To anyone, like myself, who has been involved in astronautical activities for over 30 years, there is a feeling of familiarity in some of the present arguments about the exploration of space. Like all revolutionary new ideas, the subject has had to pass through three stages which may be summed up by these reactions:

(1) "It's crazy—don't waste my time";
(2) "It's possible, but it's not worth doing";
(3) "I always said it was a good idea."

As far as orbital flights, and even journeys to the Moon, are concerned, we have made excellent progress through all of these stages, though it will be a few years yet before everyone is in category three. But where flights to the planets are involved, we are still almost where we were 30 years ago. True, there is much less complete skepticism—to that extent, history has *not* repeated itself—but there remains, despite all the events of the past decade, a widespread misunderstanding of the possible scale, importance and ultimate implications of travel to the planets.

Let us start by looking at some fundamentals, which are not as well known as they should be—even to space scientists. Forgetting all about rockets and today's astronautical techniques, consider the basic problem of lifting a man away from the Earth, purely in terms of the work done to move him against gravity. For a man of average mass, the energy requirement is about 100 kilowatt-hours, which customers with a favorable tariff can purchase for ten dollars from their electric company. What may be called the basic cost of a one-way ticket to space is thus the modest sum of ten dollars.

For the smaller planets and all satellites—Mercury, Venus, Mars, Pluto, Moon, Titan, Ganymede, etc.—the exit fee is even less; you need only 50 cents' worth of energy to escape from the Moon. Giant planets such as Jupiter, Saturn, Uranus and Neptune are

SOURCE. From *Playboy*, Vol. 15, No. 3 (March, 1969), p. 95; copyright by HMH Publishing Co., Inc., reprinted by permission of the author and the author's agents, Scott Meredith Library Agency, Inc., New York.

naturally much more expensive propositions. If you are ever stranded on Jupiter,
you'll have to buy almost $300 worth of
energy to get home. Make sure you take
enough traveler's checks!

Of course, the planetary fields are only
part of the story; work also has to be done
traveling from orbit to orbit and thus moving
up or down the enormous gravitational field
of the Sun. But, by great good luck, the Solar
System appears to have been designed for the
convenience of space travelers: All the
planets lie far out on the gentle slope of the
solar field, where it merges into the endless
plain of interstellar space. In this respect,
the conventional map of the Solar System,
showing the planets clustering round the
Sun, is wholly misleading.

We can say, in fact, that the planets are
99 percent free of the Sun's gravitational
field, so that the energy required for orbital
transfers is quite small; usually, it is considerably less than that needed to escape from
the planets themselves. In dollars and cents,
the energy cost of transferring a man from
the surface of the Earth to that of Mars is
less than $20. Even for the worst possible
case (surface of Jupiter to surface of Saturn),
the pure energy cost is less than $1000.

Hardheaded rocket engineers may well
consider that the above arguments, purporting to prove that space travel should be about
a billion times cheaper than it is, have no
relevance to the practical case—since, even
today, the cost of the fuel is trivial, compared
with the cost of the hardware. Most of the
mountainous Saturn 5 standing on the pad
can be bought for, quite literally, a few cents
a pound; kerosene and liquid oxygen come
cheap. The expensive items are the precision-
shaped pieces of high-grade metals and all
the little black boxes that are sold by the
carat.

Although this is true, it is also, to a large
extent, a consequence of our present im-
mature, no-margin-for-error technology.
Just ask yourself how expensive driving
would be if a momentary engine failure were
liable to write off your car—and yourself—
and the fuel supply were so nicely calculated

that you couldn't complete a mission if the
parking meter you'd aimed at happened to
be already occupied. This is roughly the
situation for planetary travel today.

To imagine what it may one day become,
let us look at the record of the past and see
what lessons we can draw from the early
history of aeronautics. Soon after the failure
of Samuel Langley's "aerodrome" in 1903,
the great astronomer Simon Newcomb wrote
a famous essay, well worth rereading, that
proved that heavier-than-air flight was impossible by means of known technology.
The ink was hardly dry on the paper when
a pair of bicycle mechanics irreverently
threw grave doubt on the professor's conclusions. When informed of the embarrassing
fact that the Wright brothers had just flown,
Newcomb gamely replied: "Well, maybe a
flying machine *can* be built. But it certainly
couldn't carry a passenger as well as a pilot."

Now, I am not trying to poke fun at one of
the greatest of American scientists. When you
look at the Wright biplane, hanging up
there in the Smithsonian Institution, Newcomb's attitude sums very reasonable, indeed; I wonder how many of us would have
been prepared to dispute it in 1903.

Yet—and this is the really extraordinary
point—there is a smooth line of development, without any major technological
breakthroughs, from the Wright "flier" to the
last of the great piston-engined aircraft, such
as the DC-6. All the many-orders-of-mag-
nitude improvement in performance came as
a result of engineering advances that, in
retrospect, seem completely straightforward
and sometimes even trivial. Let us list the
more important ones: variable-pitch air-
screws, slots and flaps, retractable under-
carriages, concrete runways, streamlining
and supercharging.

Not very spectacular, are they? Yet these
things, together with steady improvements
in materials and design, lifted much of the
commerce of mankind into the air. For they
had a synergistic effect on performance;
their cumulative effect was much greater
than could have been predicted by considering them individually. They did not merely

add; they multiplied. All this took about 40 years. Then there was the second technological breakthrough—the advent of the jet engine—and a new cycle of development began.

Unless the record of the past is wholly misleading, we are going to see much the same sequence of events in space. As far as can be judged at the moment, the equivalent items on the table of aerospace progress may be: refueling in orbit, air-breathing boosters, reusable boosters, refueling on (or from) the Moon and lightweight materials (e.g., composites and fibers).

Probably the exploitation of these relatively conventional ideas will take somewhat less than the 40 years needed in the case of aircraft; their full impact should be felt by the turn of the century. Well before then, the next breakthrough or quantum jump in space technology should also have occurred, with the development of new propulsion systems—presumably fission-powered but, hopefully, using fusion as well. And with these, the Solar System will become an extension of the Earth—if we wish it to be.

It is at this point, however, that all analogy with the past breaks down; we can no longer draw meaningful parallels between aeronautics and astronautics. As soon as aircraft were shown to be practical, there were obvious and immensely important uses for them: military, commercial and scientific. They could be used to provide swifter connections between already highly developed communities—a state of affairs that almost certainly does not exist in the Solar System and may not for centuries to come.

It seems, therefore, that we may be involved in a peculiarly vicious circle. Planetary exploration will not be really practical until we have developed a mature spaceship technology; but we won't have good spaceships until we have worthwhile places to send them—places, above all, with those adequate refueling and servicing facilities now sadly lacking elsewhere in the Solar System. How can we escape from this dilemma? Fortunately, there is one encouraging factor.

Almost all of the technology needed for long-range space travel will inevitably and automatically be developed during the exploration of *near* space. Even if we set our sights no higher than 1000 miles above the Earth, we would find that by the time we had perfected the high-thrust, high-performance surface-to-surface transports, the low-acceleration interorbital shuttles and the reliable, closed-cycle space-station ecologies, we would have proved out at least 90 percent of the technology needed for the exploration of the Solar System—and the most expensive 90 percent, at that.

Perhaps I had better deal here with those strange characters who think that space is the exclusive province of automatic, robot probes and that we should stay at home and watch TV, as God intended us to. This whole man-machine controversy will seem, in another couple of decades, to be a baffling mental aberration of the early space age.

I won't waste much time arguing with this viewpoint, as I hold these truths to be self-evident: (1) Unmanned spacecraft should be used whenever they can do a job more efficiently, cheaply and safely than manned vehicles: (2) Until we have automatons superior to human beings (by which time, all bets will be off), all really sophisticated space operations will demand human participation. I refer to such activities as assembling and servicing the giant applications satellites of the next decade and running orbital observatories, laboratories, hospitals and factories—projects for which there will be such obvious and overwhelming commercial and scientific benefits that no one will dispute them.

In particular, medium-sized telescopes outside the atmosphere—a mere couple of hundred miles above the Earth—will have an overwhelming impact on Solar System studies. The recent launching of OAO II— the initials stand for "Orbiting Astronomical Observatory"—was a promising beginning. Until the advent of radar and space probes, everything we knew about the planets had been painfully gathered, over a period of about a century and a half, by astronomers with inadequate instruments, hastily sketching details of a tiny, trembling disk glimpsed during moments of good sighting. Such moments—when the atmosphere is stable and the image undistorted—may add up to only a few hours in an entire lifetime of observing.

In these circumstances, it would be amazing if we had acquired any *reliable* knowledge about planetary conditions; it is safest to assume that we have not. We are still in the same position as the medieval cartographers, with their large areas of "*Terra Incognita*" and their "Here Be Dragons," except that we may have gone too far in the other direction—"Here Be *No* Dragons." Our ignorance is so great that we have no right to make either assumption.

As proof of this, let me remind you of some horrid shocks the astronomers have received recently, when things of which they were quite sure turned out to be simply not true. The most embarrassing example is the rotation of Mercury. Until a couple of years ago, everyone was perfectly certain that it always kept the same face toward the Sun, so that one side was eternally dark, the other eternally baked. But now, radar observations indicate that it turns on its axis every 59 days; it has sunrise and sunset, like any respectable world. Nature seems to have played a dirty trick on several generations of patient astronomers.

Einstein once said: "The good Lord is subtle, but He is not malicious." The case of Mercury casts some doubt on this dictum. And what about Venus? You can find, in the various reference books, rotation periods for Venus ranging all the way from 24 hours to the full value of the year, 225 days. But, as far as I know, not one astronomer ever suggested that Venus would present the extraordinary case of a planet with a day longer than its year. And, of course, it *would* be the one example we had no way of checking, until the advent of radar. Is this subtlety—or malice?

And look at the Moon. Five years ago, everyone was certain that its surface was either soft dust or hard lava. If the two schools of thought had been on speaking terms, they would at least have agreed that

there were no alternatives. But then Luna 9 and Surveyor 1 landed—and what did they find? Good honest dirt.

These are by no means the only examples of recent shocks and surprises. There are the unexpectedly high temperature beneath the clouds of Venus, the craters of Mars, the gigantic radio emissions from Jupiter, the complex organic chemicals in certain meteors, the clear signs of extensive activity on the surface of the Moon. And now Mars seems to be turning inside out. The ancient, dried-up sea beds may be as much a myth as Dejah Thoris, Princess of Helium; for it looks as if the dark *Maria* are actually highlands, not lowlands, as we had always thought.

The negative point I am making is that we really know nothing about the planets. The positive one is that a tremendous amount of reconnaissance—the essential prelude to *manned* exploration—can be carried out from Earth orbit. It is probably no exaggeration to say that a good orbiting telescope could give us a view of Mars at least as clear as did Mariner 4. And it would be a view infinitely more valuable—a continuous coverage of the whole visible face, not a signal snapshot of a small percentage of the surface.

Nevertheless, there are many tasks that can best be carried out by unmanned spacecraft. Among these is one that, though of great scientific value, is of even more profound psychological importance. I refer to the production of low-altitude oblique photographs. It is no disparagement of the wonderful Ranger, Luna and Surveyor coverage to remind you that what suddenly made the Moon a real place, and not merely an astronomical body up there in the sky, was the famous photograph of the Crater of Copernicus from Lunar Orbiter 2. When the newspapers called it the picture of the century, they were expressing a universally felt truth. This was the photograph that first proved to our emotions what our minds already knew but had never really believed—that Earth is not the only world. The first high-definition, oblique photos of Mars, Mercury and the satellites of the giant planets will have a similar impact, bringing our mental images of these places into sharp focus for the first time.

The old astronomical writers had a phrase that has gone out of fashion but that may well be revived: the plurality of worlds. Yet, of course, every world is itself a plurality. To realize this, one has only to ask: How long will it be before we have learned everything that can be known about the planet Earth? It will be quite a few centuries before terrestrial geology, oceanography and geophysics are closed, surprise-free subjects.

Consider the multitude of environments that exists here on Earth, from the summit of Everest to the depths of the Marianas Trench—from high noon in Death Valley to midnight at the South Pole. We may have equal variety on the other planets, with all that this implies for the existence of life. It is amazing how often this elementary fact is overlooked and how often a single observation or even a single extrapolation from a preliminary observation based on a provisional theory has been promptly applied to a whole world.

It is possible, of course, that the Earth has a greater variety of more complex environments than any other planet. Like a jet-age tourist "doing Europe" in a week, we may be able to wrap up Mars or Venus with a relatively small number of "landers." But I doubt it, if only for the reason that the whole history of astronomy teaches us to be cautious of any theory purporting to show that there is something special about the Earth. In their various ways, the other planets may have orders of complexity as great as ours. Even the Moon—which seemed a promising candidate for geophysical simplicity less than a decade ago—has already begun to unleash an avalanche of surprises.

The late Professor J. B. S. Haldane once remarked—and this should be called Haldane's Law—"The universe is not only stranger than we imagine, it is stranger than we *can* imagine." We will encounter the operation of this law more and more frequently as we move away from home. And as we prepare for this move, it is high time that we face up to one of the more shattering realities of the astronomical situation. For all practical pur-

poses, we are still as geocentrically minded as if Copernicus had never been born; to all of us, the Earth is the center, if not of the Universe, at least of the Solar System.

Well, I have news for you. There is really only one planet that matters; and that planet is not Earth but Jupiter. My esteemed colleague Isaac Asimov summed it up very well when he remarked: "The Solar System consists of Jupiter plus debris." Even spectacular Saturn doesn't count; it has less than a third of Jupiter's enormous mass—and Earth is a hundred times smaller than Saturn! Our planet is an unconsidered trifle, left over after the main building operations were completed. This is quite a blow to our pride, but there may be much worse to come, and it is wise to get ready for it. Jupiter may also be the *biological*, as well as the *physical*, center of gravity of the Solar System.

This, of course, represents a complete reversal of views within a couple of decades. Not long ago, it was customary to laugh at the naïve ideas of the early astronomers—Sir John Herschel, for example—who took it for granted that all the planets were teeming with life. This attitude is certainly overoptimistic; but it no longer seems as simple-minded as the opinion, to be found in the popular writings of the 1930s, that ours might be the only solar system and, hence, the only abode of life in the entire Galaxy.

The pendulum has, indeed, swung—perhaps for the last time; for in another few decades, we should know the truth. The discovery that Jupiter is quite warm and has precisely the type of atmosphere in which life is believed to have arisen on Earth may be the prelude to the most significant biological findings of this century. Carl Sagan and Jack Leonard put it well in their book *Planets*: "Recent work on the origin of life and the environment of Jupiter suggests that it may be more favorable to life than any other planet, not excepting the earth."

The extraordinary color changes in the Jovian atmosphere—in particular, the behavior of that Earth-sized, drifting apparition, the Great Red Spot—hint at the production of organic materials in enormous quantities.

Where this happens, life may follow inevitably, given a sufficient lapse of time. To quote Isaac Asimov again: "If there are seas on Jupiter . . . think of the fishing." So that may explain the mysterious disappearances and reappearances of the Great Red Spot. It is, as Polonius agreed in a slightly different context, "very like a whale."

Contrary to popular thinking, gravity on Jupiter would not pose insurmountable difficulties. The Jovian gravity is only two and a half times Earth's—a condition to which even terrestrial animals (rats in centrifuges) have managed to adapt. The Jovian equivalent of fish, of course, couldn't care less about gravity, because it has virtually no effect in a marine environment.

Dr. James Edson, late of NASA, once remarked, "Jupiter is a problem for my grandchildren." I suspect that he may have been wildly optimistic. The zoology of a world outweighing 300 Earths could be the full-time occupation of mankind of the next 1000 years.

It also appears that Venus, with its extremely dense, furnace-hot atmosphere, may be an almost equally severe yet equally promising challenge. There now seems little doubt that the planet's average temperature is around 700 degrees Fahrenheit; but this does not, as many have prematurely assumed, rule out all possibility of life—even life of the kind that exists on Earth.

There may be little mixing of the atmosphere and, hence, little exchange of heat between the poles and the equator on a planet that revolves as slowly as Venus. At high latitudes or great altitudes—and Venusian mountains have now been detected by radar—it may be cool enough for liquid water to exist. (Even on Earth, remember, the temperature difference between the hottest and the coldest points is almost 300 degrees.) What makes this more than idle speculation is the exciting discovery, by the Russian space probe Venera IV, of oxygen in the planet's atmosphere. This extremely reactive gas combines with so many materials that it cannot occur in the free state—unless it is continuously renewed by vegetation.

Free oxygen is an almost infallible indicator of life: If I may be allowed the modest cough of the minor prophet, I developed precisely this argument some years ago in a story of Venusian exploration, *Before Eden*.

On the other hand, it is also possible that we shall discover no trace of extra-terrestrial life, past or present, on any of the planets. This would be a great disappointment; but even such a negative finding would give us a much sounder understanding of the conditions in which living creatures are likely to evolve; and this, in turn, would clarify our views on the distribution of life in the Universe as a whole. However, it seems much more probable that long before we can certify the Solar System as sterile, the communications engineers will have settled this ancient question—in the affirmative.

For that is what the exploration of space is really all about; and this is why many people are afraid of it, though they may give other reasons, even to themselves. It may be just as well that there are no contemporary higher civilizations in our immediate vicinity; the cultural shock of direct contact might be too great for us to survive. But by the time we have cut our teeth on the Solar System, we should be ready for such encounters. The challenge, in the Toynbeean sense of the word, should then bring forth the appropriate response.

Do not for a moment doubt that we shall one day head out for the stars—if, of course the stars do not reach us first. I think I have read most of the arguments proving that interstellar travel is impossible. They are latter-day echoes of Professor Newcomb's paper on heavier-than-air flight. The logic and the mathematics are impeccable; the premises, wholly invalid. The more sophisticated are roughly equivalent to proving that dirigibles cannot break the sound barrier.

In the opening years of this century, the pioneers of astronautics were demonstrating that flight to the Moon and nearer planets was possible, though with great difficulty and expense, by means of chemical propellants. But even then, they were aware of the promise of nuclear energy and hoped that it would be the ultimate solution. They were right.

Today, it can likewise be shown that various conceivable, though currently quite impracticable, applications of nuclear and medical techniques could bring at least the closer stars within the range of exploration. And I would warn any skeptics who may point out the marginal nature of these techniques that, at this very moment, there are appearing simultaneously on the twin horizons of the infinitely large and the infinitely small, unmistakable signs of a breakthrough into a new order of creation. To quote some remarks made recently in my adopted country, Ceylon, by a Nobel laureate in physics, Professor C. F. Powell: "It seems to me that the evidence from astronomy and particle physics that I have described makes it possible that we are on the threshold of great and far-reaching discoveries. I have spoken of processes that, mass for mass, would be at least a thousand times more productive of energy than nuclear energy. ... It seems that there are prodigious sources of energy in the interior regions of some galaxies, and possibly in the 'quasars,' far greater than those produced by the carbon cycle occurring in the stars ... and we may one day learn how to employ them." And, if Professor Powell's surmise is correct, others may already have learned, on worlds older than ours. So it would be foolish, indeed, to assert that the stars must be forever beyond our reach.

More than half a century ago, the great Russian poineer Tsiolkovsky wrote these moving and prophetic words: "The Earth is the cradle of the mind—but you cannot live in the cradle forever." Now, as we enter the second decade of the age of space, we can look still further into the future.

The Earth is, indeed, our cradle, which we are about to leave. And the Solar System will be our kindergarten.

9.3   *Red Giants and White Dwarfs*

Robert Jastrow

The stars seem immutable, but they are not. They are born, evolve and die like living organisms. The life story of a star begins with the simplest and most abundant element in nature, which is hydrogen. The universe is filled with thin clouds of hydrogen, which surge and eddy in the space between the stars. In the swirling motions of these tenuous clouds, atoms sometimes come together by accident to form small pockets of gas. These pockets are temporary condensations in an otherwise highly rarefied medium. Normally the atoms fly apart again in a short time as a consequence of their random motions, and the pocket of gas quickly disperses to space. However, each atom exerts a small gravitational attraction on its neighbor, which counters the tendency of the atoms to fly apart. If the number of atoms in the pocket of gas is large enough, the accumulation of all these separate forces will hold it together indefinitely. It is then an independent cloud of gas, preserved by the attraction of each atom in the cloud to its neighbor.

With the passage of time, the continuing influence of gravity, pulling all the atoms closer together, causes the cloud to contract. The individual atoms "fall" toward the center of the cloud under the force of gravity; as they fall, they pick up speed and their energy increases. The increase in energy heats the gas and raises its temperature. This shrinking, continuously self-heating ball of gas is an embryonic star.

As the gas cloud contracts under the pressure of its own weight, the temperature at the center mounts steadily. When it reaches 100,000 degrees Fahrenheit, the hydrogen

SOURCE. From *Red Giants and White Dwarfs* (New York: Harper and Row. 1967); copyright by Robert Jastrow, reprinted by permission of Harper and Row Publishers, Inc.

atoms in the gas collide with sufficient violence to dislodge all electrons from their orbits around the protons. The original gas of hydrogen atoms, each consisting of an electron circling around a proton, becomes a mixture of two gases, one composed of electrons and the other of protons.

At this stage the globe of gas has contracted from its original size, which was 10 trillion miles. To understand the extent of the contraction, imagine the Hindenberg dirigible shrinking to the size of a grain of sand.

The huge ball of gas—now composed of separate protons and electrons—continues to contract under the force of its own weight, and the temperature at the center rises further. After 10 million years the temperature has risen to the critical value of 20 million degrees Fahrenheit.[1] At this time, the diameter of the ball has shrunk to one million miles, which is the size of our sun and other typical stars.

Why is 20 million degrees a critical temperature? The explanation is connected with the forces between the protons in the contracting cloud. When two protons are separated by large distances, they repel one another electrically because each proton carries a positive electric charge. But if the protons approach within a very close distance of each other, the electrical repulsion gives way to the even stronger force of nuclear attraction. The protons must be closer together than one 10-trillionth of an inch for the nuclear force to be effective. Under ordinary circumstances, the electrical repulsion serves as a barrier to prevent as close an approach as this. In a collision of exceptional violence, however, the protons may pierce the barrier which separates them, and come within a range of their nuclear attraction. Collisions of the required degree of violence first begin to occur when the temperature of the gas reaches 20 million degrees.

Once the barrier between two protons is pierced in a collision, they pick up speed as a result of their nuclear attraction and rush rapidly toward each other. In the final moment of the collision the force of nuclear attraction is so strong that it fuses the protons together into a single nucleus. At the same time the energy of their collision is released in the form of heat and light. This release of energy marks the birth of the star.

The energy passes to the surface and is radiated away in the form of light, by which we see the star in the sky. The energy release, which is one million times greater per pound than that produced in a TNT explosion, halts the further contraction of the star, which lives out the rest of its life in a balance between the outward pressures generated by the release of nuclear energy at its center and the inward pressures created by the force of gravity.

The fusion of two protons into a single nucleus is only the first step in a series of reactions by which nuclear energy is released during the life of the star. In subsequent collisions, two additional protons are joined to the first two to form a nucleus containing four particles. Two of the protons shed their positive charges to become neutrons in the course of the process. The result is a nucleus with two protons and two neutrons. This is the nucleus of the helium atom. Thus, the sequence of reactions transforms protons, or hydrogen nuclei, into helium.

The fusion of hydrogen to form helium is the first and longest stage in the history of a star, occupying about 99 percent of its lifetime. In the second stage, which takes up most of the remaining 1 percent of the star's life, three nuclei of helium combine to form the nucleus of the carbon atom. Afterwards, the nuclei of oxygen and other still heavier elements are fabricated, at an increasingly rapid pace, until all elements have been built up. In this way the elements of the

---

[1]Twenty million degrees is a very high temperature. For comparison, the temperature of the flame in the gas burner of the kitchen stove is one thousand degrees, and the temperature of the hottest steel furnace is ten thousand degrees.

universe are manufactured out of hydrogen nuclei at the center of the star during the course of its life.[2]

All stars lead similar lives to the time of their demise. Their manner of dying, however, depends on their size. The small stars shrivel up and fade away, while the large ones disappear in a gigantic explosion. The sun happens to lie just below the dividing line; we are not certain which turn it will take at the end of its life, but we suspect that it will fade away.

The size of a star also determines how long it will live. Surprisingly, the largest stars are the first to expire. In a large star, the temperature in the interior is higher and the protons collide at its center with greater violence. As a consequence, the protons fuse into helium more rapidly, and the star burns out very quickly. The sun's life span is 10 billion years; it has existed for 4.5 billion years, and it is now well into middle age, with perhaps another 5 billion years of life remaining before its fuel reserves are exhausted. A star 10 times as massive as the sun lives only 10 million years, which is the blink of an eye in the normal scale of stellar lifetimes. On the other hand, a star one-tenth as massive as the sun, which is about as small as a star can be, is cooler at the center, burns slowly, and lives for a trillion years.

The first signs of old age in a star are a swelling and reddening of its outer regions. The sun will reach this stage in another 5 billion years, at which time it will have swollen into a vast, distended sphere of gas 100 times its present radius. At the same time its color will change from yellow-white to red. This red globe of gas will cover most of the sky when viewed from the earth. Unfortunately we will not be able to linger and observe the magnificent sight, because the rays of the swollen sun will heat the surface of the earth to 4000 degrees Fahrenheit and eventually evaporate its substance. Perhaps Jupiter will be a suitable habitat for us by then. More likely, we will have fled to another part of the Galaxy.

Such distended, reddish stars are called red giants by the astronomers. An example of a red giant is Betelgeuse, a fairly bright star in the constellation Orion which appears distinctly red to the naked eye.

A star continues to live as a red giant until its reserves of hydrogen fuel are exhausted. With its fuel gone it can no longer generate the pressures needed to maintain itself against the crushing force of gravity, and it begins to collapse once more under its own weight.

At the center of the collapsing star there is a core of pure helium, which has been produced by the fusion of protons throughout its life. Helium does not fuse into heavier nuclei at the ordinary stellar temperature of 20 million degrees because the helium nucleus, with *two* protons, carries a double charge of positive electricity, and, as a consequence, the electrical repulsion between two helium nuclei is stronger than the repulsion between two protons. A temperature of 200 million degrees is required to produce collisions which will pierce the helium barrier.

As soon as the temperature reaches the critical level of 200 million degrees, helium nuclei commence to fuse in groups of three to form carbon nuclei, releasing more nuclear energy in the process and rekindling the fire at the center of the star. The release of energy halts the gravitational collapse of the star and it obtains a new lease on life,

---

[2]The transmutation of heavy hydrogen into helium and heavier elements has been duplicated on the earth for brief moments in the explosion of the hydrogen bomb. However, we have never succeeded in fusing hydrogen nuclei under controlled conditions in such a way that the energy released can be harnessed for constructive purposes. The United States, the Soviet Union and other countries have invested prodigious amounts of money and energy in the effort, for the stakes are high, but physics has not yet been equal to the task. The difficulty is that no furnace has yet been constructed on the earth whose walls can contain fire at the temperature of the millions of degrees necessary to produce nuclear fusion. The only furnace that can do this is provided by nature in the heart of a star.

burning helium nuclei to produce carbon. This stage will last for about one hundred million years in the sun. At the end of that time the reserves of fuel, composed now of helium rather than hydrogen, once again are exhausted.

Nearly all stars reach the stage in which their helium fuel has been used up. Exhaustion of the helium is followed by the inevitable collapse under the pressure of the star's own weight. From this point onward, the history of the star varies according to its size. In the case of small stars, the collapse continues until all the matter of the star is squeezed into a space the size of the earth. The density of the collapsed star is so great that a volume the size of a matchbox weighs ten tons. The collapse makes the surface of the star white-hot. These shrunken white-hot stars are called white dwarfs. Slowly the white dwarf radiates into space the last of its heat. In the end, its temperature drops, and it fades into a blackened corpse.

A very different fate awaits a large, massive star. Because the weight of the star is so great, its collapse generates an enormous amount of heat, greater than the heat generated in the creation of the white dwarf.

If the star is massive enough, the temperature produced by the collapse will reach another critical level—this time of 600 million degrees—at which the carbon nuclei at the center of the star collide with sufficient violence to stick together, forming still heavier elements than carbon, ranging from oxygen to sodium.

Eventually, the carbon fuel reserves are also exhausted; once again their exhaustion is followed by further stages of collapse, heating, and renewed nuclear burning, leading to the production of other elements.

In this way, through the alternation of collapse and nuclear burning, a massive star successively manufactures all elements up to iron. But iron is a very special element. This metal, which lies halfway between the lightest and the heaviest elements, has an exceptionally compact nucleus, so tightly packed that no energy can be squeezed from

it in any sort of nuclear reaction. When a large amount of iron accumulates at the center of the star, the fire cannot be re-kindled; it goes out for the last time, and the star commences a final collapse under the force of its own weight.

The ultimate collapse is a catastrophic event. The heat generated by it drives the central temperature up to 100 billion degrees, and every possible nuclear reaction comes into play. It is in this last gasp that the heaviest elements, those extending beyond iron to uranium, are produced.

The star rebounds from the final collapse in a great explosion which disperses to space most of the elements manufactured in its interior during its lifetime. Thus, the life story of a large star is a cycle of dust to dust.

The exploding star is called a supernova. About fifty supernovas have been photographed with telescopes, in the last 75 years, and in our own galaxy a few occur every thousand years that are bright enough to be seen by the unaided eye. One of the earliest reported supernovas was a brilliant explosion recorded by Chinese astronomers in A.D. 1054. At the position of this supernova there is today a great cloud of gas known as the Crab Nebula, expanding at the speed of 1000 miles per second, which contains the remains of the star that exploded a thousand years ago.

The supernova explosion sprays the material of the star out into space, where it mingles with fresh hydrogen to form a mixture containing all 92 elements. Later in the history of the galaxy, other stars are formed out of clouds of hydrogen which have been enriched by the products from these explosions. The sun is one of these stars; it contains the debris of countless supernova explosions dating back to the earliest years of the Galaxy. The planets also contain this debris, and the earth, in particular, is composed almost entirely of it. We owe our corporeal existence to events which took place billions of years ago, in stars that lived and died long before the solar system came into being.

9.4   *The Origin of the Elements*
                            William A. Fowler

THE ASTRONOMICAL SETTING

... Astronomers speak of the system of stars of which the sun is a part as the Galaxy, with a capital G, no less. We see part of it as the Milky Way. The Galaxy is one of seventeen galaxies making up the so-called local cluster of galaxies. The nearest group of galaxies to the local group is 40 million light years away.[1] It is called the Virgo Cluster. There are many such clusters, some containing a great number of galaxies as members and the study of these clusters has been one of the most intriguing and rewarding in astronomy. They indicate structure in the Universe on the enormous scale of many millions of light years.

Our nearest neighbor galaxy in our own cluster is the Andromeda Nebula illustrated in Figure 9.4–1. It is 2 million light years away. The Galaxy is a flattened, spiral system somewhat similar to the Andromeda Nebula, its "twin." If the sun were located in Andromeda, it would be in one of the spiral arms about two-thirds of the way out or 30,000 light years distant from the bright central core or nucleus of the galaxy. It is the spiral arm of the Galaxy in which we are embedded which we see as the Milky Way. The greater part of the Galaxy is hidden from visible observation by great clouds of obscuring gas and dust. Radio waves can penetrate these clouds and it is by radio observations that the overall structure of the Galaxy has been mainly determined.

Twelve billion years ago our Galaxy was not at all like it is at the present time,[2] it was an enormous mass of hydrogen gas hanging tenuously in space. It was roughly spherical in shape, and it was slowly rotating. In some way, not completely understood but certainly involving gravitational contraction, this mass of gas became separated from the rest of the universe. In regions of low turbulence and high density, stars formed from some of the gas, and as they condensed and contracted, their gravitational potential energy was converted into internal thermal energy and into radiant heat and light. After a few tens of millions of years, stars stop contracting and settle down for a relatively long period, constant in size like our sun, during which they emit light uniformly and steadily, again like our sun. How did this stability come about? Where did the energy come from, after contraction and the release of gravitational energy stopped?

Again for the answers we must turn to the nuclear laboratory. There we find that with particle accelerators we can cause pairs of deuterons, the heavy hydrogen nuclei with a mass of two units, to fuse into helium nuclei with a mass of four units. This process is called *fusion*. Several steps are required, but we can produce helium nuclei one at a time with our accelerators. In spite of the valiant efforts of scientists throughout the world—in Russia, in England and in the United States—no one has been able to produce a *self-sustaining* fusion process.

Why should anyone want to do so? The answer lies in the fact that all the masses we have quoted have been approximate. The two

SOURCE. From *Nuclear Astrophysics* (Philadelphia: American Philosophical Society, 1967).

[1] The light year is the distance traveled by light in one year. Since the speed of light is 3×10^{10} or-thirty billion centimeters per second, and since the year contains approximately 3×10^7 or thirty million seconds, the light year is approximately equal to 10^{18} or one billion billion centimeters.

[2] It is generally agreed that the birth of the Galaxy took place 7 to 15 billion years ago. My own best estimate is 12 billion years ago.

Figure 9.4–1. The Andromeda Nebula, the great spiral galaxy in *Andromeda*, which is the "twin" of the Galaxy in which the solar system is lacated. (Mount Wilson and Palomar Observatories.)

deuterons together have slightly greater mass than the helium nucleus they form, and when the fusion occurs, the excess mass is converted into energy. By Einstein's famous equation, one obtains the energy released by multiplying the excess mass by the velocity of light, once and then once again. Thus a large amount of energy results from the small change in mass, and if the fusion process could be made self-sustaining, as is the fission process in reactors, then we could burn as fuel the heavy hydrogen which forms a part of all sea water. We would have a source of energy sufficient for all mankind forever. The unsolved problem lies in the fact that the nuclear *burning* must take place at such high temperature that we cannot build a furnace which will confine and contain the fuel.

The fusion problem has been solved in stars. Because of their size, stars contain a sufficient amount of material so that they can confine nuclear fuel gravitationally, sustain fusion processes and shine on the energy that comes from these processes.

Most importantly, we shall see that stars can fuse protons, the nuclei of ordinary hydrogen into helium and do not require the very rare heavy hydrogen as fuel. Moreover, in the star's early stages of contraction the stellar material is heated until the nuclear processes are triggered by the high temperature. The release of the nuclear energy adds to the violent internal motions, and the contraction of the star is stopped with a delicate balance resulting between inwardly directed gravitational forces and outwardly directed thermal pressure.

As successive nuclear processes take place, the composition of a star changes and the star is said to evolve as its internal structure and external appearance vary in response to these composition changes. More about this anon. It is essential in the point of view of stellar synthesis that instabilities arise during the evolution and aging of a star that return the transmitted material to interstellar space. It is there mixed with the uncondensed hydrogen gas in the Galaxy so that it is available for condensation into second- and later-generation stars. The general state of affairs in this "equilibrium" between stars and the interstellar gas and dust is illustrated in Figure 9.4–2.

Stellar nucleosynthesis demands that there exists this interchange of material between stars and the interstellar medium of gas and dust. The stars are the nuclear furnaces; the space between is the site of the mixing and dilution which result in the average abundance distribution over fairly large astronomical regions. Observations confirm that matter is given off by stars, both slowly and explosively, and that new stars are continually forming from the interstellar material. Giant stars lose mass at a fairly substantial rate; even our sun slowly ejects matter into space. The planetary nebulae, such as that shown in Figure 9.4–3, show spherical shells moving away from a central star. Projected on our line of sight through the telescope, such a shell gives the appearance of a gigantic "smoke ring." The most spectacular instabilities in stars result in the novae and supernovae that are observed to flare up suddenly in the sky and then die away in brightness. In the case of novae, a mass loss of the order of one-tenth to one per cent can suddenly occur after which the star involved returns in most cases to approximately its original luminosity. Novae can reoccur. In supernova explosions, all or a substantial fraction of the mass of what is presumed to have been a star is ejected with high velocity into space.

In all these cases, ranging from slow to catastrophic mass loss, we see astronomical evidence that material is transferred from the star to the interstellar medium. If material from the deep interior of the star is involved in the ejection or explosion, then the new elements produced by nuclear burning in this material will be mixed into the inter-

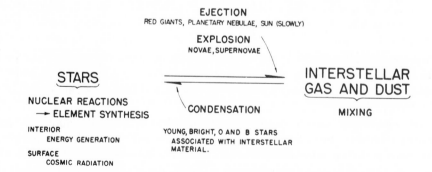

Figure 9.4–2. Transfer of material between stars and interstellar gas and dust. Synthesis of elements occurs in the stars, and mixing to yield the relative cosmic abundance of the elements occurs in interstellar space. Mechanisms for the transfer as observed astronomically are indicated.

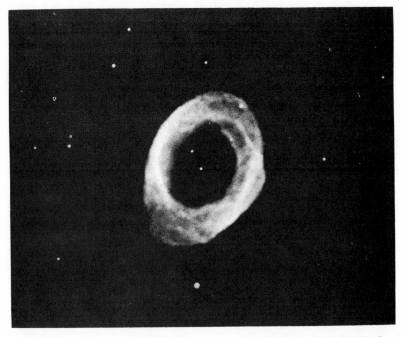

Figure 9.4–3. The "Ring" planetary nebula in *Lyra* showing the spherical shell of gas which is moving away from the central star and was presumably ejected by it. The off-center star within the ring is a field star. (Mount Wilson and Palomar Observatories.)

stellar medium. This nuclear burning could have taken place during the previous sedentary history of the star as energy was steadily generated and emitted as light or at a greatly accelerated rate during the final explosive stage. In any case the interstellar medium becomes increasingly contaminated with the debris from past stellar mass losses as the Galaxy ages. Thus there is good observational evidence to believe that new stars which continue to form from the interstellar medium will not consist solely of the primordial hydrogen but also of material enriched in other elements by the addition of nuclear debris from previous generations of stars.

The reverse process to the breakup of stars, the formation of new stars, also has substantial observational confirmation, albeit somewhat indirect. There are stars in the heavens so bright for their known mass that even nuclear processes cannot have kept them shining for more than a few million years. They are thus much younger than

the sun and the Galaxy. The bright stars are "young stars," and since they occur only in regions observed to be populated with relatively large amounts of gas, it is reasonable to assume that they condensed from the interstellar material. These "bright" stars often beautifully illuminate the remaining gas and dust near them, as seen in Figures 9.4–4, 5 and 6.

Consider the very bright star seen just above the tall column of gas and dust in the photograph of the nebulosity in the constellation *Monoceros* shown in Figure 9.4–6. The nebulosity is called the *Madonna Nebula* for obvious reasons. Note the light from the top of the column. It is scattered light originally emitted by the bright star but absorbed and re-emitted by atoms in the column. Now the very bright star is approximately 10,000 times as luminous as the sun but it is only about 10 times as massive. Thus it is consuming its nuclear fuel 1,000 times faster than the sun and its lifetime can be expected to be only a thousandth that of the

Figure 9.4–4.   The Lagoon Nebula in *Sagittarius* showing a region in the Galaxy where bright, young stars have recently formed from interstellar gas and dust. The remaining gas and dust are illuminated by these stars. (Mount Wilson and Palomar Observatories.)

sun. The sun's present age is thought to be 4.5 billion years and it is believed that it will remain in much its present state for another 5 or 6 billion years. At the age of 10 billion years the sun will change markedly. It will probably explode as a supernova but it may quietly but quickly become a dwarf star. In any case 10 billion years is a good round number for the lifetime of the sun as a normal star. This tells us then the maximum lifetime of the bright star in *Monoceros*—a mere 10 million years, very short compared to the age of the Galaxy. Thus sometime within the last 10 million years this star was formed presumably from gas and dust similar to that remaining in the nebulosity.

In regions of no gas and dust, there are few great bright stars; only old stars of low mass, low central temperatures, and low nuclear reaction rates are found there. They are the slow burners from the original and succeeding condensations that cleaned up the vicinity in which they are located. The galactic rotation has acted to concentrate the gas and dust into the spiral arms of the equatorial plane of the Galaxy. Under the influence of the rotation and the galactic gravitational attraction, both stars and the interstellar matter execute generalized Keplerian orbits in which they are carried through the galactic equatorial plane from time to time. Atomic collisions retard the motions of the gas but not of the stars, and thus the gas accumulates in the galactic plane, where it forms the spiral arms and the bright, young stars found predominantly in these arms. For example, the central stars of planetary nebulae [Figure 9.4–3] will be stripped of their gaseous, external clouds as they pass through the galactic equator. The older stars are found throughout the galaxy in what is termed the galactic "halo" and in general have much higher velocities than the stars in the spiral arms. Stars of intermediate age and velocity are formed in a region called the galactic disk which is somewhat thicker than that containing the spiral arms. Stars which populate these various regions are referred to as halo

Figure 9.4–5.   The Orion Nebula showing bright, young stars embedded in interstellar gas and dust. (Mount Wilson and Palomar Observatories.)

population (originally called population II), arm population (originally called population I), and disk population (differentiated only recently).

PURE HYDROGEN BURNING

The process of gravitational contraction of a protostar containing only hydrogen leads to a temperature and density rise in the interior and to a gradient in these quantities such that the temperature and density are highest at the center of the star and drop off rapidly to relatively low values at the stellar surface. When the central temperature reaches 10^7 degrees absolute and the density reaches 100 g/cm^3, the hydrogen begins to interact through the so-called direct proton-proton chain which consists in part of the reactions shown in Figurge 9.4–7. The reactions proceed with the emission of large amounts of energy even when the reactants have relatively low energies. The reaction products rapidly lose their large energies

Figure 9.4–6.   Bright, young stars in *Monoceros*, which illuminate the gas and dust remaining after their formation. (Mount Wilson and Palomar Observatories.)

by atomic (not nuclear) collisions, and thus the reactions are not reversed . . .

The steps in the proton-proton chain illustrated in Figure 9.4–7 can be described in the following manner. In a violent collision two protons (H^1) fuse together to form the deuteron (D^2) with the emission of a neutrino (ν) and a positron (e^+). The neutrino escapes directly from the star because it interacts only very infrequently with other forms of matter. The positron soon collides with an electron, and both are annihilated with the appearance of radiant energy. The deuteron eventually collides with a proton with the formation of a nucleus of helium with a mass of three units (He^3) and the appearance of more radiant energy. The He^3 does not interact with protons except through the process of nuclear scattering in which the He^3 and H^1 change their individual energies with no change in their combined energy. Such scatterings play no direct role in nucleosynthesis or energy

H- BURNING
THE FUSION OF ORDINARY HYDROGEN
IN MAIN SEQUENCE STARS (THE SUN)

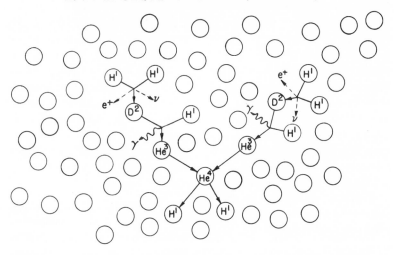

OVERALL RESULT: 4 HYDROGEN NUCLEI → HELIUM NUCLEUS

ENERGY RELEASE = 100 MILLION KILOWATT— HOURS
PER POUND CONVERTED

Figure 9.4–7. Schematic representation of the fusion of hydrogen into helium by the p–p chain, which occurs in main sequence stars of one solar mass or less. Density: 10^2 gm cm^3. Temperature: 10^7 degrees K.

generation. The deuterons are very rare since they react with protons to form He³ just as fast as they are produced and thus interactions between He³ and D² are very rare indeed. As a consequence the He³ builds up in abundance until pairs of He³ nuclei interact to form the common form of helium with a mass of four units (He⁴) plus two protons.

Thus we see that six protons interact in a fairly complicated way to form He⁴ with the reappearance of two protons. The overall result is the conversion of four protons into the He⁴ nucleus. Of the four electrons which originally balanced the charge on the four protons, two were annihilated by positrons and two remain to balance the double charge of He⁴. Two neutrinos escape for each He⁴ produced. It is important to realize that the detailed processes just described are not theoretical pipe dreams. With the exception of the initial proton-proton fusion all of these reactions have been observed in the

nuclear laboratory and the rates at which they proceed have been carefully measured. The proton-proton fusion rate is too low to be detected experimentally but it can be calculated quite accurately from the rate of proton-proton scattering with which the fusion process competes and from the rate of other beta-decay processes in which positrons and neutrinos are emitted.

Laboratory measurements also reveal the amount of energy released in the various steps in the proton-proton chain. The overall energy release corresponds to the difference in initial and final masses multiplied by the velocity of light squared as is well known from Einstein's equation $E = Mc^2$, E being the energy released when mass M is destroyed. The velocity of light is designated by c. Since c is equal to 3×10^{10} centimeter per second, its square is equal to 9×10^{20} centimeter^2 per second^2. The metric energy unit, the erg can be expressed as gram centimeter^2

per second^2 and so $c^2 = 9 \times 10^{20}$ ergs per gram. This is to say that each gram converted results in the release of 9×10^{20} ergs. Now the four protons and the two annihilated electrons have a mass 0.7 per cent greater than that of the He^4 nucleus *from laboratory measurements*. Thus the conversion of one gram of hydrogen into helium results in the release of approximately 6×10^{18} ergs of energy. Translated into practical units this means that one pound of hydrogen yields 100 million kilowatt-hours of energy when converted into 0.993 pounds of helium. This is one more example of the fact that nuclear burning is prodigal in energy release compared to chemical burning—prodigal enough to keep the sun shining for 10 billion years before its interior hydrogen will have been consumed.

From the standpoint of stellar nucleosynthesis, the proton-proton chain is important because it is a mechanism by which *pure* hydrogen can be converted into helium. The "big bang" production of helium is not needed in principle. However, the proper question is this: Has there been enough helium production in stars? The answer cannot be given at the present time. I do not believe that ordinary stellar activity will suffice but [it is possible that there is] copious production of helium in the explosion of super-massive stars.

The fusion of protons into helium can occur in stars even though protons are all positively charged and mutually repel each other. As a matter of fact, on classical Newtonian mechanics, the fusion cannot occur, because even at stellar temperatures the protons do not have sufficient relative velocities to overcome their mutual repulsion. Sir Arthur Eddington, who proposed hydrogen fusion as the source of energy in stars in 1920, gave a magnificent answer to those who criticized him on classical grounds:

> We do not argue with the critic who urges that the stars are not hot enough for this process; we tell him to go and find a hotter place.

Eddington's critics were saved from their classical fate by modern quantum mechanics, which governs the behavior of atomic particles and permits fusion to occur even when it is "impossible" on Newtonian mechanics. This matter is one of degree. Electrostatic repulsions between positively charged nuclei retard the nuclear interactions between them but do not prevent them entirely at low temperatures and low relative velocities.

The above quotation from Eddington is taken from his address to the British Association for the Advancement of Science at Cardiff, Wales. Because it is so relevant to the subject at hand and so prophetic, I cannot refrain from quoting from this address once again:

> If, indeed, the sub-atomic energy in the stars is being freely used to maintain their great furnaces, it seems to bring a little nearer to fulfillment our dream of controlling this latent power for the well-being of the human race—or for its suicide.

PURE HELIUM BURNING

Stars which live and shine from energy generated through the process $4H^1 \rightarrow He^4$ fall in a luminosity-color classification called the "main sequence." However, as the hydrogen in the central regions of the star is exhausted, the star ceases to be homogeneous in composition throughout its interior and will move, or "evolve," of the main sequence. The conversion of hydrogen "fuel" into helium "ash" occurs in the core of the star because the temperature and density are highest there. Judging from astrophysical observations, it appears that the reaction product, helium, is mixed with the outer envelope, still hydrogen, with extreme difficulty. Thus, a core of helium develops and gradually increases in size as more and more hydrogen is converted. Because of greater electrostatic repulsions, the doubly charged He^4 does not burn at 10^7 degrees or even at considerably higher temperatures, and so energy generation ceases except in a thin shell surrounding the helium core. This shell now contains the hottest hydrogen in the star. It has been estimated that the shell temperatures reach 3×10^7 degrees, while the density is of the order of 10 gm/cm³. In the central regions, the nuclear hydrogen furnace goes out for lack of fuel, and one would expect from ordinary experience with furnaces that the temperature would drop. But this is not at all the case in stars because of their great potential gravitational energy. The helium in the core begins to contract and its temperature rises as gravitational energy is converted into kinetic energy.

This "anomalous" behavior of stars is not all pure conjecture, for the sudden rise in temperature of the core also heats up the envelope, which expands enormously and increases the surface area of the star. The increased area means that energy can be radiated at a lower surface temperature, and thus the surface reddens in color. Larger in area and redder in color than main-sequence stars of the same luminosity, these stars are aptly called the "red giants" by astronomers.

Eventually the helium in the core reaches temperatures ($\sim 10^8$ degrees) and densities ($\sim 10^5$ gm/cm³) at which Coulomb repulsions should no longer critically inhibit nuclear processes between two helium nuclei. What these processes might be constituted for a long time the Gordian knot of nuclear astrophysics. Two helium nuclei, upon interacting, might be expected to form Be⁸. However, as noted previously, no nucleus of mass 8 exists in nature, and from this, early investigators inferred that it must be unstable. Shortly after World War II, this was confirmed in quantitative measurements of the Be⁸ decay at Los Alamos and the California Institute of Technology. In both laboratories it was found that when Be⁸ was produced artificially in nuclear reactions, it promptly broke up into two alpha particles. However, the energy of breakup was found to be relatively small, slightly less than 100 keV. With this last fact in mind, Salpeter of Cornell University then pointed out that, although hot interacting helium in a star will not produce a stable Be⁸ nucleus, it will produce, at 10^8 degrees and 10^5 gm/cm³, a small but real concentration of Be⁸ as a result of the equilibrium between the formation and breakup processes. Now, nuclei are

found in the laboratory to capture alpha particles with the emission of energy in the form of gamma radiation. Salpeter pointed out that the Be⁸ should behave similarly and that if, after its formation from two alpha particles, it collided with a third, the well-known stable nucleus C¹² should be formed. Because of the low equilibrium concentration of the Be⁸, about 1 part in 10 billion at 100 million degrees, Hoyle emphasized that the Be⁸ capture process had better be a very rapid one, or a "resonant" reaction in nuclear parlance. Experiments at Stanford, Brookhaven, and the California Institute of Technology have shown that this is the case. It has been possible to show that there exists an excited state of the C¹² nucleus at 7.656 MeV, with almost the exact energy of excitation and other properties which Hoyle predicted that it must have in order to serve as a thermal resonance for the formation of C¹² from Be⁸ and He⁴ in stars.

Thus, there now exists a reasonable experimental basis for the two state process by which three alpha particles in the hot dense cores of red giant stars can synthesize carbon, bypassing the intervening elements lithium, beryllium, and boron. This process is indicated schematically in Figure 9.4–8. The overall process can, in fact, be looked upon as an equilibrium between three helium nuclei and the excited carbon C¹²*, with occasional irreversible leakage out of the equilibrium to the ground state of C¹². In reaction notation, we have

$$3He^4 \rightleftarrows C^{12*} \to C^{12}.$$

The C¹² frequently captures a helium nucleus to form O¹⁶ before the helium is exhausted. In extreme cases this results in the over-all process $4He^4 \to O^{16}$. A small amount of Ne²⁰ is produced in the capture of He⁴ by O¹⁶. In stars, there is no difficulty at mass 5 and the difficulty at mass 8 has been surmounted. When the central conditions in a red giant reach 10^8 degrees and 10^5 gm/cm³, the helium begins to burn and energy is released. Because of the small fraction, 0.07 per cent, of mass converted into energy in the above process, the red giant star is not stabilized for

THE FUSION OF HELIUM IN RED GIANT STARS

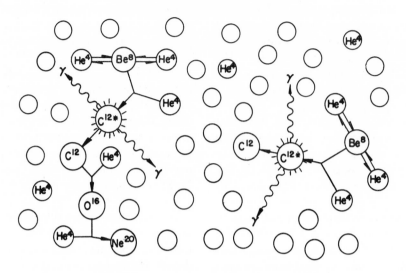

DENSITY : 10^5 GRAMS PER CM^3
TEMPERATURE : 1.3×10^8 DEGREES KELVIN

Figure 9.4–8.   Schematic representation of the fusion of helium to form C^{12} which occurs in red giant stars.

any long period after the onset of the helium burning. The major release in nuclear energy comes in the first process, $4H^1 \to He^4$. In any case, however, the astronomical evidence indicates that the trend toward catastrophic internal temperatures is stopped and the evolutionary track reversed. Stars that become unstable at this point will eject unburnt hydrogen and helium and the synthesized carbon, oxygen and neon into interstellar matter. Others which remain stable will continue the synthesis process.

.   .   .   .   .

The first essentials of stellar nucleosynthesis have now been told . . . In this lecture we have seen how it is possible in stars for hydrogen to fuse into helium and for helium to fuse into carbon, some oxygen, and a little neon. The story is not too greatly different from the ideas of the ancient Greeks . . . Consider a quotation from the Greek philosopher,

Simplicius, of the sixth century A.D. in his analysis of the ideas of another Greek philosopher, Leucippus, of the fifth century B.C., generally considered to be the originator of the theory of atomism:

They (atoms) move in the void and catching each other up jostle together, and some recoil in any direction that may chance, and others become entangled with one another in various degrees according to the symmetry of their shapes and sizes and positions and order, and they remain together and thus the coming into being of composite things is effected.

It might seem that we have not learned very much in the last 2,500 years. In this lecture, however, I have tried to emphasize one important feature of our present-day science which distinguishes it from that of the Greeks. Astronomers now make observations with powerful telescopes. Nuclear physicists perform experiments with powerful accelerators. The concepts of modern-day nuclear astrophysics are rooted in experimental and observational evidence which can be tested and retested. It is the interplay of theory with experiment on the one hand and observation on the other which sharpens and clarifies our knowledge of the world about us.

And now permit me to pass along one final thought ... My major theme has been that all of the elements heavier than helium, and perhaps the helium too, have been synthesized in stars. Let me remind you that your bodies consist for the most part of these heavier elements. Thus it is possible to say that you and your neighbor and I, each one of us and all of us, are truly and literally a little bit of stardust.

9.5 *Why is it Dark at Night*
Herman Bondi

One of the bases of modern cosmology is known as Olbers' paradox, which makes the darkness of the night sky appear as a curious phenomenon. The argument leading up to this is so simple and attractive and beautiful that it may not be out of place to consider it here in full.

When one looks at the sky at night one notices that there are some very bright stars, more medium bright ones and very large numbers of faint ones. It is easy to see that this phenomenon might be accounted for by the fact that the bright-looking stars happen to be near; the medium bright ones rather farther away, and the faint ones a good deal farther away still. In this way one would not only account for the variations in brightness but also for the fact that there are more of the faint ones than of the medium bright ones, and more of the medium bright ones than of the very bright ones, for there is more space farther away than nearby. One can now speculate about stars yet farther away, so far away, in fact, that they cannot be seen individually, not by the naked eye, nor even by the telescope. The question then arises of whether these very distant stars, though they would individually be too faint to be seen, might not be so exceedingly numerous as to provide an even background illumination of the night sky? This is the question that the German astronomer Olbers asked 130-odd years ago. The argument will now be presented in the light of the astronomical knowledge of 1826, without considering any of the phenomena discovered by modern astronomy.

OLBERS' PARADOX

Olbers then attempted to calculate what the brightness of the background of the sky should be on this basis. He immediately realized that in trying to consider effects from regions too far away to be seen in detail, he was forced to make assumptions about what the depths of the universe were like. He then made a set of assumptions which looks so plausible even nowadays that they may well serve as a model of what the beginning of a scientific investigation should be like. He first assumed, in the light of the knowledge of his day (1826), that the distant regions of the universe would be very much like our own. He expected there would be stars there, with the same average distance between them as between near stars. He expected that while each star would have an intrinsic brightness of its own, there would be an average brightness of stars very much like that in our astronomical neighborhood. In other words, he assumed that we get a typical view of the universe. This is in full accord with the ideas that have been current since the days of Copernicus, that there is nothing special,

SOURCE. From *The Universe at Large* (New York: Doubleday, 1959); copyright by the Illustration London New Sketch, Ltd., reprinted by permission of Doubleday and Co.

nothing pre-selected about our position in the scheme of things. This is a convenient assumption from a scientific of view and a very fruitful one because we can assume that what goes on around us holds elsewhere as well, if not in detail at least on the average.

Unfortunately, this assumption is not sufficient for the calculation Olbers wished to make. For light travels at a finite speed: at a high speed, it is true, but a finite one nevertheless. The light we now receive from many distant regions was sent out by the objects there a long time ago, having spent the intervening period on its journey from there to here. What is important for us, therefore, in trying to calculate the amount of light we get from the depths of the universe, is not how much the stars there radiate *now*, but how much they radiated at the time when the light which we receive now was sent out by them. We have to make a guess about the variation of astronomical conditions, not only with space, but also with time. And here, again, Olbers made the simplest of all possible assumptions, for he assumed time to matter as little as space. In other words, he supposed that not only in other parts of the universe, but also at other times, there would be stars, that their brightness would be the same as it is in our astronomical neighborhood, and similarly, their average distance apart would be the same as it is near us. Next, Olbers assumed, very naturally, that the laws of physics, as we know them from here, apply elsewhere and at other times. In particular, he assumed that the laws of the propagation of light—the way light spreads out after leaving its source—applied over these vast regions just as much as they apply in our rooms here. This, again, is the most obvious, most convenient and most fruitful assumption one can make. It would seem a stupid thing to set out on a voyage of discovery into the depths of the universe by first throwing away all the knowledge we have gained in our vicinity. Finally, Olbers made an assumption which is of the utmost importance, but he made it implicitly. He was not aware of the fact that he was making an assumption at all. Scientists know very well that this is the most

dangerous kind of assumption. This assumption was that there were no large, systematic motions in the universe; that the universe was static.

THE MATHEMATICS OF STARLIGHT

On the basis of these four assumptions, it is easy to work out the background light of the sky. Imagine a vast, spherical shell surrounding us (Figure 9.5–1). The thickness of the shell is supposed to be small compared with its radius; but the whole shell is supposed to be so enormous that there are vast numbers of stars within the shell. How many stars are there in this shell? In order to work this out we have to know the volume of the shell. If we call the radius of the shell R and its thickness H, then we see readily that the surface of the sphere on which the shell is built is $4\pi R^2$ and thus the volume of the shell is, to a sufficient approximation, $4\pi R^2 H$. If, now, N is the number of stars per unit volume, then the number of stars in the shell of volume $4\pi R^2 H$ will be $4\pi R^2 HN$. How much light will all the stars in the shell send out? If the average rate at which an individual

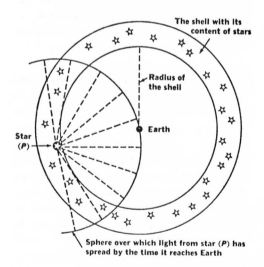

Figure 9.5–1. Each star gives off a large amount of light, but only a small fraction reaches our earth. Since there are so many millions of stars, a large amount of light should reach us. If we consider shells of equal thickness, equal amounts of light should reach us from each shell. If this were so, the sky should be bright and there would be neither night nor day nor any life on earth.

star sends out light is L, then all the stars in the shell put together will send out $4\pi R^2 HNL$. However, what interests us is not how much light all these stars send out, but how much light we receive from them. Consider the light of an individual star in the shell. By the time the light from it reaches us, it will have traveled through a distance R: and so it will have spread out over a sphere of surface $4\pi R^2$. That is to say, the light of each individual star has to be divided by $4\pi R^2$ to tell us the intensity of light from it which is received here. This is true of all the stars in the shell, and therefore, the total light we receive from all the stars in the shell is the total light they send out divided by $4\pi R^2$. This division leads to the cancellation of the factor $4\pi R^2$ and we are left with HNL.

It will be seen that this does not involve the radius of the shell at all. The amount of light we receive from any shell of equal thickness is the same irrespective of the radius of the shell. If, therefore, we add shell after shell, then, since we get the same amount of light from each shell, the amount received will go up and up without limit. On this basis, we should be receiving an infinite amount of light from all the shells stretching out to infinity. However, this argument not only leads to an absurd conclusion, but is not quite right. For each star, in addition to sending out light, obstructs the light from the stars beyond it. In other words, we will not be receiving light from stars in the very distant shells because there will generally be a star in between us and there—a nearer star—which will intercept the light. Of course, it will be realized that stars send out very much light, considering how small a surface they have. Therefore, this obscuring or shadowing effect is not very strong; it will prevent the sum from going up to infinity, but it still leads to our getting from all these shells of stars a flood of light equal to 50,000 times sunlight when the sun is in the zenith. On this basis, then, it should be incredibly bright both day and night. Everything would be burned up; it would correspond to a temperature of over 10,000 degrees Fahrenheit. Naturally, this remarkable result astonished Olbers, and he tried to find a way out. He thought that this flood of light might be stopped by obscuring clouds of matter in space between us and these distant stars. However, this way of escape does not work. For, if there were such a cloud, it would be getting hot owing to the very fact that it was absorbing light from stars; and it would go on getting hotter and hotter until it radiated by its glow as much light as it received from the stars. And then it would not be a shield worth having any longer. Other ways out have been tried, but none of them works. We are, therefore, inevitably led to the result that, on the basis of Olbers' assumptions, we should be receiving a flood of light which is not, in fact, observed.

EXPANSION AND DIMNESS

This little argument may well serve as a proto-type of scientific arguments. We start with a theory, the set of assumptions that Olbers made. We have deduced from them by a logical argument consequences that are susceptible to observation, namely, the brightness of the sky. We have found that the forecasts of the theory do not agree with observation, and thus the assumptions on which the theory is based must be wrong. We know, as a result of Olbers' work, that whatever may be going on in the depths of the universe, they cannot be constructed in accordance with his assumptions. By this method of empirical disproof, we have discovered something about the universe and so have made cosmology a science.

In order to escape from this paradox, we have to drop at least one of his assumptions. . . . It is the one that the universe is static. If the universe is expanding, then the distant stars will be moving away from us at highest speeds, and, it is well known from ordinary physics that light emitted by a receding source is reduced in intensity compared with light emitted by a source at rest. With an expanding universe, such as the one we live in, it may indeed be dark at night, for the light from the distant shells is tremendously weakened by the fact that the luminous objects in them are rushing away from us at high speed. Thus, the darkness of the night sky, the most obvious of all astronomical observations, leads us almost directly to the expansion of the universe, this remarkable and outstanding phenomenon discovered by modern astronomy.

Other changes made by modern astronomy in Olbers' assumptions are relatively minor. It is true that we know that our stars do not go on and on, but form a large stellar system, our galaxy; but we also know that beyond our galaxy there are millions and millions of other galaxies, all more or less like ours. We could, therefore, put Olbers' argument into modern language by changing the reference to stars going on and on in space to galaxies going on and on. The substance of the argument would not be affected.

9.6   The Expanding Universe: Its Origin and Fate
Ernst J. Öpik

The study of the distances and motions of extra-galactic nebulae represents one of the most spectacular chapters of astronomy in the twentieth century. It led, in the first place, to the understanding of the structure of these nebulae—these other Milky Ways, consisting of numberless millions of suns and their planets, potential carriers of life. Yet the consequences of the study are much more far-reaching than that: unexpectedly we have come close to the question of the origins of our world.

The observational fact which made possible an approach to this problem, considered insoluble until now, was provided by the astronomer's "magic eye"—the spectroscope. The spectra of the nebulae, consisting of a continuous color band from red to violet, interrupted by dark Fraunhofer lines of absorption similar to those observed in the sun and stars, indicated that the nebulae consisted, indeed, of stars or suns, as was anticipated 200 years ago by Lambert and Kant. The positions of the absorption lines, however, did not coincide exactly with those of the solar spectrum, but were found to be displaced systematically towards the red, and the greater the displacement, the fainter and more distant were the nubulae. This phenomenon is called the *red shift*. It would indicate that the nebulae are moving away from us; the farther they are, the faster they move.

Of course, the stars of our Milky Way also reveal shifts of their spectral lines. However, these shifts are taking place both ways, toward the red and blue ends of the spectrum, indicating motions of recession and approach, whereas the nebular shifts are one-sided, revealing only recession. Further, the velocities of the stars are relatively small, being of the order of those encountered within the solar system, from 10 to 200 kilometers per second, while nebular motions attain thousands of kilometers per second, the highest value so far recorded being 60,000 kilometers per second. For this reason the nebular red shifts are rightly considered to have a more profound significance than the random motions of the stars. They have become the chief clue to the interpretation of the structure and life history of the whole world, and not only of a limited part of it.

If science is not to be judged by technical achievements, but by its contribution to the understanding of the universe and our place in it, one of its greatest successes in modern times consists in this tackling of the problems of the beginning of our material world and its future fate. Not that final answers have been obtained, but the problems have been put on a concrete basis for the first time in

SOURCE. From *The Oscillating Universe* (New York: New American Library, 1960) Chapter 29; copyright by Ernst J. Öpik, reprinted by arrangement with The New American Library, Inc.

the history of human thought; in particular, a picture of the evolution of the universe for the past 4,000 or 5,000 million years can now be drawn with little ambiguity.

The fundamental question is that of the beginning and age of the universe. By the end of the nineteenth century this problem still remained outside the realm of exact science, simply for lack of observational approach. Of course, the second law of thermodynamics was already established. This stated that thermal evolution is a one-way process, like the run of a clock which is not kept wound up; heat flows from hotter towards colder bodies and is lost by radiation to space; an equalization of the temperature of different parts of the universe follows. Given sufficient intervals of time, a state of equalization of temperature would result, that of Nernst's *Wärmetod* or thermal death, where, in the absence of temperature differences, no relative motion except that of molecules could take place. The state of our present world is certainly very remote from the lethargy of equalization. The mere fact that temperature differences exist, that suns shine and planets carry life on their surface, in defiance of the insatiable immensities of heat-absorbing space, would point to the youthfulness of the world. If so, there must have been a beginning; and, inevitably, there must also be an end.

Scientists of half a century ago somehow preferred to ignore this writing on the wall. One of the reasons—or, rather, pretexts—for such an attitude was that the speed with which the universe is approaching equalization is unknown and may be very slow; thus, the time intervals required by this process may be very long. Without proper logical justification, the indefiniteness of the problem led to a widespread feeling that the universe should have neither beginning nor end, a viewpoint which was more influenced by opposition to former mythological or religious ideas of creation than by impartial reasoning.

Since the second decade of this century a number of new observational facts became known which entirely changed the outlook; they concordantly pointed to a finite age of our universe, of the order of a few thousand million years. Of these, the most revolutionary consequences were brought about by the red shift in the spectra of extragalactic nebulae. Being interpreted as a Doppler effect of motion, it indicated a recession of the nebulae; they are all moving away from us, and from one another, with velocities that increase with distance.

The discovery brought home to us a type of organized motion which hitherto was not know among cosmic systems. Over the entire hierarchy of these systems, from atoms to galaxies, we find that generally a rotational motion keeps the individual members of a system in equilibrium against the centripetal forces of attraction. Thus, the electrons revolve in atoms on their tiny orbits of less than one hundred-millionth of a centimeter in diameter; the moon and other satellites revolve around their respective planets at distances of several hundred thousand kilometers; the earth and her sister planets revolve around the sun, keeping to distances of several hundred million kilometers; the stars of the Milky Way system, a typical galaxy, whose extent is 100 million times that of the solar system, revolve around a common center of gravity; among them the sun, one of 100,000-odd million suns of the system, performs one revolution in about 200 million years. With this last-mentioned motion, called galactic rotation, ends the sequence of rotational systems in our universe.

Next comes the system of extragalactic nebulae which already behaves in a different manner. The flattened shape of these other Milky Way systems indicates rotation within each of them, similar to galactic rotation, and this is also confirmed by direct spectroscopic observations of motion. They are separated from one another by gaps of almost empty space, measured in millions of light years, and the most distant ones still observable with the largest telescopes are over 1,000 million light years away—a million million times farther than the farthest known planets of the solar system.

In all directions of space where the view is not obstructed by local Milky Way clouds of cosmic smoke, extragalactic nebulae are visible more or less uniformly. There exist groupings or clusters of nebulae which, however, do not influence the general picture of uniformity. The uniformity of the distribution, without preference for a particular plane, distinguishes the universe of the nebulae from subordinate systems in which planes of preference are prominent. The peculiarity of motion—the recession of the nebulae—explains this peculiarity in distribution; there is no systematic rotational motion among them, only a radial motion of *expansion*. The nebulae are moving away not only from the terrestrial observer, but also from one another.

A belief has been formed that the universe of extragalactic nebulae is the last step in the hierarchy of cosmic systems, or that it embodies the universe itself; no further systems of a higher order are expected to exist. Although only a belief, it has proved a fruitful working hypothesis. The peculiar distribution of motions in this supersystem, as well as the high velocities of recession encountered—60,000 kilometers per second or one fifth of the velocity of light being the observed value for the remotest nebulae, without any sign that the increase of velocity with distance would stop at that value—points to its unique position.

Thus, we choose to say that the universe itself is expanding in all directions. The word "expansion," although scientifically precise, is too modest to convey to lay ears the meaning of what is actually happening. The observed velocities of recession correspond to energies per unit mass equal to seven times those released in the explosion of a hydrogen bomb. In the unobserved range the velocities are doubtless even greater. We are witnesses to an explosion involving the whole universe, of an intensity surpassing everything hitherto encountered on earth. The fragments of the explosion—the galaxies—are flying apart with velocities which exceed a hundred thousand times the velocities in the most violent non-atomic terrestrial explosion. In a corner of one of these fragments—the Milky Way—is our sun, sheltering the planet Earth where life is nestling, where the marvel of the human mind is prying into the mysteries of the cosmos. We have just discovered our precarious position, sitting like Baron Munchausen on a flying bombshell, and are wondering what is coming next. Whence have we come, and where will we land? Is there time enough for us to fulfill our destination?

As to time, there is even too much of it. The velocities of the cosmic projectiles are great, but so also are the distances to be covered. If the present velocities of recession of the nebulae remained unaltered, their distances would be doubled in 6,700 million years. Actually, restrained by gravitation, the velocities will decrease with time, and the time of expansion will be even greater than the above-mentioned figure. We can be sure that doubling of the nebular distances would not mean fundamental changes in the properties of our universe. It follows that there are many thousand million years ahead for the happenings inside the subordinate systems, in the suns and on the surfaces of their planets, to pursue their normal course without being influenced by the universal expansion. It took "only" 500 million years for life on earth to develop from the primitive forms of the early Cambrian to the present-day level and diversity of forms; man entered the stage less than a million years ago, and his civilization has lasted for less than 10,000 years. The time of doubling the nebular distances is one million times longer than the duration of our civilization, and exceeds twenty times the post-Cambrian span of organic evolution. There is plenty of time ahead. Our position on a fragment of a cosmic superexplosion is not so precarious, after all. As to the consequences of the explosion, we can feel at present perfectly safe and comfortable and make plans for the future, not only for our immediate descendants, but also for those creatures of a higher order which may succeed man after millions of centuries. This does not mean that there are no cosmic dangers in the future; for instance, our sun may play a bad trick on life on earth within the coming 1,000 million years by becoming so hot that the oceans will boil. However, this and other menaces are in no relation to the phenomenon of universal expansion, whose influence upon the fate of individual worlds will not come into play for, say, 25,000 million years, and then only if the present expansion stops, and is followed by a contraction.

So much for the future; but what about the past? Tracing the expansion back in time, we find that the universe was in a more compact state, the distances between the nebulae having been the smaller the farther into the past we go. The velocities of recession were greater than now, because since then gravitational attraction has been slowing them down; this means that the recession of the nebulae never stopped in the past, and that there was a moment when the nebulae were in contact. The contact must have taken place simultaneously for all nebulae, because their velocities of recession are proportional to their mutual distances, and the time required to cover any distance is the same for all. Modern astronomical data for the velocities of recession, the distances of the nebulae and the average amount of gravitating matter per unit volume in the universe, interpreted on the basis of the Friedmann-Einstein "cosmological model" (gravitation versus expansion), set the time of contact as 4,500 million years ago. This figure can be called the age of the universe in a restricted sense.[1] Actually, it is the duration of the present stage of the universe, without prejudice to what was, or whether there was anything at all, before. When in contact, the whole matter of the universe must have been dissolved into a uniform substance; all the classes of celestial bodies which are familiar to us now must have been completely absent.

Thus, the universe exploded 4,500 million years ago front a superdense state: this was the beginning of our present universe. The elastic forces which sent the fragments flying

with velocities of at least one fifth of that of light could not have originated anywhere but in the most condensed state of matter known to the physicist, the so-called *nuclear fluid*. This is the substance that fills the interior of atomic nuclei; one cubic centimeter of nuclear fluid weighs 250 million tons. At that density the whole universe—which is believed, rightly or not, to be of finite mass—would have been squeezed into a volume equal to that of a sphere of 220 million kilometers radius, or about that of the orbit of Mars. Lemaître was the first to visualize such an initial state of the universe; he pointed out that, at the density of nuclear fluid, this state would correspond to one giant atom. This "primeval atom," "the egg from which the universe hatched," must have been unstable, in the same manner as are radioactively unstable all atoms heavier than bismuth. It must have decayed spontaneously, exploding the instant it was formed. Our present universe would thus consist of the debris of radioactive decay of one single atom!

How did the primeval atom come into being? If it had no predecessor, it must have been created on the spot. It would seem that here, in the absence of positive evidence, the acceptance of an Act of Creation could be left to the taste and esthetic judgment of those concerned with the problem.

The alternative would be contained in the assumption of a previous state, a *collapsing* universe which crashed from all directions into one spot. The collapse was stopped at the maximum possible density of matter, that of nuclear fluid, whose elastic forces not only broke the impact, but sent the universe back expanding, like a rubber ball rebounding from a wall. The primeval atom would in this case be deprived of its unique significance; it would represent only a transient stage of maximum compression.

The question arises: Where did the previous collapsing universe come from? To do away with the necessity of assuming the

[1]At present the age of the universe is thought to be somewhat greater—about 12×10^9 years. [Ed.]

spontaneous creation of a collapsing universe, which would appear much less plausible than the creation of the primeval atom, we are compelled to assume its origin from an expanding universe whose expansion was somewhere stopped by gravitation and converted into contraction, or collapse. We arrive thus at the picture of an oscillating universe, the "cosmic pendulum," which expands to a certain maximum volume, then contracts and rebounds from the almost pointlike state of maximum density to start the next stage of expansion, and so forth.

Theoreticians have, of course, foreseen the possibility of an oscillating universe. Some observational data which are now available seem to give support to this model. An expanding universe may develop along two different lines, according to the actual amount of matter in it. If the amount of matter is below a certain limit, gravitation will be unable to halt expansion; the universe will continue expanding indefinitely and irreversibly. On the contrary, if there is more matter than the limit, gravitation will ultimately prevail—expansion will cease, to be succeeded by contraction; this is the case of the oscillating universe.

Whether or not the real universe is of the oscillating type depends thus upon the "world density," or the average amount of matter per unit volume of space. The limiting, or critical, value of world density, for the observed rate of recession of the nebulae, is 39 grams of matter within a cube of 100,000 kilometer edge (not very much, indeed); an estimate of the actual world density by the writer gave 25 grams, or 64 per cent of the critical value. The result could easily be in error by as much as a factor of 2, or even more. If this result is taken literally, it would mean that the universe is irreversibly expanding. However, the opposite conclusion appears to be warranted here. We note that a relatively small increase, by about 60 per cent of the estimated value, would bring it above the critical limit. The estimate, based upon the gravitational attraction which a cluster of galaxies (the Virgo cluster) exerts upon its members, takes into account only the mass of the cluster, neglecting the diffuse matter (gas, dust, and stray stars) which may be present in greater amounts in "open space" outside the cluster. Indeed, Zwicky at Mount Wilson has found indications of great amounts of diffuse matter in intergalactic space and arrives at a world density 25 times our value; although his estimate is based only upon the outer appearance of the patches of diffuse matter, and is therefore less reliable than ours, it definitely points to a higher value of the world density. Therefore, it is highly probable that the world density exceeds the critical value, although perhaps only slightly, and that the universe is of the oscillating type. Let us consider some consequences of this hypothesis.

The timetable of the "cosmic pendulum" could then be roughly as follows. The excess of world density over the critical value being slight, it would mean that the expansion is still in full swing, and that its climax is far ahead, perhaps 10,000 million years from now. With the 5,000-odd million years in the past, this gives 15,000 million years as the total duration of expansion, and an equal interval of time for the ensuing contraction. The total duration of a cycle of oscillation is thus of the order of 30,000 million years, or nearly seven times the age of the solar system or the duration of past expansion. During this interval of time, which could be called the "cosmic year" ("the Day of Brahma"), galaxies, stars and planets are formed, so to speak, in flight; suns are born and become extinct, or explode; life develops on the surface of planets, and perishes.

At present we are in the beginning of a cycle of cosmic oscillation. The "age of the universe," the 4,500 million years of past expansion, is not a true age but only the time elapsed since the last rebirth of the world, which took place in the state of greatest compression into which the universe had previously crashed. Our sun with its planetary system came into being soon after the start of the present expansion, and can now look back upon more than one half of its lifetime; from what we know about stellar structure and evolution, it will not last for more than a further 3,000 to 4,000 million years. Life may disappear from the surface of our planet long before that. There will be other suns, some newly born from interstellar clouds, some those fainter ones which do not spend their energy so profusely and therefore last longer; many of them will shelter life in corners of the innumerable galaxies of the universe, long after our sun has gone.

Some 25,000 million years from now the Day of Reckoning will come. The whole universe— all galaxies with their suns shining or extinct, their planets dead or still carrying life on their surface—will precipitate itself into a narrow space, almost a point. Everything will perish in a fiery chaos well before the point of greatest compression is reached. All bodies and all atoms of the world will dissolve into the nuclear fluid of the primeval atom—which in this case is not truly primeval—and a new expanding world will surge from it, like Phoenix out of the ashes, rejuvenated and full of creative vigor. No traces of the previous cycle will remain in the new world, which, free of traditions, will follow its course in producing galaxies, stars, living and thinking beings, guided only by its own laws—the laws of nature and God.

In each new oscillation, all the structural phases of the previous ones will repeat themselves, without, however, an "eternal recurrence of all things" in Nietzsche's sense. The exact repetition of individual phenomena will be practically impossible, being forbidden by the law of chance. Therefore, the individual celestial bodies in successive oscillations would not be identical, nor would their inhabitants. On the contrary, an unlimited variety of combinations and of prospects of evolution would be possible during each phase of the oscillation.

Disapproval, on purely esthetic grounds, of the idea of an oscillating universe repeating its general features has been voiced in some quarters; the prospect has been described as dull and uninviting. However, human esthetic considerations are here absolutely irrelevant. The only pertinent question is: What is the Plan of our world? We cannot insist upon the correctness or finality of our interpretation, which only claims a certain degree of probability. But, if the interpretation turns out to be correct, we have to accept it without demur, even though the Plan was laid down without our being consulted beforehand. Besides, we cannot see why the Great Repetition should claim a lesser esthetic value than, for example, the annual succession of seasons so praised by poet and layman. Would we call the coming spring dull because there has been springtime before? Or is today boring because sunrise and sunset have been recurring for eons in the past?